Frontiers in Biomaterials

(Volume 3)

(Chitosan Based Materials and its Applications)

Editors

Guilherme Luiz Dotto
Chemical Engineering Department, Federal University of Santa Maria (UFSM), Roraima Avenue, 1000, Santa Maria, RS, Brazil

Sergio Paulo Campana-Filho
Instituto de Química de São Carlos – Universidade de São Paulo, Av. Trabalhador são-carlense, 400 - CEP, 13566-590 São Carlos/SP, Brazil

Luiz Antonio de Almeida Pinto
Industrial Technology Laboratory, School of Chemistry and Food, Federal University of Rio Grande – FURG, Rio Grande – RS, Brazil

Frontiers in Biomaterials

Volume # 3

Chitosan Based Materials and its Applications

Editors: Guilherme Luiz Dotto, Sergio Paulo Campana-Filho & Luiz Antonio de Almeida Pinto

eISSN (Online): 2352-3921

ISSN (Print): 2468-0168

eISBN (Online): 978-1-68108-485-5

ISBN (Print): 978-1-68108-486-2

General:

1. Any dispute or claim arising out of or in connection with this License Agreement or the Work (including non-contractual disputes or claims) will be governed by and construed in accordance with the laws of the U.A.E. as applied in the Emirate of Dubai. Each party agrees that the courts of the Emirate of Dubai shall have exclusive jurisdiction to settle any dispute or claim arising out of or in connection with this License Agreement or the Work (including non-contractual disputes or claims).
2. Your rights under this License Agreement will automatically terminate without notice and without the need for a court order if at any point you breach any terms of this License Agreement. In no event will any delay or failure by Bentham Science Publishers in enforcing your compliance with this License Agreement constitute a waiver of any of its rights.
3. You acknowledge that you have read this License Agreement, and agree to be bound by its terms and conditions. To the extent that any other terms and conditions presented on any website of Bentham Science Publishers conflict with, or are inconsistent with, the terms and conditions set out in this License Agreement, you acknowledge that the terms and conditions set out in this License Agreement shall prevail.

Bentham Science Publishers Ltd.
Executive Suite Y - 2
PO Box 7917, Saif Zone
Sharjah, U.A.E.
Email: subscriptions@benthamscience.org

BENTHAM SCIENCE

CONTENTS

FOREWORD

Petroleum is an important source of energy and materials in our world and many products we use every day are derived from it. However, environmental and economic considerations lead to consider new sources to allow less dependency on fossil fuels. Natural polymers may be one of these alternatives, and among them, polysaccharides form a large class containing a number of industrially-important ones. Chitin, despite constituting the second most abundant natural polymer has not achieved the same commercial importance as for, cellulose, starch or alginate. Academic as well as industrial scientists have developed strong energy to find new and potential applications for this material and most preferentially for chitosan, being its major derivative. The latter presents very specific properties particularly due to its capacity to be polycationic in acidic medium but also a great interest considering the multiple forms in which it can be processed. As a consequence, possible applications cover a wide domain, and this book in a comprehensive manner summarizes many of the most recent technical and research accomplishments in chitosan covering its acquisition processes, characterization, chemical modifications and specific applications such as dyes and metals removal, drug delivery, coatings (especially for food applications), and permeation processes.

The contents of this book highlight the most important challenges and information needed to understand chitosan science. Chitosan is derived from chitin and many processes were designed for deacetylation reaction leading to acquire this derivative. For a few years, microwave-assisted as well as ultrasound processes have been developed and found to be of great interest as it is developed in this book. The characterization of obtained chitosans is of major importance and still remains a challenge for users. The used techniques are discussed in this book and this will strongly help the chitosan users. But chitosan as well as other polysaccharides, do not exhibit all the required properties necessary for specific applications. As a consequence, it is necessary to carry out modifications and the presence of amino groups in chitosan structure allows selective reactions which are described in this book. It is possible to prepare cationic, anionic, amphiphilic or crosslinked materials demonstrating the versatility of this polysaccharide. All of these actions allow the chitosan to be used in applications, covering wide domains as biomedical or pharmaceutics, cosmetics, food, agriculture, water treatment and permeation.

This book is a very useful addition to literature on chitin and chitosan, particularly for those working on the development of new products for specific applications. The authors have to be thanked for their successful contribution to chitosan development demonstrating its strong interest and unlimited potential of applications and it will stimulate researchers in order to continue their efforts so that chitosan finds the place within the polysaccharide world.

Dr. Jacques Desbrières
Professeur Directeur Adjoint UFR Sciences et Techniques de Pau,
Hélioparc Pau Pyrénées,
2 Avenue P. Angot,
64053 PAU cedex 9, France
Tél: 05 59 40 76 02
E-mail: jacques.desbrieres@univ-pau.fr

PREFACE

Chitosan is an alternative, low cost and available biomaterial, which can be obtained from natural and renewable resources. Chitosan can be obtained from chitin deacetylation and, its properties are dependent on the used method and the experimental conditions. Chitosan has several interesting properties from industrial and technological viewpoints, such as, biocompatibility, bioavailability, adhesivity, film forming properties and is polycationic in acid mean. Chitosan can be processed in the form of films, membranes, hollow fibers, nanofibers, nanoparticles and scaffolds. On the basis in these interesting properties, chitosan has an unlimited potential of application, for example, in agriculture, medicine, odontology, pharmacy, engineering and other fields. Studies involving chitosan increased significantly in the last 15 years. Consequently, many frontiers were overcome and new research fields were opened. Based on the above mentioned facts, the traditional and new trends about chitosan based materials were organized in a book: "Chitosan Based Materials and its Applications". It is expected that these information will assist future research in the chitosan field.

This book consists of 10 contributions or chapters. The chapters were developed by eminent researchers from Brazil, Portugal, the United Kingdom and Poland, which worked on chitosan and chitosan based materials. Chapter 1 shows general considerations about chitosan and its based materials, while chapter 2 is focused on the obtention processes and characteristics. In chapters 3 to 5, chitosan based materials like nanofibers, nanoparticles, nanocapsules and chemically modified chitosans are addressed. The characterization techniques for chitosan and its based materials are presented in chapter 6. Chapters 7 to 10 are focused on the applications such as, water treatment, drug delivery, edible films and pervaporation membranes. Based on this interesting material, we believe that this book continue to inspire the development and implementation of technologies involving chitosan based materials in the academic and industrial fields.

Dr. Guilherme Luiz Dotto
Chemical Engineering Department,
Federal University of Santa Maria (UFSM),
Roraima Avenue, 1000, Santa Maria, RS, Brazil

Dr. Sergio Paulo Campana-Filho
Instituto de Química de São Carlos – Universidade de São Paulo,
Av. Trabalhador são-carlense, 400 - CEP,
13566-590 São Carlos/SP, Brazil

Dr. Luiz Antonio de Almeida Pinto
Industrial Technology Laboratory, School of Chemistry and Food,
Federal University of Rio Grande – FURG,
Rio Grande – RS, Brazil

List of Contributors

Adalberto E. Junior	Chemistry Department, Universidade Federal de Santa Catarina, Trindade - Florianópolis, Santa Catarina-Brazil
Adriana Pavinatto	Instituto de Química de São Carlos , Universidade de São Paulo, Av. Trabalhador são-carlense, 400 – CEP, 13566-590 São Carlos/SP, Brazil
Andrea de Lacerda Bukzem	Instituto de Química de São Carlos, Universidade de São Paulo, Av. Trabalhador são-carlense, 400 – CEP, 13566-590 São Carlos/SP, Brazil
Alexandre L. Parize	Chemistry Department, Universidade Federal de Santa Catarina, Trindade - Florianópolis, Santa Catarina-Brazil
Alisson S. Camara	Industrial Technology Laboratory, School of Chemistry and Food, Federal University of Rio Grande - FURG, Rio Grande-RS, Brazil
Ana Luísa Daniel-da-Silva	CICECO-Aveiro Institute of Materials, Department of Chemistry, University of Aveiro, Aveiro, Portugal
Anderson Fiamingo	Instituto de Química de São Carlos, Universidade de São Paulo, Av. Trabalhador são-carlense, 400 - CEP, 13566-590 São Carlos/SP, Brazil
Bárbara J. Leite Ferreira	CICECO-Aveiro Institute of Materials, Department of Chemistry, University of Aveiro, Aveiro, Portugal
Daliomar Lourenço de Oliveira Júnior	Chemical Engineering Department, Environmental Processes Laboratory, Federal University of Santa Maria-UFSM, Santa Maria-RS, Brazil
Daniella de Souza e Silva	Instituto de Química de São Carlos , Universidade de São Paulo, Av. Trabalhador são-carlense, 400 – CEP, 13566-590 São Carlos/SP, Brazil
Daniel Assumpção Bertuol	Chemical Engineering Department, Environmental Processes Laboratory, Federal University of Santa Maria-UFSM, Santa Maria-RS, Brazil
Danilo Martins dos Santos	Instituto de Química de São Carlos , Universidade de São Paulo, Av. Trabalhador são-carlense, 400 – CEP, 13566-590 São Carlos/SP, Brazil
Denys Alberto S. Rodrigues	Industrial Technology Laboratory, School of Chemistry and Food, Federal University of Rio Grande – FURG, Rio Grande – RS, Brazil
Eduardo Hiromitsu Tanabe	Chemical Engineering Department, Environmental Processes Laboratory, Federal University of Santa Maria-UFSM, Santa Maria-RS, Brazil
Gabriela Dudek	Department of Physical Chemistry and Technology of Polymers, Faculty of Chemistry, Silesian University of Technology, Strzody 9, 44100 Gliwice, Poland
Guilherme Luiz Dotto	Chemical Engineering Department, Federal University of Santa Maria-UFSM, Santa Maria-RS, Brazil
Jorge Augusto de Moura Delezuk	Instituto de Química de São Carlos, Universidade de São Paulo, Av. Trabalhador são-carlense, 400 - CEP, 13566-590 São Carlos/SP, Brazil
Luiz Antonio de Almeida Pinto	Industrial Technology Laboratory, School of Chemistry and Food, Federal University of Rio Grande – FURG, Rio Grande – RS, Brazil
Mery Luiza Garcia Vieira	Industrial Technology Laboratory, School of Chemistry and Food, Federal University of Rio Grande - FURG, Rio Grande-RS, Brazil

Nathalie B. Barroca	School of Mechanical and Aerospace Engineering, Queen's University Belfast, United Kingdom
Natália B. Caon Branco	Chemistry Department, Universidade Federal de Santa Catarina, Trindade - Florianópolis, Santa Catarina-Brazil
Nauro da Silveira	Industrial Technology Laboratory, School of Chemistry and Food, Federal University of Rio Grande – FURG, Rio Grande – RS, Brazil
Poliana P. Lopes	Chemical Engineering Department, Federal University of Santa Maria, Santa Maria, Brazil
Regiane Cristina Corrêa	Instituto de Química de São Carlos, Universidade de São Paulo, Av. Trabalhador são-carlense, 400 - CEP, 13566-590 São Carlos/SP, Brazil
Roman Turczyn	Department of Physical Chemistry and Technology of Polymers, Faculty of Chemistry, Silesian University of Technology, Strzody 9, 44100 Gliwice, Poland
Sérgio Paulo Campana-Filho	Instituto de Química de São Carlos, Universidade de São Paulo (IQSC/USP), São Carlos, São Paulo, Brazil
Tito Roberto Sant'Anna Cadaval	Industrial Technology Laboratory, School of Chemistry and Food, Federal University of Rio Grande – FURG, Rio Grande – RS, Brazil
Tonimar Arrighi Domiciano Senra	Instituto de Química de São Carlos , Universidade de São Paulo, Av. Trabalhador são-carlense, 400 – CEP, 13566-590 São Carlos/SP, Brazil
Vanderlei C. Souza	Mechanical Engineering Department, Laboratory of Porous Media and Thermophysical Properties, Federal University of Santa Catarina-UFSC, Florianópolis-SC, Brazil
William Marcondes Facchinatto	Instituto de Química de São Carlos , Universidade de São Paulo, Av. Trabalhador são-carlense, 400 – CEP, 13566-590 São Carlos/SP, Brazil

Frontiers in Biomaterials

CHAPTER 1

General Considerations about Chitosan

Guilherme L. Dotto[1,*] and **Luiz Antonio A. Pinto**[2]

[1] *Chemical Engineering Department, Federal University of Santa Maria-UFSM, Santa Maria-RS, Brazil*

[2] *Industrial Technology Laboratory, School of Chemistry and Food, Federal University of Rio Grande – FURG, Rio Grande-RS, Brazil*

Abstract: Chitosan is a polysaccharide composed of repeated units of N-acetyl-2-amino-2-D-glucopyranose and 2-amino-2-deoxy-D-glucopyranose, which are linked by β-(1→4)-glycosidic bonds. This biopolymer was discovered in 1859, and the industrial scale production started from 1971. During the last 20 years, a considerable interest for chitosan based materials has been observed. Chitosan is mainly obtained from crustacean shells, but other sources are also possible. In general, the shells are submitted to sequential steps of demineralization, deproteinization and deodorization to obtain chitin. Chitosan is then obtained by alkaline deacetylation of the chitin. The quality of chitosan is evaluated taking into account the characteristics such as, molecular weight, deacetylation degree and crystallinity. These characteristics are responsible for properties like biocompatibility, bioadhesivity, solubility and polycationic character. The properties of chitosan make this biopolymer an excellent and attractive material for several chemical and physical modifications, aimed at diverse applications. This chapter presents some general considerations about the biopolymer chitosan, including, definitions, history, main sources, obtention processes, characteristics, properties, chitosan-based materials and their unlimited potential applications.

Keywords: Applications, Biopolymers, Characteristics, Chitin, Chitosan, Production processes, Properties.

INTRODUCTION

Chitosan is a polycationic biopolymer composed of 2-acetamido-2-deoxy-β-D-glucopyranose and 2-amino-2-deoxy-β-D-glucopyranose residues [1] (Fig. **1**). It is known that the NH_2 and OH groups in its structure are mainly responsible for its properties and unlimited potential applications [2]. Chitosan was discovered

* **Corresponding author Guilherme L. Dotto:** Environmental Processes Laboratory (LAPAM), Chemical Engineering Department, Federal University of Santa Maria-UFSM, Santa Maria-RS, Brazil; Tel/Fax: 55 (55) 32208448; E-mail: guilherme_dotto@yahoo.com.br

Guilherme Luiz Dotto, Sergio Paulo Campana-Filho and Luiz Antonio de Almeida Pinto (Eds)

by Rouget [3] in 1859, when studying deacetylated forms of chitin. Production in industrial scale was first registered in 1971 in Japan. The "1st International Conference of Chitin and Chitosan" occurred in 1977 (Boston, USA). From this conference, the scientific and industrial communities demonstrated an exponential interest in chitin and chitosan [4]. From 2010 to 2015, more than 15,000 articles and about 20 books on chitosan were published worldwide.

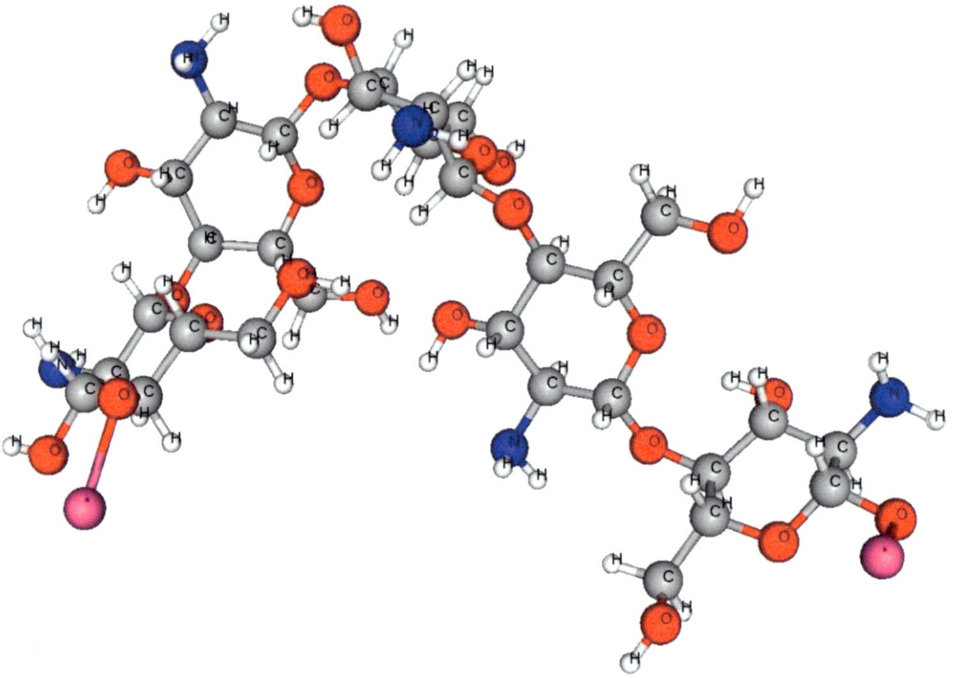

Fig. (1). Three-dimensional chemical structure of the chitosan biopolymer (from the authors).

Crustaceans, insects, mollusks and fungi are the main sources of chitosan [5]. Industrially, chitosan is obtained from shrimp and crab shells, which are wastes from the seafood industries [6]. First, the crustacean wastes are submitted to demineralization, deproteinization, deodorization and drying steps, in order to obtain chitin [5 - 8]. Afterwards, the chitin is deacetylated, purified and dried to obtain chitosan [9, 10]. This process is economically feasible since the raw material has zero cost and can include the recovery of astaxanthin [6]. Nowadays, chitosan is commercially produced in Japan, India, China, Italy, Canada, Chile, Poland, Norway, USA and Brazil [5, 7].

The price of powdered chitosan depends on the range of application, characteristics and purity [5]. For example, the price of chitosan for agricultural use is about US$ 20/kg; for water treatment is about US$ 30/kg; for food applications is US$ 150/kg; for general use in laboratory is US$ 500/kg. The price of high purity 99% chitosan (1 kg) varies from US$ 1000 to 1300.

The main characteristics that define the chitosan biopolymer are its molecular weight and deacetylation degree [11]. Other features such as, crystallinity, surface area, particle size, moisture and ash contents are important [12]. These characteristics are fundamental to ensure the chitosan properties, including, solubility, polycationic character, antimicrobial, biocompatibility and bioadhesivity [7]. Furthermore, the abovementioned characteristics are responsible for determining what chitosan based material could be prepared (films, nanofibers, nanoparticles, nanocapsules, scaffolds, hydrogels) and what the possible application of the biopolymer will be (water treatment, food industry, cosmetics, agriculture, biomedicine) [11 - 13].

Taking into account the importance of chitosan for the academic, scientific and industrial communities, this chapter intends to present some general considerations about this biopolymer. The definitions, main sources, obtention processes, characteristics, properties, chitosan based materials and their unlimited potential of application are presented in general lines. Specific discussions about chitosan based materials, their characteristics, properties and applications are presented in the other chapters of this book.

MAIN SOURCES

Crustaceans

Shrimp and crab wastes are the main sources of the industrial production of chitosan [5, 14]. Chitosan is obtained from α-chitin, which is a component of the exoskeleton of these crustaceans. The exoskeletons are wastes from the seafood industries. It is estimated that the total global annual generation of seafood wastes is around 5.1×10^6 metric tons [15]. The chitin content in the dried exoskeletons varies from 5% to 42%, depending on the crustacean species [5]. Nouri and coworkers [16] used Persian Gulf shrimp wastes to produce chitosan and, obtained a good product with a deacetylation degree of 89% and molecular weight of 806 Da. Yen and coworkers [17] prepared chitosan from crab shells, obtaining a biopolymer with a deacetylation degree higher than 80% and molecular weight of about 500 kDa. In brief, chitosan obtained from shrimp and crab wastes is very attractive, since these sources are available, renewable and have zero cost. Furthermore, it is an alternative form of appropriate management of solid wastes in the seafood industries.

Insects

Insects are an alternative source of chitosan, but, until recently, only laboratory scale studies were performed. Kaya and coworkers [18] obtained chitosan from the Colorado potato beetle (adult and larvae). They found that the dry weight chitin contents of the adult beetles and larvae were 20% and 7%, respectively. Furthermore, the chitin produced chitosan yields of 72% from the adult beetles and 67% from the larvae. Chitosan from beetles presented a molecular weight of around 2.600 kDa. Liu *et al.* [19] also obtained chitosan from beetles. Other insects used to produce chitosan are blowflies [20], silkworm chrysalides [21] and houseflies [22]. The use of insects to obtain chitosan is based on their biodiversity, since they represent 95% of the animal kingdom. Furthermore, the insect cuticles have lower levels of inorganic material compared to crustacean shells, which makes their demineralization treatment more convenient [19]. Another advantage is the insect control in agricultural areas [18].

Mollusks

Mollusks are another source of chitosan in laboratory scale. Species such as *Sepia kobiensis*, *Sepia spp.*, *Loligo lessoniana* and *Loligo formosana* have been used for this purpose [23 - 25]. Ramasamy *et al.* [23] prepared chitosan from *S. kobiensis* cuttlebone and obtained a product with a deacetylation degree of 85.55%, molecular weight of 322.04 kDa and good antioxidant properties. Al Sagheer *et al.* [25] prepared chitosan from cuttlefish, obtaining chitosan from α-chitin (from crustaceans) and β-chitin (from cuttlefish). During the process, they found that the deacetylation reaction was faster when β-chitin (from cuttlefish) was used. In general, the preparation of chitosan from mollusks is suitable. An example is the squid pen, in which chitin consists of about 35% of the dry weight [24]. The deacetylation process is also facilitated because chitin is in the β form is more reactive [23].

Fungi

Chitosan can be also prepared from fungi (fungal chitosan), since its precursor, chitin, is a characteristic component of the taxonomical groups Zygo-, Asco-, Basidio- and Deuteromycetes [26]. In this case, fermentative routes are used. Vendruscolo and Ninow [27] prepared chitosan from *Gongronella butleri* using apple pomace as the substrate for fermentation. Tayel *et al.* [28] obtained chitosan from *Aspergillus brasiliensis* and demonstrated that the biopolymer can be used as a biopreservative for minced meat. Dhillon *et al.* [29] studied the preparation of chitosan from fungal waste mycelium as co-product during citric acid fermentation. They concluded that the co-extraction of chitosan from fungal waste mycelium resulting from citric acid production is an eco-friendly alternative to the chitosan derived from marine shell wastes. The production of fungal chitosan is justified by the high worldwide production of mushrooms [26] and high content of chitin in the fungal cell walls (up to 50% of dry biomass) [29].

CONVENTIONAL OBTENTION PROCESS

In general, two main processes are used to obtain chitosan: the conventional process and the fermentative process [5, 6]. In the conventional process, the sources are crustaceans, insects or mollusks [2]. Firstly, chitin is isolated from the raw materials by several chemical treatments, which remove ashes, proteins, color and flavors [30]. The chitin is then converted into chitosan by alkaline deacetylation [31]. The purification can be performed by several steps of dissolution/precipitation/centrifugation, and finally, the drying step is employed [12]. In the fermentative process, chitosan is obtained from fungi [26]. Initially, fungal biomass containing glucan complexes with chitin or chitosan is obtained by fermentation [29]. After, the biomass is separated from the fermentation media by filtration, it is dried and sequential steps are performed to extract chitosan [26, 29]. It is known that each process has advantages and drawbacks, but, in this chapter, only the conventional process will be addressed since this process is the most common and is used industrially. More details about the fermentative process are presented in chapter 2 of this book. Fig. (**2**) shows the main steps of the conventional process used to obtain chitosan.

The conventional process generally starts with the demineralization step (Fig. **2**). The main objective is the reduction of the ash content (mineral content) in the crustacean wastes. The ashes are mainly composed by carbonates, phosphates and other mineral salts [5, 31]. Demineralization is mainly performed using diluted solutions of HCl (1-10%) under stirring and ambient temperature (20-30 °C), for short time periods (1-3 h) [32 - 34]. This treatment ensures the complete removal of the mineral content, without degradation of the chitin polymeric chains. Other acids, such as, HNO_3, H_2SO_4, HCOOH, H_3CCOOH and EDTA can be used [5, 6, 32 - 34]. These acids are efficient, but can cause damages in the chitin structure, for example, depolymerization.

The second step is deproteinization (Fig. **2**). In this step, alkalis, including, NaOH, Na_2CO_3, $NaHCO_3$, KOH, K_2CO_3 or $Ca(OH)_2$ are used to remove the protein content adherent on the crustaceans wastes [5, 15, 35]. The most common option is the use of NaOH (1-10%) under stirring. The temperature used in this operation varies from 20 to 100 °C and the time period varies from 2 h to 72 h [15, 35 - 37]. Deproteinization requires the effective control of time and temperature, since high temperatures and prolonged time periods can provoke depolymerization and deacetylation [38]. An alternative form of deproteinization is the enzymatic treatment with proteases (pepsin, trypsin) [39]. The enzymatic alternative is milder, but is less efficient than the alkali treatment [5, 6].

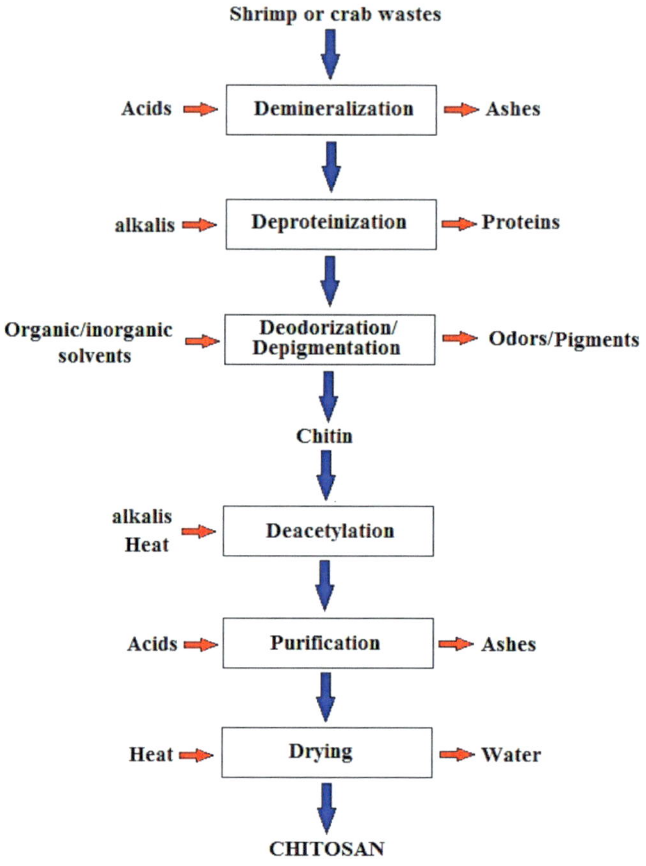

Fig. (2). Conventional process to obtain chitosan (from authors).

The third step to isolate chitin is deodorization/depigmentation (Fig. **2**). This operation is responsible for removing pigments and odors [6]. Ethanol, acetone, $KMnO_4$, NaOCl or H_2O_2 are commonly used for this purpose [5, 7]. If solvent extraction is employed in this step, the recovery of astaxanthin is also possible [5]. It is important to highlight that the aforementioned steps (demineralization, deproteinization and deodorization/depigmentation) are followed by several consecutive washings until neutral pH [30, 31]. The product of demineralization, deproteinization and deodorization/depigmentation is chitin. It can then be dried for further use or immediately deacetylated.

The deacetylation reaction is the most important step in the process presented in Fig. (**2**). In this step, chitin is converted into chitosan. During the course of deacetylation, the chitin acetamido groups in position C-2 are converted into amino groups (NH_2) [2, 6, 8, 9]. When the percentage of amino groups attains around 50% or more, the polymer becomes soluble in aqueous acidic media and is called chitosan [11]. Alkaline hydrolysis with NaOH (40-50%) at high temperatures (80-120 °C) is the most common form of obtaining chitosan from chitin [5, 8, 11, 30 - 33]. Other alkalis are also used instead of NaOH [40]. The main factors that affect the deacetylation reaction are alkali concentration, reaction time, solid/liquid ratio, temperature and chitin particle size [5 - 9, 30 - 33]. Some alternative techniques are being investigated to improve or substitute the conventional alkaline hydrolysis, these include enzymatic means [41], steam explosion [42], microwave assisted deacetylation [43] and ultrasound assisted deacetylation [44]. Independent of the deacetylation type, the reaction parameters should be carefully controlled since they directly affect the chitosan characteristics such as molecular weight, deacetylation degree and crystallinity. These characteristics, in turn, affect the properties and applications of chitosan.

The purification step (Fig. **2**) can be performed to obtain a high purity chitosan. In this step, the product of deacetylation is dissolved in acid medium generating a viscous solution with an insoluble fraction (composed of impurities, mainly ashes which were not completely removed in the demineralization step). This viscous solution is then centrifuged or filtered and the insoluble fraction is discarded. The alkali is then added into the viscous solution, forming a suspension with precipitated chitosan. Subsequently, the suspension is centrifuged again, generating pure chitosan (90-95% chitosan) in paste form. The abovementioned steps (dissolution/centrifugation/precipitation/centrifugation) can be repeated many times aiming to obtain a high purity chitosan (purity higher than 99.9%) [4, 30 - 33].

Drying is the last step necessary to obtain chitosan. This operation is important in chitosan production, to guarantee the necessary moisture content for product storage, without causing alterations in the material [12]. In general, after drying, the desired product should contain moisture content lower than 10% (wet basis), to ensure good physicochemical and microbiological aspects during prolonged storage. Polymerization and Maillard reactions are the main alterations that should be avoided during the drying operation. Some techniques such as convective tray drying [10, 45], spouted bed drying [12], spray drying [46], sun drying [47], oven drying, infrared drying [48], lyophilization [49] and low-pressure superheated steam drying [50] have been used to obtain dried chitosan. All these techniques have advantages and drawbacks, but it is common sense that factors such as air temperature and residence time are key factors, which should be controlled to obtain a good quality product.

CHITOSAN CHARACTERISTICS

Chitosan characteristics are strictly dependent on the source and obtention process [2]. The knowledge and control of chitosan characteristics is fundamental, since these affect the properties and applications of the biopolymer [6]. The main characteristics that define the quality and application range of chitosan are molecular weight, deacetylation degree, crystallinity, surface area and particle size [2, 5, 6, 8, 9, 11, 51]. Details about these characteristics are presented in this section.

Molecular Weight

Chitosan molecular weight (MW) is a characteristic associated with the number of monomeric units of the biopolymer [9, 11]. The control, evaluation and modification of this characteristic are fundamental, since MW affects the properties (viscosity, solubility) and applications of chitosan [6]. In general, chitosan MW ranges from 20 to 1200 kDa [6, 11]. Regarding the molecular weight, chitosan can be classified as: low molecular weight chitosan (LMWC), medium molecular weight chitosan (MMWC) and high molecular weight chitosan (HMWC) [9]. Chitosan MW can be determined by HPLC and light scattering, but the viscosimetric method is the most common and simple way [2]. Generally, chitosan MW can be modified starting from a HMWC. So, depolymerization techniques are employed aiming to obtain LMWC [9].

Some researchers have verified the effect of molecular weight on the properties and applications of chitosan. According to Dotto [52] and Fernandez-Pan [53], chitosan edible films from LMWC are more efficient than HMWC in preserving papaya fruits and fishes during the storage period. Bekale and co-workers [54] studied the effect of polymer molecular weight on chitosan-protein interaction. They found that the increase in the chitosan molecular weight from 15 to 200 kDa improved the binding constant of the complexes between bovine serum albumin and chitosan. Bof and co-workers [55] studied the effect of chitosan molecular weight on starch-composite film properties. It was verified that the filmogenic solutions based on LMWC exhibited a Newtonian rheological behavior while those containing MMWC or HMWC exhibited a pseudoplastic one, increasing both consistency index and apparent viscosity with polymer molecular weight. They also concluded that the physicochemical properties of the chitosan based film were strongly affected by its molecular weight.

Deacetylation Degree

The deacetylation degree (DD) is one of the most important characteristics that defines the properties and applications of chitosan. DD is the relation between 2-

acetamido-2-deoxy-β-D-glucopyranose and 2-amino-2-deoxy-β-D-glucopyranose units. If all monomeric units are 2-amino-2-deoxy-β-D-glucopyranose, the biopolymer is fully deacetylated and DD is equal to 100%. In general, when the percentage of 2-amino-2-deoxy-β-D-glucopyranose units attains 50% or more, the polymer becomes soluble in aqueous acidic media and is called chitosan [6, 9, 11]. In the same way as MW, DD affects the properties and applications of chitosan. For example, increases in DD lead to an increase in the free amino groups on the chitosan polymeric chain. As consequence, the chitosan solubility and polycationic character is increased [8]. DD can be estimated by spectroscopic methods, including Fourier transform infrared spectroscopy (FTIR), X-ray diffraction (XRD) and magnetic nuclear resonance [56, 57]. Potentiometric titration can also be used [31]. Chitosans with different DD can be obtained by varying the conditions of the deacetylation reaction [31, 33].

Piccin and co-workers [32] studied the adsorption of FD&C Red 40 dye by chitosan with different deacetylation degrees. It was verified that the DD increase from 42 to 84% caused an increase from 266 to 373 mg/g in the adsorption capacity. Gonçalves and others [58] obtained similar results in the adsorption of food dyes by chitosan in the binary system. Literature shows that the DD influences the mechanical properties (elongation and tensile strength) of chitosan films [33] and chitosan/gelatin films [59]. According to Li and Xia [60], chitosan-emulsifying activity is also dependent on DD.

Crystallinity

Crystallinity is a characteristic of chitosan that represents the ratio between the crystalline and amorphous fractions of the biopolymer. This feature is quantified as crystallinity index (CI) [7, 11]. In solid state, chitosan is a semi-crystalline biopolymer and can present polymorphism [3]. The unit cell of chitosan is orthorhombic and contains two antiparallel chains without water molecules [7]. The chitosan crystallinity is dependent on the source and preparation procedures. Crystallinity is maximum for both chitin (0% deacetylated) and fully deacetylated chitosan (100%). Commercial chitosans are semi-crystalline polymers and the CI is a function of the DD [51]. The quantification of CI is important since this characteristic affects the swelling properties, porosity, hydration and sorption properties of chitosan [11, 51]. CI can be found from the relation between the X-ray diffraction (XRD) characteristic peaks [57].

Thakhiew and others [50] studied how the physical and mechanical properties of chitosan films are affected by the different drying methods. They used hot air drying (HD) and low-pressure superheated steam drying (LPSSD). They found that LPSSD provided films with lower CI (3-5%) than the films obtained from

HD (10-16%). Consequently, the swelling degree was 48-100% (LSSPD) and 75-110% (HD), respectively.

Surface Area and Particle Size

Surface area and particle size are important characteristics of chitosan. These characteristics are dependent on the source and obtention process, being related to the porosity, pore volume and pore size distribution of the chitosan particle [51]. It is known that chitosan powder or flakes have low surface area (lower than 10 m^2/g), being non-porous materials. Particles lower than 1 mm are commonly used in most of the applications [32, 61]. Surface area and particle size are fundamental for applications such as adsorption and enzyme immobilization, since in these fields, many accessible sites and a porous structure are required [14, 62]. Since chitosan is a non-porous material, some modifications are performed to increase its surface area [63, 64]. Indeed, the correct determination of surface area and particle size is fundamental. Surface area is normally obtained by N_2 adsorption-desorption isotherms using the BET method [62]. Particle size can be obtained by scanning electron microscopy (SEM), sieving test or, using a particle analyzer [51, 62 - 65].

Piccin and others [62] investigated the effects of particle size, surface area and pore volume of chitosan on the adsorption of FD&C Red 40. The particle sizes used were 0.10, 0.18 and 0.26 mm, with surface areas of 4.2, 3.4 and 1.6 m^2/g, respectively. The results showed that, the increase in the surface area and a decrease in particle size doubled the adsorption capacity. Dotto and others [61] compared chitin and chitosan as adsorbents for Tartrazine yellow dye. They concluded that chitosan was a better adsorbent than chitin due to its higher deacetylation degree and higher surface area, pore volume and average pore radius.

CHITOSAN PROPERTIES

Chitosan proprieties are dependent on its characteristics and are consequently responsible for the quality and application range of chitosan. Some important properties are solubility, polycationic character, film forming, antimicrobial and biocompatibility [2, 5, 6, 8, 9]. Details about these properties are presented in this section.

Solubility and Polycationic Character

The solubility and polycationic character of chitosan are fundamental properties for the preparation of derivates and application in several fields. As a rule, chitosan is a strong base with primary amino groups, which have a pKa of 6.3.

Thus, this biopolymer is soluble in dilute acid solutions below pH 6.0. At low pH values, the H^+ ions in the solution protonate the NH_2 groups of chitosan leading to the formation of $(NH_3)^+$. Thus, chitosan becomes a polycationic biopolymer. On the other hand, at pH higher than 6.0, deprotonation occurs and chitosan becomes insoluble [9]. The soluble-insoluble transition occurs at its pKa value with the pH being between 6.0 and 6.5. Formic acid is the most appropriate solvent for chitosan, while acetic acid is the most commonly used. Other alternatives are hydrochloric acid and nitric acid. Chitosan is insoluble in sulfuric and phosphoric acids. Concentrated acids are not indicated since they can break the chitosan polymeric chains leading to depolymerization [8]. The solubility of chitosan and chitosan based materials is dependent on several important factors including; time, temperature, DD, MW, particle size, type and concentration of acid, and obtention process [6, 8, 13].

Film Forming Property

The ability of chitosan to form films is one of the most important properties of chitosan. This property makes it possible to form a solid/gel film or a membrane with interesting mechanical, chemical and textural characteristics. These films can also be mixed with other materials, such as essential oils, proteins and others in order to improve its characteristics and functionalities [33, 66]. The ability of chitosan to form films is attributed to its mucoadhesive and gelling properties. The polycationic character and the possibility of hydrogen and hydrophobic bonds between the polymeric chains confer these properties for chitosan [67].

Antimicrobial Properties

Chitosan has intrinsic antimicrobial properties seen in fungi, viruses and bacteria (gram-positive and gram-negative) [68]. This is a key property of chitosan, which allows its application in areas like food preservation, pharmacy and biomedicine [9, 13, 68]. According to Harish Prashanth and Tharanathan [9], the antibacterial property of chitosan is related to its polycationic character. The polycationic chitosan molecule strongly interacts with the microbial cell surface, leading to gradual shrinkage of cell membrane and finally death of the cell. Elsabee and Abdou [13] stated that the bactericidal activity of chitosan is caused by the electrostatic interaction between $(NH_3)^+$ groups of chitosan and the phosphoryl groups of the phospholipid component of the cell membrane. In general, it is reasonable that polycationic chitosan molecule interacts with the anionic cell wall components (lipopolysaccharides and proteins) of the microorganism, resulting in the leakage of intracellular components due to changes in the permeability barrier; preventing nutrients from entering the cell; upon entry into the cell, binding to DNA, and thus inhibiting RNA and protein synthesis [9, 13, 68]. The antimicrobial properties of chitosan are dependent on its characteristics, such as deacetylation degree and molecular weight [9, 13, 52].

Biocompatibility

Biocompatibility can be defined as "compatibility with living tissue or a living system by not being toxic, injurious, or physiologically reactive and not causing immunological rejection" [69]. One of the most important properties of chitosan is its biocompatibility. This property confers to chitosan the possibility to be applied in tissue engineering, wound dressing, bone regeneration and other related areas, being a potential biomaterial [70]. The biocompatibility of chitosan and chitosan-based materials can be evaluated by *in vitro* and *in vivo* tests [70]. Regarding *in vitro* evaluation, the cytotoxicity test is the most common [71]. For *in vivo* evaluation, the characterization is dependent on the application. Genetic toxicity test, intramuscular implantation test and others are possible [70].

CHITOSAN BASED MATERIALS AND THEIR APPLICATION

One of the main advantages of chitosan biopolymer is the possibility of a wide range of modifications, from chemical to physical viewpoints, providing different materials for several applications. Some examples of chitosan based materials are: films, membranes, nanofibers, nanoparticles, nanocapsules, scaffolds, hydrogels and various chemically modified chitosans. In this section, some of these chitosan-based materials are addressed. In this book, some applications are discussed, including water treatment, drug delivery, edible films and coatings, and permeation/pervaporation membranes. The application of chitosan based materials in the textile sector is also an important area. Detailed information on the application of chitosan based materials in the textile sector can be verified in the works of the Ferrero group [72 - 80].

Chitosan Films

Due to its characteristics, chitosan can be used as an efficient filmogenic matrix. It has been researched mainly in the application in food preservation [13, 52] and adsorption process [81 - 84]. Many techniques have been used to prepare chitosan films [13], but the simplest is the casting technique [33, 66]. This technique entails of the mixture of chitosan with an adequate solvent under stirring, until complete dissolution. The resulting filmogenic solution is dried for solvent evaporation, forming a solid film [33]. In some cases of food preservation, the filmogenic solution can be directly applied to the product and dry under ambient conditions [52].

Thin chitosan films are normally characterized according to their mechanical properties, thickness, water vapor permeability, color parameters, texture, functional groups and crystallinity [10, 66, 81]. As presented by Moura and others [33, 66], these characteristics are dependent on the chitosan molecular weight and deacetylation degree, as well as on the operational conditions used in the chitosan preparation [10].

Recently, Dotto and co-workers [52] investigated the use of chitosan filmogenic solutions for the extension of the microbiological shelf life of papaya fruits during storage at room temperature. Chitosans with molecular weights of 150 kDa and 300 kDa were employed. They found that after 10 days of storage, the Log (CFU/g) of mesophilic bacteria and yeasts and molds were, respectively, 1.3 and 2 times lower for chitosan coated fruits. It was demonstrated that the 150 kDa chitosan solution was more adequate to preserve the papaya fruits. The use of 150 kDa chitosan solutions extended the shelf life of papaya fruits by 4-7 days, during the storage at room temperature. In order to illustrate the good results, Dotto and co-workers [52] presented the papaya fruits after 10 days of storage: (a) coated with chitosan 150 kDa and (b) without chitosan coating (Fig. **3**). Furthermore, in this field, chitosan filmogenic solutions have been adequate in extending the shelf life of some food products, including, mangoes [85], water caltrop [86], Chinese water chestnut [87], eggs [88], fruit-based salad [89] and Eksotika II papaya [90].

Fig. (3). Papaya fruits after 10 days of storage: (**a**) coated with chitosan 150 kDa and (**b**) without chitosan coating (from reference [52]).

Chitosan thin solid films have been used for water treatment aiming to remove dyes and metallic ions from aqueous effluents [81 - 85]. Dotto and Pinto group [81] investigated the application of chitosan films for the removal of food dyes from aqueous solutions by adsorption. The results showed that the maximum experimental adsorption capacities were 194.6 mg/g and 154.8 mg/g for acid red

18 and FD&C blue 2, respectively. It was concluded that chitosan films maintained their structure and were easily separated from the liquid phase after the adsorption process, thus being a potential adsorbent. In another research [82], chitosan films were used to adsorb erythrosine B and indigo carmine from aqueous media. In this case, a diffusional mass transfer model was developed and this model revealed that the surface diffusion was the predominant intraparticle mass transfer mechanism. In the same way, Rêgo and co-workers [83] applied chitosan films to remove azo dyes (tartrazine and amaranth) from aqueous solutions by adsorption.

They found adsorption capacities of 413.8 mg/g and 278.3 mg/g, respectively. The interactions between the protonated amino groups of chitosan films and anionic form of the dyes at pH 2 were confirmed. It was also found that chitosan films could be reused for two cycles. Chitosan films were recently used to remove metals such as, vanadium [84] and chromium [91] from aqueous solutions, presenting good results. Based on the above studies, it should be stated that chitosan films are good adsorbents for the removal of pollutants from aqueous media, presenting advantages such as, good swelling, high adsorption capacity, high efficiency, potential for reuse and easy phase separation. These interesting characteristics are presented in Fig. (**4**).

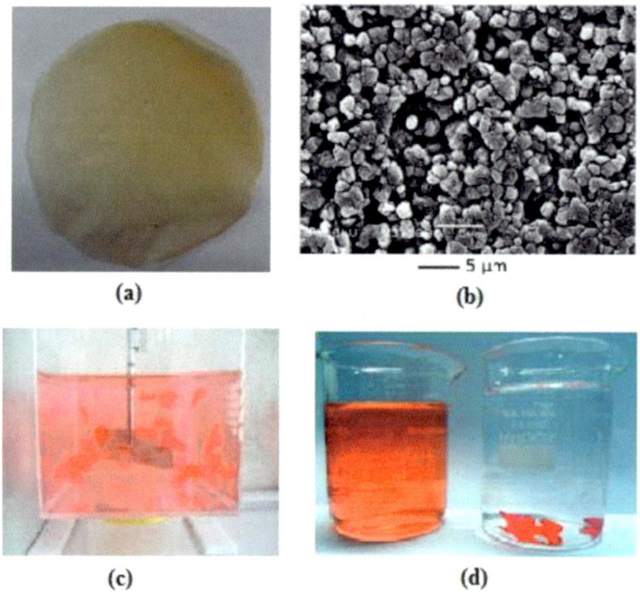

Fig. (4). (**a**) chitosan film produced by the casting technique, (**b**) SEM image of chitosan film (from ref [66].), (**c**) batch adsorption experiment using chitosan films and (**d**) dye solution before and after the adsorption with chitosan film (images (**a**), (**c**) and (**d**) are courtesy of the Industrial Technology Laboratory, LTI, FURG).

Chitosan Nanofibers and Nanoparticles

Chitosan can be processed in the form of nanomaterials like nanofibers and nanoparticles [3, 92]. Chitosan nanofibers are defined as fibers with cross-sectional diameter in the range of 1-1000 nm [93]. These nanofibers can be obtained by the sol-gel method, chemical vapor deposition, electrospinning, thermal oxidation, forcespinning among others, with electrospinning being the most common. The formation of nanofibers through electrospinning is based on the uniaxial stretching of a viscoelastic solution [92, 93]. The preparation of chitosan nanofibers by electrospinning depends on several factors including, solution and process conditions, as detailed in chapter 3. Forcespinning is an innovative technique, which uses the centrifugal force used to prepare chitosan nanofibers [94]. Independent of the preparation technique, the quality of nanofibers is evaluated by analytical techniques like scanning electron microscopy (SEM), transmission electron microscopy (TEM), X-ray diffraction (XRD), thermal profile, zeta potential and tensile properties [93]. Fig. (**5**) shows the images of chitosan nanofibers (SEM) and nanoparticles (TEM).

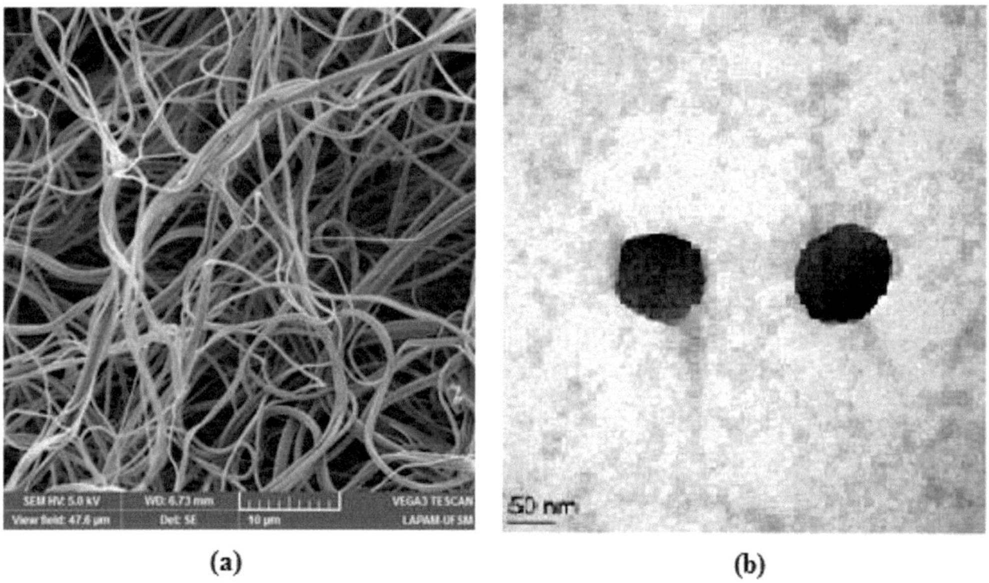

Fig. (5). (**a**) SEM image of chitosan/nylon nanofibers prepared by forcespinning (courtesy of the Environmental Processes Laboratory, LAPAM, UFSM) and (**b**) TEM image of chitosan nanoparticles (from ref [95]).

Chitosan nanofibers can be used in a wide range of applications. Marysková and others [96] reported the preparation of polyamide/chitosan nanofibers for laccase immobilization. Nanofibers with an average diameter of 185±19 nm and a surface density of 3 g/m^2 were obtained. The authors concluded that nanofibers proved to be a suitable carrier for immobilized and modified laccase. Furthermore, they were also efficient in the removal of a mixture of endocrine disrupting chemicals. Razzaz and others [97] used chitosan nanofibers functionalized by TiO_2 nanoparticles for the removal of heavy metal ions. They found that the maximum adsorption capacities for Cu(II) and Pb(II) were 710.3 and 579.1 mg/g. Zhao and others [98] prepared chitosan/sericin nanofibers for wound dressing applications. The nanofibers presented a good morphology with diameters between 240 nm and 380 nm. The authors demonstrated that the chitosan/sericin composite nanofibers were biocompatible and promote cell proliferation. The nanofibers also showed good bactericidal activity against both gram-positive and gram-negative bacteria. Toskas and co-workers [99] prepared chitosan (PEO)/silica hybrid nanofibers for bone regeneration. Nanofibers with an average diameter of 182±16 nm were obtained. The results revealed that the nanofiber matrices were cytocompatible when seeded with bone-forming 7F2-cells, promoting attachment and proliferation over 7 days.

Chitosan based nanoparticles are defined as particulate dispersion of solid particles with sizes ranging from 1nm to 1000 nm. These nanoparticles can be obtained by self-assembling, emulsion, ionic gelation and reverse micellar methods, among others [3, 92, 93]. The nanoparticles are characterized by scanning electron microscopy (SEM), polydispersity index (PDI), atomic force microscopy (AFM), transmission electron microscopy (TEM), X-ray diffraction (XRD), zeta potential and thermal profile [93]. It is clear that nanoparticles and nanofibers are also characterized by specific techniques, depending on their application. Detailed information about the preparation and characterization methods can be found in chapters 3, 4 and 6. Chitosan based nanoparticles can be used in a wide range of applications. In terms of application, the majority of studies prefer the preparation and use of chitosan nanoparticles along with other material, instead of pure chitosan.

O'Callaghan and Kerry [100] prepared chitosan nanoparticles by ionic gelation, and verified their antimicrobial activity against cheese-derived cultures, as well as gram-positive and gram-negative microorganisms. They found nanoparticles with mean diameters from 132 to 202 nm, with an intermediary PDI. It was concluded that nanoparticles in an acidic medium are promising as antimicrobial agents in the area of food packaging, particularly for use with cheese products. Wang and co-workers [101] developed red fluorescent chitosan nanoparticles grafted with poly (2-methacryloyloxyethyl phosphorylcholine) for live cell imaging. This study demonstrated that the nanoparticles presented excellent water dispersibility, biocompatibility and photostability, which made them potential for long-term tracing applications. The development of chitosan nanoparticles loaded with essential oil for antimicrobial and antioxidant applications was studied by Feyzioglu and Tornuk [102]. From their study, it was possible to obtain thermally stable chitosan nanoparticles with mean diameters ranging from 140.25 to 237.60 nm. In brief, the authors concluded that chitosan nanoparticles loaded with essential oils were most adapted to environmental factors, such as pH and temperature, maintaining good bioactive properties.

Water treatment is an important field for the application of chitosan based nanoparticles. Tanhaei and others [103] prepared chitosan/Al_2O_3/magnetite nanoparticles to remove methyl orange from aqueous solutions. The nanoparticles presented a good adsorbent potential, with removal percentage of 93% and adsorption capacity of 417 mg/g. The reuse was possible for five consecutive adsorption/desorption cycles. Shaker [104] studied the adsorption of Co(II), Ni(II) and Cu(II) ions onto chitosan-modified poly(methacrylate) nanoparticles. The nanoparticles presented diameters of 100-120 nm, with adsorption capacities in the range of 195-340 mg/g and recovery percentages from 85 to 90%. Zhou and co-workers [105] investigated the adsorption of food dyes from aqueous solution by glutaraldehyde cross-linked magnetic chitosan nanoparticles. The size distribution ranged from 60 to 95 nm. The adsorption capacities were of 475.61 and 292.07 mg/g, for FD&C Blue 1 and FD&C Yellow 5 respectively.

Chitosan Nanocapsules

Nanocapsules are colloidal systems, in nanometric dimensions, which contain an active agent inside of a wall material. Due its biocompatibility, biodegradability, mucoadhessiveness and longer *in vivo* circulation time, chitosan has been used as wall material to prepare nanocapsules. In this context, chitosan is responsible for protecting and distributing the active agent. Chitosan nanocapsules can be obtained by high pressure homogenization, ultrasonication, spontaneous emulsification or phase inversion methods [106]. The nanocapsules quality is verified by analytical techniques such as scanning electron microscopy (SEM), transmission electron microscopy (TEM), X-ray diffraction (XRD), thermal profile, zeta potential, droplet size distribution and polydispersity [107]. Depending of the active agent, *in vivo* and *in vitro* characterizations are necessary. Characterizations pertinent to the active agent have also been realized in order to verify the performance of chitosan as an encapsulating agent [108]. To illustrate, Fig. (**6**) shows the SEM images of nanocapsules prepared from chitosan and unsaturated fatty acids concentrates.

Esquerdo and co-workers [109] prepared nanocapsules containing unsaturated fatty acids concentrate (UFAC) using chitosan as wall material (UFAC-chitosan nanocapsules). The nanocapsules were prepared to avoid the primary oxidation of

UFAC. The results showed that the nanocapsules presented a spherical shape with diameter of 332 nm and polydispersity index of 0.214. The peroxide values demonstrated that the nanocapsules were able to protect the UFAC from primary oxidation. The encapsulation efficiency was 74.1%, showing that chitosan has potential to be used as an encapsulating agent for unsaturated fatty acids concentrate (UFAC).

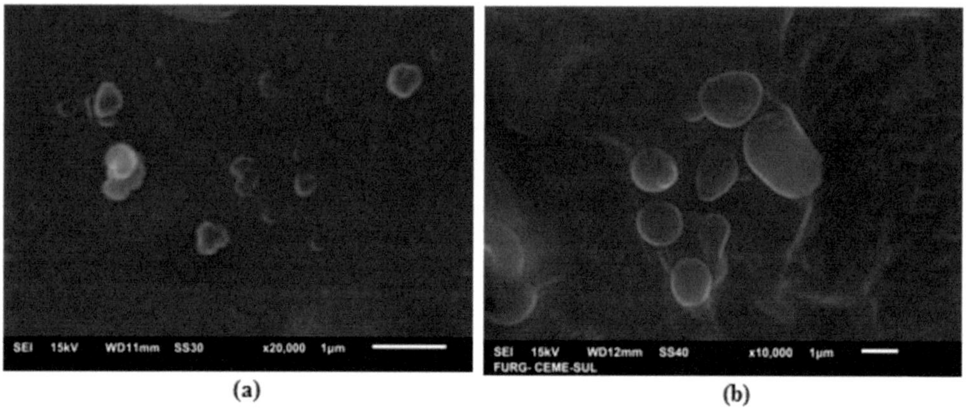

Fig. (6). SEM images of nanocapsules prepared from chitosan and unsaturated fatty acids concentrates (courtesy of the Industrial Technology Laboratory, LTI, FURG).

The development and evaluation of alginate-chitosan nanocapsules for controlled release of acetamiprid pesticide was studied by Kumar and others [110]. According to the authors, the nanoformulation was superior to commercial formulation in terms of controlled release. Therefore, this formulation can be helpful in the reduction of consumption of pesticide. Thus, there are dual benefits for humans and the environment; first, non-toxic, biodegradable and inexpensive polymer for encapsulation and other is encapsulation of hazardous pesticide.

Rivera and others [111] presented the development of biodegradable nanocapsules as carriers of two bioactive compounds, 5-aminosalycilic acid and glycomacropeptide. Nanocapsules were produced through layer-by-layer deposition of chitosan and alginate layers on polystyrene nanoparticles. The bioactive compounds were incorporated on the third layer of the nanocapsules evaluating its encapsulation efficiency and release behavior. The results demonstrated that the synthesized nanocapsules presented spherical shape and a good capacity to encapsulate different bioactive compounds, with the best results obtained for the system containing 5-aminosalycilic acid. The multilayer nanocapsules could be a promising carrier of bioactive compounds for applications in food and pharmaceutical industries.

Chitosan Scaffolds

The biopolymer chitosan can be prepared in the form of different scaffolds to be applied in many fields. In this sense, chitosan can be pure, but it is generally mixed with other compounds to attain the desired properties and applications. For example, the chitosan based scaffolds can be biocompatible, biodegradable, antibacterial, osteoconductive, osteoinductive, osteoproductive, optical, electrical, bioactive and photoactive. These proprieties make the scaffold applicable in many fields, like tissue engineering, biosensors, wound healing, drug and gene delivery, electronics, photocatalysis, adsorption, chromatography and others [63, 112, 113]. Several methods are used to prepare chitosan based scaffolds, including high speed homogenization followed by lyophilization or freeze drying, supercritical CO_2 foaming, unidirectional freeze drying, salt-leaching and solvent casting-particulate leaching. The quality of chitosan based scaffolds is evaluated by several analytical techniques, in the same way as chitosan films, nanocapsules or nanoparticles [114, 115]. There are also specific characteristics, which should be evaluated, depending on the application. For example, if the scaffold is used for tissue/bone regeneration or wound healing, then *in vivo* and *in vitro* tests are required to verify the effectiveness [110]. If the scaffold is used as an adsorbent, the surface area and adsorption capacity should be evaluated [63]. Fig. (7) shows the images of chitosan based scaffolds, which were used for different applications.

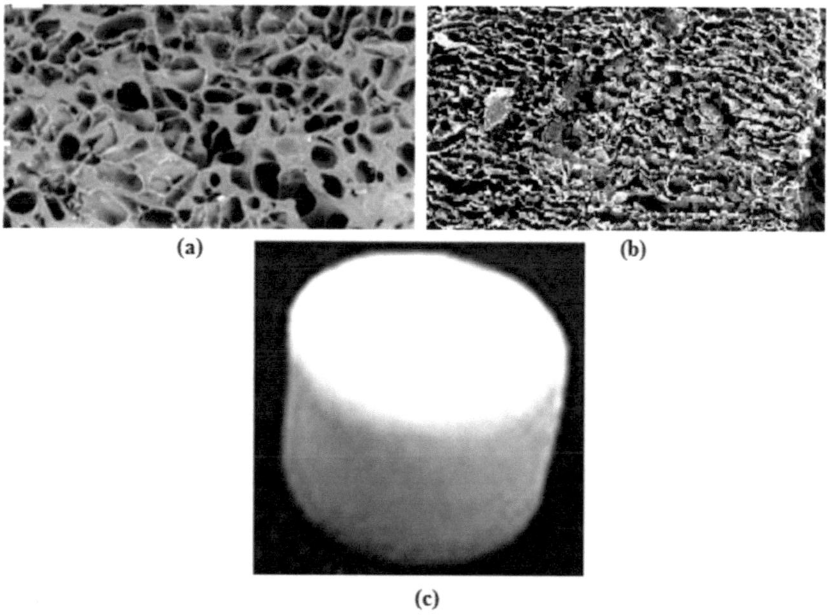

Fig. (7). Chitosan based scaffolds used for (**a**) skin reconstruction [114], (**b**) dye removal [63] and (**c**) biomedical applications [112].

Mahmoud and Salama [114] prepared norfloxacin-loaded collagen/chitosan scaffolds for skin reconstruction and evaluated the *in vivo* wound healing assessment. Regarding the scaffold characteristics, the authors verified that all investigated scaffolds displayed a buff color with round tablet-like shape and spongy consistency. The cross-sectional morphology of lyophilized scaffold revealed a highly interconnected porous sponge-like structure. The porosity was higher than 90% and the maximum swelling index was 306%. The *in vivo* tests revealed that the scaffolds enhanced the rate of wound healing.

Esquerdo and co-workers [63] developed a chitosan scaffold to be used as an alternative adsorbent for the removal of hazardous food dyes from aqueous solutions. The chitosan scaffold presented pore sizes from 50 to 200 μm, porosity of 92.2% and specific surface area of 1135 m^2/g. The two-step Langmuir model was suitable to represent the adsorption equilibrium data. The adsorption was a spontaneous, favorable, exothermic and enthalpy-controlled process. Electrostatic interactions occurred between the chitosan scaffold and the dyes. The chitosan megaporous scaffold showed good structural characteristics and high adsorption capacities, ranging from 788 to 3316 mg/g.

Tylingo and others [112], investigated the preparation and characterization of porous scaffolds from chitosan-collagen-gelatin composite. They concluded that the use of biopolymer mixtures of fish collagen, fish gelatin and chitosan had a significant impact on the rheological properties of obtaining scaffold materials. It was demonstrated that the physicochemical properties of the biopolymer mixtures were affected by many factors and depending on the origin of the biopolymers, it is possible to create materials with specific properties. The authors indicated that future works should be conducted to assess the suitability of the obtained preparations as a wound dressing material, attractive for biomedical uses, notably in tissue engineering or as a specific active compounds or drug delivery system.

Chitosan Hydrogels

Hydrogels can be described as a three dimensional network of polymers made of natural or synthetic materials possessing a high degree of flexibility due to large water content. They are able to retain a large amount of water and are characterized by a soft rubbery consistency similar to living tissues, making them an ideal substance for a variety of applications [116]. Chitosan can be processed in the form of hydrogels for several applications. The chitosan hydrogels can be prepared using only chitosan, but in this case, the mechanical properties are fragile. Therefore, a combination of chitosan with other materials is normally used to prepare hydrogels [117, 118]. Some methods used to prepare chitosan hydrogels include radiation crosslinking technology, freeze-thawing, dissolution/

lyophilization, dissolution/freeze drying and others [119]. The quality of hydrogels is addressed using SEM, FTIR, XRD, DSC/TGA in the same way as other chitosan based materials. Furthermore, specific characterizations are performed, like, swelling, elastic modulus, gel strength, mechanical properties among others depending on the application [120].

Cheng and others [120] developed thermo sensitive chitosan-based hydrogel as a topical ocular drug delivery system of latanoprost for glaucoma treatment. The hydrogel was characterized according its rheological properties, *in vitro* drug release, *in vivo* and *in vitro* biocompatibility. The results showed that latanoprost release from the developed hydrogel displayed a sustained-release profile. In a rabbit model of glaucoma, TA-induced elevated IOP was significantly decreased within 7 days and was maintained within a normal range by a weekly topical administration of 250 μg/ml of latanoprost-loaded hydrogel for the remainder of the study period. According to the authors, the hydrogel possesses promising properties that contribute to its sustained release of latanoprost. This newly developed chitosan based formulation may provide a non-invasive alternative to traditional eye drops for long-term IOP control in the near future.

Fan and others [121] prepared chitosan/gelatin/PVA hydrogel for wound dressings. Gelatin combined with polyvinyl alcohol (PVA) and chitosan (CS) were mixed to form a CS/gelatin/PVA hydrogel using γ-Ray irradiation as the free radical initiator. Several experimental conditions were varied and, the hydrogels were extensively characterized. The highest tensile strength reached 2.2 Mpa. All hydrogels presented a good coagulation effect, meaning that the hemostatic effect of hydrogels was optimal. The hydrogels also showed good pH-sensitivity, swelling ability and water evaporation rate. According to the authors, this hydrogel showed a promising potential to be applied as wound dressing.

Chemically Modified Chitosan

The structure of chitosan allows several chemical modifications to improve its ability in different fields. Since chitosan has a long polysaccharide chain, which is longer than most of the functional molecules introduced to chitosan, the modified chitosan derivatives are usually prepared by grafting [8]. According to Wang and co-workers [122], the chemical modifications on the chitosan structure can be classified as N-substitution, O-substitution, free radical graft copolymerization or other modification methods. The high reactivity of chitosan amino groups makes N-substitution a more convenient and promising method to modify chitosan. Since the reactivity of hydroxyl groups is lower than that of amino groups, most reagents for O-substitution of chitosan can react with the amino groups of chitosan as well [123]. The free radical graft copolymerization employs different initiators and later, different molecules can be inserted in the polymeric chain [124]. Each modification performed on the chitosan structure aims to improve a specific property and consequently a specific application. In general, these modifications should be proved by analytical techniques, including, Fourier transform infrared spectroscopy (FTIR), X-ray diffraction (XRD), nuclear magnetic resonance (NMR), thermal analyses (DSC/TGA) and X-ray dispersive spectroscopy (EDX) [122 - 124].

Chemically modified chitosan has been used for water treatment. Gonçalves and others [125] prepared cyanoguanidine-crosslinked chitosan for the adsorption of food dyes in an aqueous binary system. The FTIR confirmed the insertion of cyanoguanidine into the amino group of chitosan. It was concluded that the crosslinked chitosan was a promising adsorbent to remove dyes from aqueous binary systems. In other study, Kahu and co-workers [126] prepared an ionic solid (ethylhexadecyldimethylammoniumbromide) impregnated phosphated chitosan (ISPC) to improve the adsorption of hexavalent chromium from an industrial effluent. The authors concluded that the dual modification in terms of structural and functional modification in the amino group as well as the hydroxyl group of chitosan makes it a wonderful material for enhanced adsorption of Cr(VI). The ISPC adsorbent presented an adsorption capacity of 266.67 mg/g, which is much higher compared to most of the reported adsorbents.

Chemically modified chitosan is also applied in enzyme immobilization, drug release and odontology. Chitosan crosslinked with genipin was used as support matrix for β-d-galactosidase immobilization [127]. It was found that β--galactosidase immobilized on this support showed to have an activity higher than the activity of the biocatalyst prepared with glutaraldehyde. Chen and others [128] prepared calcium-crosslinked modified chitosan as potential matrix material for theophylline release. The drug loading content and encapsulation efficiency reached 53.40% and 69.61%, respectively. Song and others [129] added chitosan quaternary ammonium salt to the denture resin material in order to promote good oral health by improving the oral environment. The authors concluded that chitosan quaternary ammonium salt-modified resin denture base material has the potential to become a new generation oral denture composite material.

CONCLUDING REMARKS

Chitosan and chitosan-based materials present a series of technological applications, which have already been investigated and much more, that are possible. For successful research of chitosan based materials, it should be considered that all the applications are dependent on the properties of the biopolymer. The properties in turn depend on the production process, which depends on the raw material. In this way, from the raw material until the application, a series of well-defined steps should be used. Firstly, the raw material (shrimp, crab or others) should be carefully selected. Secondly, based on the characteristics of this raw material, the process should be defined in order to obtain a high quality chitosan. Then, chitosan can be modified by different ways, to acquire special properties to be applied in different fields. During all of these steps, conventional and advanced analytical techniques should be employed to prove the main findings.

CONFLICT OF INTEREST

The authors confirm that they have no conflict of interest to declare for this publication.

ACKNOWLEDGEMENTS

The authors would like to express their thanks to CAPES, CNPq, FAPERGS, Federal University of Santa Maria (UFSM) and Federal University of Rio Grande (FURG) for supporting this work.

REFERENCES

[1] Pontoni L, Fabbricino M. Use of chitosan and chitosan-derivatives to remove arsenic from aqueous solutions a mini review. Carbohydr Res 2012; 356: 86-92. [http://dx.doi.org/10.1016/j.carres.2012.03.042] [PMID: 22537862]

[2] Kurita K, Sannan T, Iwakura Y. Studies on chitin 4: Evidence for formation of block and random copolymers of N-Acetyl-D-glucosamine and D-glucosamine by hetero and homogeneous hydrolyses. Makromol Chem 1977; 178: 3197-202. [http://dx.doi.org/10.1002/macp.1977.021781203]

[3] Shukla SK, Mishra AK, Arotiba OA, Mamba BB. Chitosan-based nanomaterials: a state-of-the-art review. Int J Biol Macromol 2013; 59: 46-58. [http://dx.doi.org/10.1016/j.ijbiomac.2013.04.043] [PMID: 23608103]

[4] Hirano S. Production and application of chitin and chitosan in Japan. In: Skjåk-Bræk G, Anthonsen T, Stanford P, Eds. Chitin and Chitosan: Sources, chemistry, biochemistry, physical properties and applications. New York: Elsevier Applied Science 1989; pp. 37-44

[5] Campana-Filho SP, Britto D, Curti E, *et al.* Extraction, structures and properties of α- and β-chitin. Quim Nova 2007; 30: 644-50. [http://dx.doi.org/10.1590/S0100-40422007000300026]

[6] Ravi Kumar MN. A review of chitin and chitosan applications. React Funct Polym 2000; 46: 1-27. [http://dx.doi.org/10.1016/S1381-5148(00)00038-9]

[7] Azevedo VV, Chaves SA, Bezerra DC, *et al.* Chitin and chitosan: Applications as biomaterials. Rev Elet Mat Proc 2007; 2: 27-34.

[8] Pillai CK, Paul W, Sharma CP. Chitin and chitosan polymers: Chemistry, solubility and fiber formation. Prog Polym Sci 2009; 34: 641-78. [http://dx.doi.org/10.1016/j.progpolymsci.2009.04.001]

[9] Harish Prashanth KV, Tharanathan RN. Chitin/chitosan: modifications and their unlimited application potential: an overview. Trends Food Sci Technol 2007; 18: 117-31.

[http://dx.doi.org/10.1016/j.tifs.2006.10.022]

[10] Dotto GL, Souza VC, Moura JM, *et al.* Influence of drying techniques on the characteristics of chitosan and the quality of biopolymer films. Drying Technol 2011; 29: 1784-91. [http://dx.doi.org/10.1080/07373937.2011.602812]

[11] Rinaudo M. Chitin and chitosan: properties and applications. Prog Polym Sci 2006; 31: 603-32. [http://dx.doi.org/10.1016/j.progpolymsci.2006.06.001]

[12] Dotto GL, Souza VC, Pinto LA. Drying of chitosan in a spouted bed: The influences of temperature and equipment geometry in powder quality. LWT -. Food Sci Technol (Campinas) 2011; 44: 1786-92.

[13] Elsabee MZ, Abdou ES. Chitosan based edible films and coatings: a review. Mater Sci Eng C Mater Biol Appl 2013; 33(4): 1819-41. [http://dx.doi.org/10.1016/j.msec.2013.01.010] [PMID: 23498203]

[14] Mendes AA, Oliveira PC, Castro HF, Giordano RL. Application of chitosan as support for immobilization of enzymes of industrial interest. Quim Nova 2011; 34: 831-40.

[15] Shahidi F, Arachchi JK, Jeon YJ. Food applications of chitin and chitosans. Trends Food Sci Technol 1999; 10: 37-51. [http://dx.doi.org/10.1016/S0924-2244(99)00017-5]

[16] Nouri M, Khodaiyan F, Razavi SH, Mousavi M. Improvement of chitosan production from Persian Gulf shrimp waste by response surface methodology. Food Hydro 2015; proof: 1-9.

[17] Yen MT, Yang JH, Mau JL. Physicochemical characterization of chitin and chitosan from crab shells. Carbohydr Polym 2009; 75: 15-21. [http://dx.doi.org/10.1016/j.carbpol.2008.06.006]

[18] Kaya M, Baran T, Erdoğan S, Menteş A, Özüsağlam MA, Çakmak YS. Physicochemical comparison of chitin and chitosan obtained from larvae and adult Colorado potato beetle (*Leptinotarsa decemlineata*). Mater Sci Eng C Mater Biol Appl 2014; 45: 72-81. [http://dx.doi.org/10.1016/j.msec.2014.09.004] [PMID: 25491803]

[19] Liu S, Sun J, Yu L, *et al.* Extraction and characterization of chitin from the beetle *Holotrichia parallela* Motschulsky. Molecules 2012; 17(4): 4604-11. [http://dx.doi.org/10.3390/molecules17044604] [PMID: 22510609]

[20] Song C, Yu H, Zhang M, Yang Y, Zhang G. Physicochemical properties and antioxidant activity of chitosan from the blowfly *Chrysomya megacephala* larvae. Int J Biol Macromol 2013; 60: 347-54. [http://dx.doi.org/10.1016/j.ijbiomac.2013.05.039] [PMID: 23792633]

[21] Paulino AT, Simionato JI, Garcia JC, Nozaki J. Characterization of chitosan and chitin produced from silkworm crysalides. Carbohydr Polym 2006; 64: 98-103. [http://dx.doi.org/10.1016/j.carbpol.2005.10.032]

[22] Ai H, Wang F, Yang Q, *et al.* Preparation and biological activities of chitosan from the larvae of housefly, *Musca domestica.* Carbohydr Polym 2008; 72: 419-23. [http://dx.doi.org/10.1016/j.carbpol.2007.09.010]

[23] Ramasamy P, Subhapradha N, Shanmugam V, Shanmugam A. Extraction, characterization and antioxidant property of chitosan from cuttlebone *Sepia kobiensis* (Hoyle 1885). Int J Biol Macromol 2014; 64: 202-12. [http://dx.doi.org/10.1016/j.ijbiomac.2013.12.008] [PMID: 24333227]

[24] Chandumpai A, Singhpibulporn N, Faroongsarng D, Sornprasit P. Preparation and physico-chemical characterization of chitin and chitosan from the pens of the squid species, *Loligo lessoniana* and *Loligo formosana.* Carbohydr Polym 2004; 58: 467-74. [http://dx.doi.org/10.1016/j.carbpol.2004.08.015]

[25] Al Sagheer FA, Al Sughayer MA, Muslim S, Elsabee MZ. Extraction and characterization of chitin and chitosan from marine sources in Arabian Gulf. Carbohydr Polym 2009; 77: 410-9.

[http://dx.doi.org/10.1016/j.carbpol.2009.01.032]

[26] Muzzarelli RA, Boudrant J, Meyer D, *et al.* Current views on fungal chitin/chitosan, human chitinases, food preservation, glucans, pectins and inulin: A tribute to Henri Braconnot, precursor of the carbohydrate polymers science, on the chitin bicentennial. Carbohydr Polym 2012; 87: 995-1012. [http://dx.doi.org/10.1016/j.carbpol.2011.09.063]

[27] Vendruscolo F, Ninow JL. Apple pomace as a substrate for fungal chitosan production in an air lift bioreactor. Biocat Agric Biotech 2014; 3: 338-42.

[28] Tayel AA, Ibrahim SI, Al-Saman MA, Moussa SH. Production of fungal chitosan from date wastes and its application as a biopreservative for minced meat. Int J Biol Macromol 2014; 69: 471-5. [http://dx.doi.org/10.1016/j.ijbiomac.2014.05.072] [PMID: 24942991]

[29] Dhillon GS, Kaur S, Sarma SJ, Brar SK. Integrated process for fungal citric acid fermentation using apple processing wastes and sequential extraction of chitosan from waste stream. Ind Crops Prod 2013; 50: 346-51.
[http://dx.doi.org/10.1016/j.indcrop.2013.08.010]

[30] Dotto GL, Rosa GS, Moraes MA, Weska RF, Pinto LA. Treatment of chitin effluents by coagulation-flocculation with chitin and aluminum sulfate. J Environ Chem Eng 2013; 1: 50-5. [http://dx.doi.org/10.1016/j.jece.2013.03.006]

[31] Weska RF, Moura JM, Batista LM, Rizzi J, Pinto LA. Optimization of deacetylation in the production of chitosan from shrimp wastes: Use of response surface methodology. J Food Eng 2007; 80: 749-53. [http://dx.doi.org/10.1016/j.jfoodeng.2006.02.006]

[32] Piccin JS, Vieira ML, Gonçalves JO, Dotto GL, Pinto LA. Adsorption of FD&C Red No. 40 by chitosan: Isotherms analysis. J Food Eng 2009; 95: 16-20. [http://dx.doi.org/10.1016/j.jfoodeng.2009.03.017]

[33] Moura CM, Moura JM, Soares NM, Pinto LA. Evaluation of molar weight and deacetylation degree of chitosan during chitin deacetylation reaction: Used to produce biofilm. Chem Eng Prog 2011; 50: 351-5.
[http://dx.doi.org/10.1016/j.cep.2011.03.003]

[34] No HK, Meyers SP. Preparation of chitin and chitosan. In: Muzzarelli RAA, Peter MG, Eds. Chitin Handbook. European chitin Society 1997; pp. 475-789.

[35] Synowiecki J, Al-Khateeb NA. Production, properties, and some new applications of chitin and its derivatives. Crit Rev Food Sci Nutr 2003; 43(2): 145-71. [http://dx.doi.org/10.1080/10408690390826473] [PMID: 12705640]

[36] Synowiecki J, Sikorski ZE, Naczk M. The activity of immobilized enzymes on different krill chitin preparations. Biotechnol Bioeng 2004; 23: 2211-5. [http://dx.doi.org/10.1002/bit.260231005]

[37] Gamage A, Shahidi F. Use of chitosan for the removal of metal ion contaminants and proteins from water. Food Chem 2007; 104: 989-96. [http://dx.doi.org/10.1016/j.foodchem.2007.01.004]

[38] Galed G, Miralles B, Paños I, Santiago A, Heras A. N-Deacetylation and depolymerization reactions of chitin/chitosan: Influence of the source of chitin. Carbohydr Polym 2005; 62: 316-20. [http://dx.doi.org/10.1016/j.carbpol.2005.03.019]

[39] Synowiecki J, Al-Khateeb N. The recovery of protein hydrolysate during enzymatic isolation of chitin from shrimp *Cragon cragon* processing discards. Food Chem 2000; 68: 147-52. [http://dx.doi.org/10.1016/S0308-8146(99)00165-X]

[40] Younes I, Ghorbel-Bellaaj O, Chaabouni M, *et al.* Use of a fractional factorial design to study the effects of experimental factors on the chitin deacetylation. Int J Biol Macromol 2014; 70: 385-90. [http://dx.doi.org/10.1016/j.ijbiomac.2014.06.051] [PMID: 24997447]

[41] Martinou A, Kafetzopoulos D, Bouriotis V. Chitin deacetylation by enzymatic means: monitoring of

deacetylation processes. Carbohydr Res 1995; 273: 235-42.
[http://dx.doi.org/10.1016/0008-6215(95)00111-6]

[42] Tan TS, Chin HY, Tsai ML, Liu CL. Structural alterations, pore generation, and deacetylation of α- and β-chitin submitted to steam explosion. Carbohydr Polym 2015; 122: 321-8.
[http://dx.doi.org/10.1016/j.carbpol.2015.01.016] [PMID: 25817675]

[43] Lertwattanaseri T, Ichikawa N, Mizoguchi T, Tanaka Y, Chirachanchai S. Microwave technique for efficient deacetylation of chitin nanowhiskers to a chitosan nanoscaffold. Carbohydr Res 2009; 344(3): 331-5.
[http://dx.doi.org/10.1016/j.carres.2008.10.018] [PMID: 19111284]

[44] Delezuk JA, Cardozo MB, Domard A, Campana-Filho SP. Ultrasound-assisted deacetylation of beta□chitin: influence of processing parameters. Polym Int 2011; 60: 903-9.
[http://dx.doi.org/10.1002/pi.3037]

[45] Batista LM, Rosa CA, Pinto LA. Diffusive model with variable effective diffusivity considering shrinkage in thin layer drying of chitosan. J Food Eng 2007; 81: 127-32.
[http://dx.doi.org/10.1016/j.jfoodeng.2006.10.014]

[46] Muzzarelli C, Stanic V, Gobbic L, Tosib G, Muzzarelli RA. Spray drying of solutions containing chitosan together with polyuronans and characterisation of the microspheres. Carbohydr Polym 2004; 57: 73-82.
[http://dx.doi.org/10.1016/j.carbpol.2004.04.002]

[47] Youn DK, No HK, Prinyawiwatkul W. Physicochemical and functional properties of chitosans affected by sun drying time during decoloration. LWT-Food Sci Technol 2009; 42: 1553-6.
[http://dx.doi.org/10.1016/j.lwt.2009.05.004]

[48] Srinivasa PC, Ramesh MN, Kumar KR, Tharanathan RN. Properties of chitosan films prepared under different drying conditions. J Food Eng 2004; 63: 79-85.
[http://dx.doi.org/10.1016/S0260-8774(03)00285-1]

[49] García-Bermejo AB, Cardelle-Cobas A, Ruiz-Matute AI, Montañés F, Olano A, Corzo N. Effect of drying methods on the reactivity of chitosan towards Maillard reaction. Food Hydro 2012; 29: 27-34.
[http://dx.doi.org/10.1016/j.foodhyd.2012.01.014]

[50] Thakhiew W, Devahastin S, Soponronnarit S. Physical and mechanical properties of chitosan films as affected by drying methods and addition of antimicrobial agent. J Food Eng 2013; 119: 140-9.
[http://dx.doi.org/10.1016/j.jfoodeng.2013.05.020]

[51] Crini G, Badot PM. Application of chitosan, a natural aminopolysaccharide, for dye removal from aqueous solutions by adsorption processes using batch studies: a review of recent literature. Prog Polym Sci 2008; 33: 399-447.
[http://dx.doi.org/10.1016/j.progpolymsci.2007.11.001]

[52] Dotto GL, Vieira ML, Pinto LA. Use of chitosan solutions for the microbiological shelf life extension of papaya fruits during storage at room temperature. LWT-Food Sci Technol 2015; 64: 126-30.
[http://dx.doi.org/10.1016/j.lwt.2015.05.042]

[53] Fernández-Pan I, Maté JI, Gardrat C, Coma V. Effect of chitosan molecular weight on the antimicrobial activity and release rate of carvacrol-enriched films. Food Hydro 2015; 51: 60-8.
[http://dx.doi.org/10.1016/j.foodhyd.2015.04.033]

[54] Bekale L, Agudelo D, Tajmir-Riahi HA. Effect of polymer molecular weight on chitosan-protein interaction. Colloids Surf B Biointerfaces 2015; 125: 309-17.
[http://dx.doi.org/10.1016/j.colsurfb.2014.11.037] [PMID: 25524222]

[55] Bof MJ, Bordagaray VC, Locaso DE, García MA. Chitosan molecular weight effect on starch-composite film properties. Food Hydro 2015; 51: 281-94.
[http://dx.doi.org/10.1016/j.foodhyd.2015.05.018]

[56] Kasaai MR. Determination of the degree of N-acetylation for chitin and chitosan by various NMR

spectroscopy techniques: A review. Carbohydr Polym 2010; 79: 801-10. [http://dx.doi.org/10.1016/j.carbpol.2009.10.051]

[57] Kumirska J, Czerwicka M, Kaczyński Z, *et al.* Application of spectroscopic methods for structural analysis of chitin and chitosan. Mar Drugs 2010; 8(5): 1567-636. [http://dx.doi.org/10.3390/md8051567] [PMID: 20559489]

[58] Gonçalves JO, Duarte DA, Dotto GL, Pinto LA. Use of chitosan with different deacetylation degrees for the adsorption of food dyes in a binary system. Clean-Soil, Air, Water 2014; 42: 767-74.

[59] Liu Z, Ge X, Lu Y, Dong S, Zhao Y, Zeng M. Effects of chitosan molecular weight and degree of deacetylation on the properties of gelatine-based films. Food Hydro 2012; 26: 311-7. [http://dx.doi.org/10.1016/j.foodhyd.2011.06.008]

[60] Li X, Xia W. Effects of concentration, degree of deacetylation and molecular weight on emulsifying properties of chitosan. Int J Biol Macromol 2011; 48(5): 768-72. [http://dx.doi.org/10.1016/j.ijbiomac.2011.02.016] [PMID: 21382402]

[61] Dotto GL, Vieira ML, Pinto LA. Kinetics and mechanism of tartrazine adsorption onto chitin and chitosan. Ind Eng Chem Eng 2012; 51: 6862-8. [http://dx.doi.org/10.1021/ie2030757]

[62] Piccin JS, Dotto GL, Vieira ML, Pinto LA. Kinetics and mechanism of the food dye FD&C red 40 adsorption onto chitosan. J Chem Eng Data 2011; 56: 3759-65. [http://dx.doi.org/10.1021/je200388s]

[63] Esquerdo VM, Cadaval TR Jr, Dotto GL, Pinto LA. Chitosan scaffold as an alternative adsorbent for the removal of hazardous food dyes from aqueous solutions. J Colloid Interface Sci 2014; 424: 7-15. [http://dx.doi.org/10.1016/j.jcis.2014.02.028] [PMID: 24767491]

[64] Mirmohseni A, Seyed Dorraji MS, Figoli A, Tasselli F. Chitosan hollow fibers as effective biosorbent toward dye: preparation and modeling. Bioresour Technol 2012; 121: 212-20. [http://dx.doi.org/10.1016/j.biortech.2012.06.067] [PMID: 22858488]

[65] Dotto GL, Pinto LA. Adsorption of food dyes onto chitosan: Optimization process and kinetic. Carbohydr Polym 2011; 84: 231-8. [http://dx.doi.org/10.1016/j.carbpol.2010.11.028]

[66] Moura JM, Farias BS, Rodrigues DA, Moura CM, Dotto GL, Pinto LA. Preparation of chitosan with different characteristics and its application for biofilms production. J Polym Environ 2015; 23: 470-7. [http://dx.doi.org/10.1007/s10924-015-0730-y]

[67] Elgadir MA, Uddin MS, Ferdosh S, Adam A, Chowdhury AJ, Sarker MZ. Impact of chitosan composites and chitosan nanoparticle composites on various drug delivery systems: A review. J Food Drug Anal 2015; 23: 619-29. [http://dx.doi.org/10.1016/j.jfda.2014.10.008]

[68] Thakur VK, Thakur MK. Recent advances in graft copolymerization and applications of chitosan: A review. ACS Sustain Chem & Eng 2014; 2: 2637-52. [http://dx.doi.org/10.1021/sc500634p]

[69] Balan V, Verestiuc L. Strategies to improve chitosan hemocompatibility: A review. Eur Polym J 2014; 53: 171-88. [http://dx.doi.org/10.1016/j.eurpolymj.2014.01.033]

[70] Chen Y, Zhou Y, Yang S, *et al.* Novel bone substitute composed of chitosan and strontium-doped α-calcium sulfate hemihydrate: Fabrication, characterisation and evaluation of biocompatibility. Mater Sci Eng C Mater Biol Appl 2016; 66: 84-91. [http://dx.doi.org/10.1016/j.msec.2016.04.070] [PMID: 27207041]

[71] Sarhan WA, Azzazy HM. High concentration honey chitosan electrospun nanofibers: biocompatibility and antibacterial effects. Carbohydr Polym 2015; 122: 135-43. [http://dx.doi.org/10.1016/j.carbpol.2014.12.051] [PMID: 25817652]

[72] Ferrero F, Periolatto M. Ultraviolet curing for surface modification of textile fabrics. J Nanosci Nanotechnol 2011; 11(10): 8663-9.
[http://dx.doi.org/10.1166/jnn.2011.3447] [PMID: 22400240]

[73] Ferrero F, Periolatto M. Antimicrobial finish of textiles by chitosan UV-curing. J Nanosci Nanotechnol 2012; 12(6): 4803-10.
[http://dx.doi.org/10.1166/jnn.2012.4902] [PMID: 22905533]

[74] Ferrero F, Periolatto M, Burelli S, Carletto RA. Silk grafting with chitosan and crosslinking agents. Fiber Polym 2010; 11: 185-92.
[http://dx.doi.org/10.1007/s12221-010-0185-7]

[75] Ferrero F, Periolatto M, Ferrario S. Sustainable antimicrobial finishing of cotton fabrics by chitosan UV-grafting: from laboratory experiments to semi industrial scale-up. J Clean Prod 2015; 96: 244-52.
[http://dx.doi.org/10.1016/j.jclepro.2013.12.044]

[76] Periolatto M, Ferrero F, Vineis C. Antimicrobial chitosan finish of cotton and silk fabrics by UV-curing with 2-hydroxy-2-methylphenylpropane-1-one. Carbohydr Polym 2012; 88: 201-5.
[http://dx.doi.org/10.1016/j.carbpol.2011.11.093]

[77] Periolatto M, Ferrero F, Vineis C, Rombaldoni F. Multifunctional finishing of wool fabrics by chitosan UV-grafting: an approach. Carbohydr Polym 2013; 98(1): 624-9.
[http://dx.doi.org/10.1016/j.carbpol.2013.06.054] [PMID: 23987391]

[78] Ferrero F, Periolatto M, Vineis C, Varesano A. Chitosan coated cotton gauze for antibacterial water filtration. Carbohydr Polym 2014; 103: 207-12.
[http://dx.doi.org/10.1016/j.carbpol.2013.12.037] [PMID: 24528721]

[79] Periolatto M, Ferrero F. Cotton filter fabrics functionalization by chitosan UV-grafting for removal of dyes. Chem Eng Trans 2013; 32: 85-90.

[80] Ferrero F, Tonetti C, Periolatto M. Adsorption of chromate and cupric ions onto chitosan-coated cotton gauze. Carbohydr Polym 2014; 110: 367-73.
[http://dx.doi.org/10.1016/j.carbpol.2014.04.016] [PMID: 24906768]

[81] Dotto GL, Moura JM, Cadaval TR Jr, Pinto LA. Application of chitosan films for the removal of food dyes from aqueous solutions by adsorption. Chem Eng J 2013; 214: 8-16.
[http://dx.doi.org/10.1016/j.cej.2012.10.027]

[82] Dotto GL, Buriol C, Pinto LA. Diffusional mass transfer model for the adsorption of food dyes on chitosan films. Chem Eng Res Des 2014; 92: 2324-32.
[http://dx.doi.org/10.1016/j.cherd.2014.03.013]

[83] Rêgo TV, Cadaval TR Jr, Dotto GL, Pinto LA. Statistical optimization, interaction analysis and desorption studies for the azo dyes adsorption onto chitosan films. J Colloid Interface Sci 2013; 411: 27-33.
[http://dx.doi.org/10.1016/j.jcis.2013.08.051] [PMID: 24112836]

[84] Cadaval TR Jr, Dotto GL, Seus ER, Mirlean N, Pinto LA. Vanadium removal from aqueous solutions by onto chitosan films. Des Water Treat 2016; 57: 16583-91.
[http://dx.doi.org/10.1080/19443994.2015.1079741]

[85] Djioua T, Charles F, Freire M, Filgueiras H, Collin MN, Sallanon H. Combined effects of postharvest heat treatment and chitosan coating on quality of fresh-cut mangoes *(Mangifera indica L.)*. Int J Food Sci Technol 2010; 45: 849-55.
[http://dx.doi.org/10.1111/j.1365-2621.2010.02209.x]

[86] Zhan L, Hu J, Zhu Z. Shelf life extension of minimally processed water caltrop (*Trapa acornis Nakano*) fruits coated with chitosan. Int J Food Sci Technol 2011; 46: 2634-40.
[http://dx.doi.org/10.1111/j.1365-2621.2011.02794.x]

[87] Peng Y, Li Y, Yin L. Effects of edible coatings on some quality parameters of Chinese water chestnut

(*Eleocharis tuberosa*) during storage. Int J Food Sci Technol 2013; 48: 1404-9. [http://dx.doi.org/10.1111/ijfs.12102]

[88] Pujols KD, Osorio L, Carrillo EP, Wardy W, Torrico DD, No HK. Comparing effects of α *vs.* β chitosan coating and emulsion coatings on egg quality during room temperature storage. Int J Food Sci Technol 2014; 49: 1383-90. [http://dx.doi.org/10.1111/ijfs.12468]

[89] D'Amato D, Sinigaglia M, Corbo MR. Use of chitosan, honey and pineapple juice as filling liquids for increasing the microbiological shelf life of a fruit-based salad. Int J Food Sci Technol 2010; 45: 1033-41. [http://dx.doi.org/10.1111/j.1365-2621.2010.02233.x]

[90] Ali A, Muhammad MT, Sijam K, Siddiqui Y. Potential of chitosan coating in delaying the postharvest anthracnose (*Colletotrichum gloeosporioides Penz.*) of Eksotika II papaya. Int J Food Sci Technol 2010; 45: 2134-40. [http://dx.doi.org/10.1111/j.1365-2621.2010.02389.x]

[91] Cadaval TRS Jr, Câmara AS, Dotto GL, Pinto LAA. Adsorption of Cr (VI) by chitosan with different deacetylation degrees. 2013; 51: 7690-99.

[92] Elsabee MZ, Naguib HF, Morsi RE. Chitosan based nanofibers. Mater Sci Eng C 2012; 32: 1711-26. [review]. [http://dx.doi.org/10.1016/j.msec.2012.05.009]

[93] Zhao LM, Shi LE, Zhang ZL, *et al.* Preparation and application of chitosan nanoparticles and nanofibers. Braz J Chem Eng 2011; 28: 353-62. [http://dx.doi.org/10.1590/S0104-66322011000300001]

[94] Sarkar K, Gomez C, Zambrano S, *et al.* Electrospinning to Forcespinning™. Mater Today 2010; 13: 12-4. [http://dx.doi.org/10.1016/S1369-7021(10)70199-1]

[95] Miladi K, Sfar S, Fessi H, Elaissari A. Enhancement of alendronate encapsulation in chitosan nanoparticles. J Drug Deliv Sci Technol 2015; 30: 391-6. [http://dx.doi.org/10.1016/j.jddst.2015.04.007]

[96] Maryšková M, Ardao I, García-González CA, Martinová L, Rotková J, Ševců A. Polyamide 6/chitosan nanofibers as support for the immobilization of *Trametes versicolor* laccase for the elimination of endocrine disrupting chemicals. Enzyme Microb Technol 2016; 89: 31-8. [http://dx.doi.org/10.1016/j.enzmictec.2016.03.001] [PMID: 27233125]

[97] Razzaz A, Ghorban S, Hosayni L, Irani M, Aliabadi M. Chitosan nanofibers functionalized by TiO^2 nanoparticles for the removal of heavy metal ions. J Taiwan Inst Chem Eng 2016; 58: 333-43. [http://dx.doi.org/10.1016/j.jtice.2015.06.003]

[98] Zhao R, Li X, Sun B, *et al.* Electrospun chitosan/sericin composite nanofibers with antibacterial property as potential wound dressings. Int J Biol Macromol 2014; 68: 92-7. [http://dx.doi.org/10.1016/j.ijbiomac.2014.04.029] [PMID: 24769088]

[99] Toskas G, Cherif C, Hund RD, *et al.* Chitosan(PEO)/silica hybrid nanofibers as a potential biomaterial for bone regeneration. Carbohydr Polym 2013; 94(2): 713-22. [http://dx.doi.org/10.1016/j.carbpol.2013.01.068] [PMID: 23544625]

[100] O'Callaghan KA, Kerry JP. Preparation of low and medium molecular weight chitosan nanoparticles and their antimicrobial evaluation against a panel of microorganisms, including cheese-derived cultures. Food Contr 2016; 69: 256-61. [http://dx.doi.org/10.1016/j.foodcont.2016.05.005]

[101] Wang K, Fan X, Zhang X, Zhang X, Chen Y, Wei Y. Red fluorescent chitosan nanoparticles grafted with poly(2-methacryloyloxyethyl phosphorylcholine) for live cell imaging. Colloids Surf B Biointerfaces 2016; 144: 188-95.

[http://dx.doi.org/10.1016/j.colsurfb.2016.04.009] [PMID: 27088188]

[102] Feyzioglu GC, Tornuk F. Development of chitosan nanoparticles loaded with summer savory (*Satureja hortensis L.*) essential oil for antimicrobial and antioxidant delivery applications. LWT-Food Sci Technol 2016; 70: 104-10. [http://dx.doi.org/10.1016/j.lwt.2016.02.037]

[103] Tanhaei B, Ayati A, Lahtinen M, Sillanpaa M. Preparation and characterization of a novel chitosan/Al_2O_3/magnetite nanoparticles composite adsorbent for kinetic, thermodynamic and isotherm studies of Methyl Orange adsorption. Chem Eng J 2015; 259: 1-10. [http://dx.doi.org/10.1016/j.cej.2014.07.109]

[104] Shaker MA. Adsorption of Co(II), Ni(II) and Cu(II) ions onto chitosan modified poly(methacrylate) nanoparticles: Dynamics, equilibrium and thermodynamics studies. J Taiwan Inst Chem Eng 2015; 57: 111-22. [http://dx.doi.org/10.1016/j.jtice.2015.05.027]

[105] Zhou Z, Lin S, Yue T, Lee TC. Adsorption of food dyes from aqueous solution by glutaraldehyde cross-linked magnetic chitosan nanoparticles. J Food Eng 2014; 126: 133-41. [http://dx.doi.org/10.1016/j.jfoodeng.2013.11.014]

[106] Esquerdo VM, Dotto GL, Pinto LA. Nanoemulsions containing unsaturated fatty acid concentrates. In: Grumezescu AM, Ed. Emulsions. Amsterdam: Elsevier 2016; pp. 71-106.

[107] McClements DJ. Critical review of techniques and methodologies for characterization of emulsion stability. Crit Rev Food Sci Nutr 2007; 47(7): 611-49. [http://dx.doi.org/10.1080/10408390701289292] [PMID: 17943495]

[108] Comuniain TA, Favaro-Trindade CS. Microencapsulation using biopolymers as an alternative to produce food enhanced with phytosterols and omega-3 fatty acids: A review. Food Hydrocoll 2016; 61: 442-57. [http://dx.doi.org/10.1016/j.foodhyd.2016.06.003]

[109] Esquerdo VM, Dotto GL, Pinto LA. Preparation of nanoemulsions containing unsaturated fatty acid concentrate-chitosan capsules. J Colloid Interface Sci 2015; 445: 137-42. [http://dx.doi.org/10.1016/j.jcis.2014.12.094] [PMID: 25617613]

[110] Kumar S, Chauhan N, Gopal M, Kumar R, Dilbaghi N. Development and evaluation of alginate-chitosan nanocapsules for controlled release of acetamiprid. Int J Biol Macromol 2015; 81: 631-7. [http://dx.doi.org/10.1016/j.ijbiomac.2015.08.062] [PMID: 26321424]

[111] Rivera MC, Pinheiro AC, Bourbon AI, Cerqueira MA, Vicente AA. Hollow chitosan/alginate nanocapsules for bioactive compound delivery. Int J Biol Macromol 2015; 79: 95-102. [http://dx.doi.org/10.1016/j.ijbiomac.2015.03.003] [PMID: 25907011]

[112] Tylingo R, Gorczyca G, Mania S, Szweda P, Milewski S. Preparation and characterization of porous scaffolds from chitosan-collagen-gelatin composite. React Funct Polym 2016; 103: 131-40. [http://dx.doi.org/10.1016/j.reactfunctpolym.2016.04.008]

[113] Deepthi S, Venkatesan J, Kim SK, Bumgardner JD, Jayakumar R. An overview of chitin or chitosan/nano ceramic composite scaffolds for bone tissue engineering. Int J Biol Macromol 2016; 93(Pt B): 1338-53. [http://dx.doi.org/10.1016/j.ijbiomac.2016.03.041] [PMID: 27012892]

[114] Mahmoud AA, Salama AH. Norfloxacin-loaded collagen/chitosan scaffolds for skin reconstruction: Preparation, evaluation and in-vivo wound healing assessment. Eur J Pharm Sci 2016; 83: 155-65. [http://dx.doi.org/10.1016/j.ejps.2015.12.026] [PMID: 26733072]

[115] Fereshteh Z, Fathi M, Bagri A, Boccaccini AR. Preparation and characterization of aligned porous PCL/zein scaffolds as drug delivery systems *via* improved unidirectional freeze-drying method. Mater Sci Eng C Mater Biol Appl 2016; 68: 613-22. [http://dx.doi.org/10.1016/j.msec.2016.06.009] [PMID: 27524061]

[116] Ullah F, Othman MB, Javed F, Ahmad Z, Md Akil H. Classification, processing and application of hydrogels: A review. Mater Sci Eng C Mater Biol Appl 2015; 57: 414-33. [http://dx.doi.org/10.1016/j.msec.2015.07.053] [PMID: 26354282]

[117] Guilherme MR, Aouada FA, Fajardo AR, *et al.* Superabsorbent hydrogels based on polysaccharides for application in agriculture as soil conditioner and nutrient carrier: A review. Eur Polym J 2015; 72: 365-85. [http://dx.doi.org/10.1016/j.eurpolymj.2015.04.017]

[118] Jiang Y, Meng X, Wu Z, Qi X. Modified chitosan thermosensitive hydrogel enables sustained and efficient anti-tumor therapy *via* intratumoral injection. Carbohydr Polym 2016; 144: 245-53. [http://dx.doi.org/10.1016/j.carbpol.2016.02.059] [PMID: 27083815]

[119] Afshari MJ, Sheikh N, Afarideh H. PVA/CM-chitosan/honey hydrogels prepared by using the combined technique of irradiation followed by freeze-thawing. Radiat Phys Chem 2015; 113: 28-35. [http://dx.doi.org/10.1016/j.radphyschem.2015.04.023]

[120] Cheng YH, Tsai TH, Jhan YY, *et al.* Thermosensitive chitosan-based hydrogel as a topical ocular drug delivery system of latanoprost for glaucoma treatment. Carbohydr Polym 2016; 144: 390-9. [http://dx.doi.org/10.1016/j.carbpol.2016.02.080] [PMID: 27083831]

[121] Fan L, Yang H, Yang J, Peng M, Hu J. Preparation and characterization of chitosan/gelatin/PVA hydrogel for wound dressings. Carbohydr Polym 2016; 146: 427-34. [http://dx.doi.org/10.1016/j.carbpol.2016.03.002] [PMID: 27112893]

[122] Wang J, Wang L, Yu H, *et al.* Recent progress on synthesis, property and application of modified chitosan: An overview. Int J Biol Macromol 2016; 88: 333-44. [http://dx.doi.org/10.1016/j.ijbiomac.2016.04.002] [PMID: 27044349]

[123] Prabaharan M, Mano JF. Chitosan derivatives bearing cyclodextrin cavities as novel adsorbent matrices. Carbohydr Polym 2006; 63: 153-66. [http://dx.doi.org/10.1016/j.carbpol.2005.08.051]

[124] Zhang K, Zhuang P, Wang Z, *et al.* One-pot synthesis of chitosan-*g*-(PEO-PLLA-PEO) *via* click chemistry and SET-NRC reaction. Carbohydr Polym 2012; 90(4): 1515-21. [http://dx.doi.org/10.1016/j.carbpol.2012.07.023] [PMID: 22944410]

[125] Gonçalves JO, Dotto GL, Pinto LA. Cyanoguanidine-crosslinked chitosan to adsorption of food dyes in the aqueous binary system. J Mol Liq 2015; 211: 425-30. [http://dx.doi.org/10.1016/j.molliq.2015.07.061]

[126] Kahu SS, Shekhawat A, Saravanan D, Jugade RM. Two fold modified chitosan for enhanced adsorption of hexavalent chromium from simulated wastewater and industrial effluents. Carbohydr Polym 2016; 146: 264-73. [http://dx.doi.org/10.1016/j.carbpol.2016.03.041] [PMID: 27112874]

[127] Klein MP, Hackenhaar CR, Lorenzoni AS, *et al.* Chitosan crosslinked with genipin as support matrix for application in food process: Support characterization and β-D-galactosidase immobilization. Carbohydr Polym 2016; 137: 184-90. [http://dx.doi.org/10.1016/j.carbpol.2015.10.069] [PMID: 26686119]

[128] Chen C, Zhong M, Li G, *et al.* Optimization on preparation conditions of calcium-crosslinked modified chitosan as potential matrix material for theophylline sustained release beads and its evaluation of release kinetics. J Alloys Compd 2016; 658: 348-55. [http://dx.doi.org/10.1016/j.jallcom.2015.10.219]

[129] Song R, Zhong Z, Lin L. Evaluation of chitosan quaternary ammonium salt-modified resin denture base material. Int J Biol Macromol 2016; 85: 102-10. [http://dx.doi.org/10.1016/j.ijbiomac.2015.12.052] [PMID: 26718869]

CHAPTER 2

Obtention Processes and Main Characteristics

Anderson Fiamingo, Jorge Augusto de Moura Delezuk, Regiane Cristina Corrêa and **Sergio Paulo Campana-Filho***

Instituto de Química de São Carlos – Universidade de São Paulo, Av. Trabalhador são-carlense, 400 - CEP, 13566-590 São Carlos/SP, Brazil

Abstract: The occurrence of chitin and chitosan is discussed with emphasis on the main biomasses used as industrial raw material to produce the former polymer. Some potential sources which nowadays are not important for sustaining large scale production are highlighted as well. The main procedures used to extract alpha-chitin from crustaceans shells and beta-chitin from squid pens are presented and its advantages and drawbacks are discussed. The production of chitosan *via* *N*-deacetylation of chitin is addressed with emphasis on thermochemical processes based on the alkaline treatment of chitin. Some unconventional processes are focused as well, mainly the ultrasound-assisted deacetylation, which allows the efficient conversion of beta-chitin into extensively deacetylated chitosan of high molar weight. The main characteristics of chitin and chitosan, namely; the average degree of acetylation ($\overline{DA}$), the pattern of acetylation (PA), the average degree of polymerization ($\overline{DP}$) and dispersity (Đ) are addressed and its influence on some polymer properties are also discussed.

Keywords: Alpha-chitin, Beta-chitin, Crustaceans shells, Chitosan, Freeze - pump out - thaw deacetylation, FPT process, Squid pens, Sponges, Thermochemical deacetylation, Ultrasound-assisted deacetylation, USAD process.

INTRODUCTION

The literature on chitin and chitosan in the latter three decades is abundant and mainly focused on applications of these polymers and their derivatives as well. Indeed, some biological activities displayed by such polymers, especially by chitosan and its derivatives, such as biocompatibility, biodegradability, low toxicity, mucoadhessiveness and antimicrobial activity, have attracted the interest

* **Corresponding author Sérgio P. Campana-Filho:** Instituto de Química de São Carlos, Universidade de São Paulo (IQSC/USP), São Carlos, São Paulo, Brazil; Tel: +55-1633739929; Fax: +55-1633739952; E-mail: scampana@iqsc.usp.br

Guilherme Luiz Dotto, Sergio Paulo Campana-Filho and Luiz Antonio de Almeida Pinto (Eds)

of academics and industrials as well, mainly for applications in Biomedicine, Pharmacy, Dentistry and other health-related fields. Thus, several chitosan-based materials, such as nano and microparticles, gels, sponges, films and membranes, have been developed for applications in drug delivery and wound healing, and to compose scaffolds for tissue regeneration, among others. However, despite the success of several studies which showed the potential of such chitosan-based materials in many of the above-mentioned applications, and although the huge number of scientific documents and patents, there are few registered products and yet less commercially available products based on chitin, chitosan and derivatives. Indeed, chitosan has been considered as a very promising biopolymer for a very long time but several initial promises have not been achieved, mainly due to poor properties reproducibility, severely precluding the development of commercial products. Currently, it is well known that the structure and microstructure of chitosan, mainly the content of 2-acetamido-2-deoxy-D-glucopyranose (GlcNAc), expressed as the average degree of acetylation ($\overline{DA}$) and 2-amino-2-deoxy-D-glucopyranose (GlcN) units and its distribution along the chains, as well as its average molecular weight, strongly affect its physical chemical properties, functionalities and biological activities. Therefore, elucidating the structure-function relationship at a molecular level is a main breakthrough toward the precise control of chitosan characteristics and properties, allowing the production of tailored chitosan according to the desired application. As chitin extracted from biomass is the main raw material for producing chitosan, it is also important to address the natural sources and the processes used to extract chitin, as they affect the characteristics of chitosan. Additionally, the reaction conditions used to promote the chemical conversion of chitin into chitosan *via* *N*-deacetylation of GlcNAc units strongly affect the characteristics of chitosan and such a theme will also be discussed in the following sections.

Chitin Structure and Occurrence

Native chitin is idealized as a linear homopolymer composed by 2-acetamido-2-deoxy-D-glucopyranose (GlcNAc) units (Fig. **1a**) and linked by β(1 → 4) glycosidic bonds but some 2-amino-2-deoxy-D-glucopyranose (GlcN) units (Fig. **1b**) can occur depending on the biomass and on the procedures used to extract chitin from it. Indeed, as chitin generally occurs strongly associated to proteins and carbonates, such as in the exoskeletons of crustaceans, its extraction from the biomass calls for chemical treatments that can provoke side reactions, such as partial *N*-deacetylation of GlcNAc units, oxidation and depolymerization, depending on the severity of the treatments.

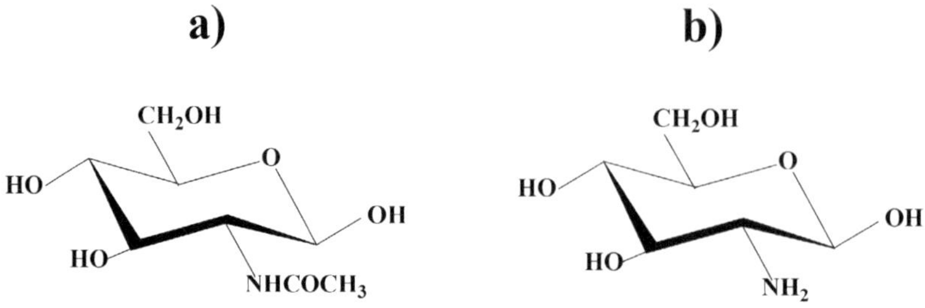

Fig. (1). Structures of (**a**) 2-acetamido-2-deoxy-D-glucopyranose (GlcNAc) and (**b**) 2-amino-2-deoxy-D-glucopyranose (GlcN).

Chitin is a semi-crystalline polymer and depending on the organism in which it occurs and to the function it plays, the chitin chains adopt different arrangements in the solid state, resulting in three different allomorphs, namely alpha-chitin, beta-chitin and gamma-chitin [1]. Alpha-chitin, the more stable and widely spread allomorph, occurs where rigidity and mechanical resistance are required, such as in the shells of crustaceans, and it adopts a dense packing due to numerous inter, intra-chains and interlamellar hydrogen bonds, strongly favored by the anti-parallel orientation of the polymeric chains in the ordered domains. In beta-chitin, which occurs where flexibility and toughness are important, such as in squid pens, the polymer chains are disposed in parallel orientation, disfavoring the establishment of hydrogen bonds involving chains pertaining to different lamellae and resulting in a less dense packing as compared to alpha-chitin. The allomorph gamma-chitin, much rarer as compared to alpha- and beta-chitin, is less studied and some authors claim that it has characteristics from both, alpha- and beta-chitin [2]. The characterization of the chitin allomorphs is usually carried out by using infrared and solid state NMR spectroscopy and X-rays diffraction [3].

Chitin is present in the exoskeletons of insects [4, 5], spiders [6] and crustaceans, including shrimps, crabs, crayfish, lobsters and krill [1, 7], in squid pens [3], in bony fish [8], in the cell wall of some fungi [9], such as Ascomycetes, Zygomycetes, Basidiomycetes and Deuteromycetes, and in algae such as diatoms [10]. Additionally, the occurrence of chitin in marine [11] and freshwater sponges [12] has been reported. Depending on the source, chitin occurs combined with proteins, carbonates, silicates, lipids and pigments in different proportions, the chitin content ranging in a relatively ample interval (10% - 40%) as a consequence (Table **1**) [1, 7, 13 - 15].

According to some authors [13, 14, 16, 17], chitin is the second most naturally abundant polysaccharide after cellulose and although it is not fully accepted, it is recognized that the replenishment rate of chitin is higher (≈2x) as compared to cellulose [18]. Some structural changes are observed in chitin depending on its source, including the content of GlcNAc and GlcN units, average molecular weight and dispersity, but the major differences concern its association with accompanying substances and its occurrence as a fibrous material, as it is found in shells of crustaceans and squid pens, or as a 2D or 3D scaffold, as found in marine and freshwater sponges.

Table 1. Mean content of chitin in some organisms (Adapted from Química Nova, 2007) [21].

Crustaceans	**Chitin (%)**
Cancer	72.1
Crangon	11.6
Macrobrachium rosenbergii	24.4
Metapenaeus	32.4
Homarus	60.8-77.0
Krill	40.2-42.0
Insects	
Blattela	10.0
Bombyx	44.2
Pieris	64.0
Molluscs	
Oysters shells	3.6
Squid pens	40.0
Fungi	
Aspergillus niger	42.0
Lactarius vellerus	19.0
Mucor roxii	44.5
Penicillium notatum	18.5
Saccharomyces cerevisiae	2.9

Chitin Extraction and Applications

Firstly, it is important to highlight that the industrial production of chitin from the residues of the fishery industry is a well-established economic activity mainly directed to the production of chitosan. Thus, the chitin/chitosan industries are spread over the entire world but they are more strongly present in the countries possessing strong sea-food industries, such as Canada, China, Chile, Germany, India, Italy, Japan, Norway, Poland and USA. Additionally, as the main industrial sources for producing chitin are the shells of shrimps and carapaces of crabs, the large majority of industries produces alpha-chitin while very few ones produce and commercialize beta-chitin, generally extracted from squid pens.

The shells of shrimps and the carapaces of crabs, used as the main industrial sources for extracting alpha-chitin, are also rich in proteins and due to their high content of organic matter, they exhibit relatively slow biodegradation [19]. Therefore, its indiscriminate disposal in water bodies increases the biochemical oxygen demand (BOD), causing the imbalance of aerobic life and endangering all living beings that are part of such ecosystems. Indeed, besides the direct environmental impact due to inappropriate discard, there are also some important harmful socio-economic consequences as the community of fishermen, who depends on these ecosystems for subsistence, is also affected by the environmental impacts already mentioned. On the contrary, extracting chitin from the residues of the fishery industry is an interesting alternative to mitigate the environmental impact due to the disposal of the residues from the sea-food industry and to add value to a chitin-rich raw material which otherwise would be discarded. However, regardless of the biomass selected as the raw material to extract chitin, the whole process should avoid the use of corrosive reactants and energy-consuming steps, the generation of low volumes of residues and its proper disposal being also very important to avoid the environment contamination.

Several procedures have been proposed for the extraction of chitin from shells of crustaceans by using chemical and enzymatic methods [18, 20 - 29]. Considering the composition of these raw materials in terms of chitin (15% - 20%), proteins (25% - 40%), calcium and magnesium carbonates (40% - 55%), and a low content (< 1%) of pigments and lipids, a sequence of physical and chemical treatments must be carried out to extract chitin from these biomasses. Besides washing and grinding steps, the extraction of alpha-chitin from shells of crustaceans calls for the execution of demineralization, deproteinization and depigmentation procedures aiming the removal of carbonates, proteins and pigments, respectively. The demineralization is usually carried out by treating the shells with aqueous hydrochloric acid to result in soluble calcium and magnesium chlorides while alkaline solutions are used to promote the hydrolysis of proteins during the deproteinization step, the soluble fragments being washed out in both steps. The depigmentation aims the removal of astaxanthin, which is done *via* extraction with organic solvents or by using oxidizing agents. The sequence of demineralization / deproteinization is usually employed but it can be inversed, first generating calcareous chitin, an organic/inorganic material that has its own applications, being then followed by demineralization to result in alpha-chitin. Regardless of the chosen sequence, it is very important to avoid harsh reaction conditions to prevent as much as possible the occurrence of side reactions, mainly those affecting chitin structure, such as *N*-deacetylation and oxidation, and resulting in properties degradation, such as chain depolymerization. Although relatively unimportant at bench scale, the consumption of reagents (acid and alkali) and mainly of water is very important at an industrial plant designed to produce chitin from shells of crustaceans. If squid pens are the raw material, the demineralization step is not necessary owing its very low mineral content (< 1%) and beta-chitin is extracted after carrying out the deproteinization step, usually by treating the biomass with dilute aqueous alkali at room temperature [30]. As already mentioned, there are few industrial suppliers of beta-chitin as compared to alpha-chitin but owing the abundance of squid pens, its high content of chitin and lower cost of extraction, one may expect that the market for beta-chitin suppliers tends to grow. Additionally, beta-chitin is less crystalline as compared to alpha-chitin and it is more able to swell and to dissolve as well, favoring its processing and conversion to chitosan too.

Due to its limited solubility, chitin is mainly applied in solid state, as in adsorption processes designed to remove pollutants, such as metal ions and dyes, from industrial effluents. Chitin can be processed as fibers, enabling its application as a biomaterial for dressing and suturing or as solid support in chromatography [7]. Chitin wyskers can be produced by using similar conditions as used to produce it from cellulose, the proper adjustment of reaction conditions being needed depending also on which chitin allomorph is used as raw material [31]. Ultrasound-based processing is promising and it has allowed the production of stable aqueous suspensions containing micro and nanoparticles of beta-chitin, which were used to produce casted films potentially useful in wound healing [32]. The production of sponge-like 3D nanoporous chitin-based materials for applications in Tissue Engineering relies on frezee-drying chitin gel which allows the production of biomaterials useful for culturing chondrocytes and in veterinary surgical practice as well [33]. Recently, the isolation of 3D chitin-based scaffolds from demosponges has been reported, its unique composition and architecture, as well as the facile and cost effective processing, being highlighted as important advantages as compared to other chitin-based scaffolds [34].

Chitin can be chemically modified at hydroxyl and amino groups to render it more soluble in aqueous media, *via* *O*-alkylation with ethylene oxide for instance, or in organic solvents by carrying out *O*-acylation, *O*-tosylation or *N*-ftaloylation reactions, some of the resulting derivatives being important intermediates to allow further chemical modifications. The controlled hydrolysis of chitin allows the production of chitin oligosaccharides (COSs) which exhibits tumoricidal, antimicrobial, antihypertensive and antioxidative stress activities, and is used in biomedical applications as it is claimed to stimulate the human immune system [35].

Chitosan

The most common chitin reaction is its *N*-deacetylation aiming the production of chitosan, copolymers of GlcNAc and GlcN units which exhibit different physical chemical properties [24, 25] depending on its average degree of acetylation ($\overline{DA}$), pattern of acetylation (PA) and average degree of polymerization ($\overline{DP}$), such parameters also affecting its applications in different fields [13, 36]. Owing to the protonation of most amino groups (pka ≈ 6.3) pertaining to its GlcN units, chitosan is soluble in moderately acidic media (pH < 6). However, the polycationic character of chitosan is due to the presence of ammonium groups and its neutralization in alkaline media results in insolubility, the average degree of deacetylation ($\overline{DA}$) and average degree of polymerization ($\overline{DP}$), strongly affecting the solubility of chitosan [37].

The hydrolysis of acetamido groups of most GlcNAc units of chitin ($\overline{DA}$ > 60%) results in chitosan ($\overline{DA}$ < 40%), the *N*-deacetylation of the parent chitin being usually carried out in concentrated alkaline medium. The alkaline *N*-deacetylation of chitin is a heterogeneous thermochemical process that is mainly affected by the alkali concentration, reaction temperature and time [38, 39]. Owing to the semi-crystallinity character of chitin and to the heterogeneous nature of the reaction, amorphous domains are more accessible and effective swelling results in more efficient reaction. During the first 30 - 50 min, the average degree of deacetylation ($\overline{DD}$ = 100 - $\overline{DA}$) increases quickly but a plateau is reached after 1 h regardless of the reaction conditions [1, 38, 40]. Thus, the reaction rate is much lower from then on while depolymerization is favored the higher the temperature and the longer the reaction.

In the thermochemical *N*-deacetylation of chitin, a large excess of concentrated alkali solution is used to assure effective swelling and to establish pseudo-first order conditions, the reaction kinetics being governed by eq. (1) [38].

$$-\frac{d[GlcNAc]}{dt} = k[NaOH][GlcAc] = k'[GlcNAc] \quad \textbf{(1)}$$

The alkaline thermochemical *N*-deacetylation of chitin, is the most commonly used process to produce chitosan in research laboratories and in industries as well. Due to the use of harsh reaction conditions, mainly highly concentrated alkali, high temperature and long reaction time, severe depolymerization can occur while only partial *N*-deacetylation is achieved. Additionally, as the reaction proceeds heterogeneously, the deacetylated products exhibit heterogeneity in terms of structural characteristics and significant variations in their properties are observed. Thus, several methods have been proposed to increase the reaction efficiency and to minimize the occurrence of depolymerization, such as the execution of the reaction in inert atmosphere, the use of less concentrated alkaline solutions, the dilution with solvents [41], the introduction of changes in processing conditions with the use of reactive extrusion [42], vapor explosion [43], ultrasound [44, 45] and microwave irradiation [46], the use of reducing agents [47] and the execution of successive deacetylation treatments [38, 39].

Lamarque and co-workers [48] developed the freeze - pump out - thaw (FPT) process, in which chitin suspended in concentrated sodium hydroxide aqueous solution is subjected to successive cycles of freezing in liquid nitrogen, suction with a pump vacuum and slow heating to the room temperature. At least six cycles must be executed and then the chitin suspension is subjected to deacetylation at 80 °C - 110 °C for short periods (up to 60 minutes). Thus, the execution of further 3 - 4 FTP processings results in extensively deacetylated chitosan ($\overline{DD}$ > 98.0%) exhibiting high weight average molar weight ($\overline{Mw}$ ≈ 450,000 g mol^{-1}). According to the authors of this study, the FPT process promotes the destruction of the crystalline regions of chitin, increasing the accessibility to reactive sites and renewing the concentration of the reactive species of sodium hydroxide (mono- and dihydrate) inside the particles of chitin, and it favors the exclusion of molecular oxygen from the reaction medium, minimizing oxidative depolymerization [48, 49].

Delezuk and co-workers [45] reported a new process known as ultrasound-assisted deacetylation (USAD), to produce chitosan by treating the alkaline suspension of beta-chitin with high intensity ultrasound irradiation. In such a study, single-step USAD procedures were carried out in non-isothermal and isothermal conditions leading to the production of chitosans presenting variable average degree of acetylation (74% < $\overline{DD}$ < 94%) and viscosity average molecular weight (115,000 g mol^{-1} < $\overline{Mv}$ < 720,000 g mol^{-1}) depending on the processing conditions, including ultrasound irradiation parameters, reaction time and temperature. Fiamingo and co-workers [50] applied multistep ultrasound-assisted deacetylation of beta-chitin to produce high molecular weight chitosan of variable degree of acetylation. Thus, after carrying out only one, two or three consecutive USAD processing at low temperatures (60 °C) and for short periods (50 min/step), randomly deacetylated chitosans possessing variable degree of acetylation (63% < $\overline{DD}$ < 96%), high weight average molecular weight (900,000 g mol^{-1} < $\overline{Mw}$ < 1,200,000 g mol^{-1}) and low dispersity (1.3 < Đ < 1.4) were produced.

Physicochemical Properties

The average degree of polymerization ($\overline{DP}$), average degree of acetylation ($\overline{DA}$) and pattern of acetylation (PA) affect the physicochemical properties of chitosan, such as solubility, viscosity, and its biological activities, such as cytotoxicity, antimicrobial activity, biodegradability and biocompatibility [51].

While chitin is limited soluble only in few solvents, such as *N,N*-dimethylacetamide/LiCl [52], chitosan is soluble in aqueous solutions of diluted acids due to protonation of amino groups pertaining to GlcN units, as the strong electrostatic repulsions involving the positive charges of ammonium groups promotes the polymer dissolution [7]. However, the molecular weight and the pattern of acetylation also affect the solubility of chitosan.

The *N*-deacetylation of chitin is rarely complete, *i.e.* some GlcNAc units remain unhydrolyzed, and chitosans are copolymers with varying average acetylation with different distributions of GlcNAc and GlcN units, except those extensively deacetylated (DD > 95%). Although enzymatic and chemical processes promote the *N*-deacetylation of chitin, the chemical processes are the most commonly used, predominantly carried out in heterogeneous conditions, generating chitosans with non-uniform characteristics and not reproducible properties. Indeed, the alkaline *N*-deacetylation of chitin *via* thermochemical heterogeneous processes conditions occurs preferentially in the amorphous regions, resulting in block-like copolymers (Fig. **2a**), *i.e.* chitosans with sequences of GlcNAc units intercalated with sequences of GlcN units. In these conditions, if the reaction is extended, the crystalline domains are partially accessible and the final product will contain deacetylated polymer chains of different extents, therefore, exhibiting different solubility. When the heterogeneous *N*-deacetylation of chitin results in chitosan with $\overline{DD} \approx 50\%$, the polymer is only partially soluble in acid medium. However, if the extensively deacetylated chitosan (Fig. **2b**) is *N*-acetylated in homogeneous conditions to attain $\overline{DD} \approx 50\%$, a random distribution of GlcNAc and GlcN units is observed (Fig. **2c**) and this chitosan is also soluble in neutral aqueous medium [53].

The distribution of GlcNAc and GlcN units along the chitosan chains is characterized by the acetylation pattern (PA) using the Bernoulli statistical model [54]. The value of PA can be determined by nuclear magnetic resonance spectroscopy of hydrogen (^{1}H NMR) or carbon (^{13}C NMR) [55], allowing the distinction among ideally alternated (PA = 2), random (PA = 1) or block-like (PA = 0) distributions.

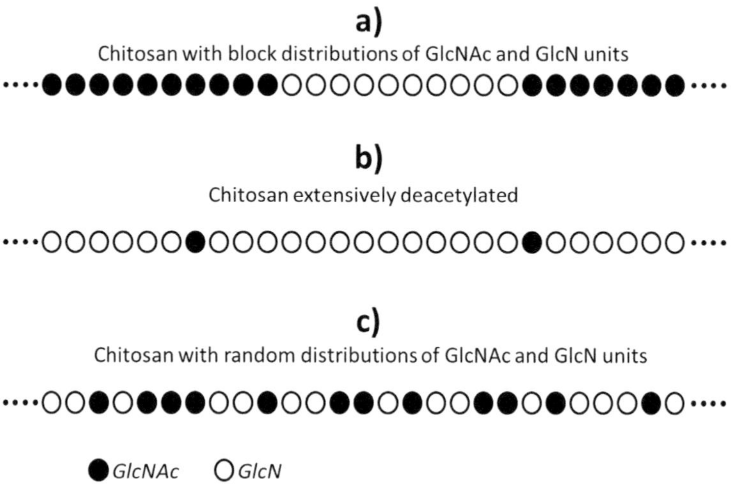

Fig. (2). Idealized representation of the distribution of GlcNAc and GlcN units in the chitosan chain (**a**) block, (**b**) extensively deacetylated chitosan and (**c**) random.

The degree of acetylation and the pattern of acetylation also affect the degradation of chitosan *in vivo*, which occurs mainly by the action of lysozyme present in body tissues and fluids, resulting in the release of oligomers and monomers. According to Nordtveit and co-workers [56] and Aiba [57], lysozyme has a specific action in *N*-acetylated units (A) and the rate of degradation is proportional to the frequency of acetylated diads (F_{AA}, percentage of GlcNAc neighboring units to another GlcNAc unit). Thus, extensively deacetylated chitosan is very slowly degraded by lysozyme while chitosans presenting $\overline{DA}$ > 30% are rapidly degraded, probably due to its high content of GlcNAc/GlcNAc diads.

The molecular weight and dispersity also affect chitosan properties and biological activities. Accordingly, many studies indicate that the antibacterial activity of chitosan depends on its molecular weight [58]. Thus, Jeon and co-workers [59] observed that the molecular weight of chitosan is an important factor governing the inhibition of Gram-negative and Gram-positive bacteria, the effective inhibition occurring when the molecular weight is higher than 10,000 g mol^{-1}. Additionally, chitosan showed a tendency to be more effective against pathogen microorganisms associated with human diseases as compared to non-pathogens.

FINAL REMARKS

The natural occurrence of chitin was addressed and it was shown that besides the shells of crustaceans, other biomasses from the seafood industry, as squid pens, can be used to extract chitin. The use of mild reaction conditions to extract chitin from shells of crustaceans and squid pens was highlighted to avoid structure modification and degradation of properties. The isolation of 3D chitin-based scaffolds from sponges seems to be an interesting alternative aiming its application as a biomaterial in Biomedicine and to develop biosensors and catalysts. The conversion of chitin into chitosan by carrying out thermochemical alkaline *N*-deacetylation of chitin was discussed as well as unconventional processes such as the freeze - pump out - thaw (FPT) process and ultrasound-assisted deacetylation (USAD) which enable the production of high molecular weight chitosan.

CONFLICT OF INTEREST

The authors confirm that they have no conflict of interest to declare for this publication.

ACKNOWLEDGEMENTS

Authors thank to Brazilian agencies, namely CAPES, CNPQ and FAPESP for financial support.

REFERENCES

[1] Roberts GA. Chitin chemistry. London: The MacMillan 1992. [http://dx.doi.org/10.1007/978-1-349-11545-7]

[2] Jang Mi-Kyeong, Kong Byeong-Gi, Jeong Young-Il, *et al.* Physicochemical characterization of alpha-chitin, beta-chitin and gamma-chitin separated from natural resources. J Polym Sci A Polym Chem 2004; 42: 3423-32. [http://dx.doi.org/10.1002/pola.20176]

[3] Lavall RL, Assis OBG, Campana-Filho SP. Beta-Chitin from the pens of *Loligo* sp.: extraction and characterization. Biores Technol 2007; 98(13): 2465-72.

[4] Neville AC, Parry DA, Woodhead-Galloway J. The chitin crystallite in arthropod cuticle. J Cell Sci 1976; 21(1): 73-82. [PMID: 932111]

[5] Liu S, Sun J, Yu L, *et al.* Extraction and characterization of chitin from the beetle *Holotrichia parallela* Motschulsky. Molecules 2012; 17(4): 4604-11. [http://dx.doi.org/10.3390/molecules17044604] [PMID: 22510609]

[6] Kaya M, Seyyar O, Baran T, Erdoğan S, Kar M. A physicochemical characterization of fully acetylated chitin structure isolated from two spider species: with new surface morphology. Int J Biol Macromol 2014; 65: 553-8. [http://dx.doi.org/10.1016/j.ijbiomac.2014.02.010] [PMID: 24530368]

[7] Agulló E, Peniche C, Tapia C, *et al.* Quitina y Quitosano. Perú: Pontificia Universidad Católica Del Perú/ Fondo Editorial 2004.

[8] Wagner GP, Lo J, Laine R, Almeder M, *et al.* Chitin in the epidermal cuticle of vertebrate (*Paralipophorys trigloids*, Blenniidae, Teleostei). Experientia 1993; 49: 317-9. [http://dx.doi.org/10.1007/BF01923410]

[9] Peter MG. Chitin and chitosan in fungi Biopolymers Online. Weinheim, UK: John Wiley & sons 2005; pp. 123-32.

[10] Durkin CA, Mock T, Armbrust EV. Chitin in diatoms and its association with the cell wall. Eukaryot Cell 2009; 8(7): 1038-50. [http://dx.doi.org/10.1128/EC.00079-09] [PMID: 19429777]

[11] Ehrlich H, Maldonado M, Spindler KD, *et al.* First evidence of chitin as a component of the skeletal fibers of marine sponges. Part I. Verongidae (demospongia: Porifera). J Exp Zoolog B Mol Dev Evol 2007; 308(4): 347-56. [http://dx.doi.org/10.1002/jez.b.21156] [PMID: 17285638]

[12] Ehrlich H, Kaluzhnaya OV, Tsurkan MV, *et al.* First report on chitinous holdfast in sponges (Porifera). Proc R Soc B 2013; 280(1762): 20130339. [http://dx.doi.org/10.1098/rspb.2013.0339]

[13] Kumar MN. A review of chitin and chitosan applications. React Funct Polym 2000; 46(1): 1-27. [http://dx.doi.org/10.1016/S1381-5148(00)00038-9]

[14] Rinaudo M. Chitin and chitosan: properties and applications. Prog Polym Sci 2006; 31(7): 603-32. [http://dx.doi.org/10.1016/j.progpolymsci.2006.06.001]

[15] Domard A. A perspective on 30 years research on chitin and chitosan. Carbohydr Polym 2011; 84(2): 696-703. [http://dx.doi.org/10.1016/j.carbpol.2010.04.083]

[16] Ruiz-Hereira J, Ruiz-Medrano R. Chitin biosynthesis in fungi. In: Dilip KA, Ed. Handbook of fungal biotechnology. London: Taylor & Francis 2003; pp. 435-47. [http://dx.doi.org/10.1201/9780203027356.ch24]

[17] Zeng JB, He YS, Li SL, Wang YZ. Chitin whiskers: an overview. Biomacromolecules. Research Support, Non-US Gov't 2012 Jan 9; 13(1): 1-11. [http://dx.doi.org/10.1021/bm201564a]

[18] Peter MG. Applications and environmental aspects of chitin and chitosan. Journal of Macromolecular Science -. Pure Appl Chem 1995; A32(4): 629-40.

[19] Shahidi F, Synowiecki J. Isolation and characterization of nutrients and value-added products from snow crab (Chinoecetes opilio) and shrimp (Pandalus borealis). Processing Discards Journal of Agricultural and Food Chemistry 1991; 39(8): 1527-32. [http://dx.doi.org/10.1021/jf00008a032]

[20] Mathur NK, Narang CK. Chitin and chitosan, versatile polysaccharides from marine animals. J Chem Educ 1990; 67(11): 938-42. [http://dx.doi.org/10.1021/ed067p938]

[21] Campana Filho SP, Britto D, Curti E, Abreu FR, Cardoso MB, Battisti MV, *et al.* Extração, estruturas e propriedades de a- e b-quitina. Quim Nova 2007; 30(3): 644-50. [http://dx.doi.org/10.1590/S0100-40422007000300026]

[22] Hirano S. Chitin and chitosan. In: GERHARTS W, Ed. Ullmann´s encyclopedia of industrial chemistry. 6th ed. New York: VCH 1986; pp. 231-2.

[23] Chaussard G, Domard A. New aspects of the extraction of chitin from squid pens. Biomacromolecules 2004; 5(2): 559-64. [http://dx.doi.org/10.1021/bm034401t] [PMID: 15003021]

[24] Sini TK, Santhosh S, Mathew PT. Study on the production of chitin and chitosan from shrimp shell by using Bacillus subtilis fermentation. Carbohydr Res 2007; 342(16): 2423-9. [http://dx.doi.org/10.1016/j.carres.2007.06.028] [PMID: 17707781]

[25] No HK, Meyers SP, Lee KS. Isolation and characterization of chitin from crawfish shell waste. J Agric Food Chem 1989; 37(3): 575-9. [http://dx.doi.org/10.1021/jf00087a001]

[26] No HK, Meyers SP. Preparation and characterization of chitin and chitosan a review. J Aquat Food Prod Technol 1995; 4(2): 27-52. [http://dx.doi.org/10.1300/J030v04n02_03]

[27] Wang SL, Chio SH. Deproteinization of shrimp and crab shell with the protease of Pseudomonas aeruginosa K-187. Enzyme Microb Technol 1998; 22(7): 629-33. [http://dx.doi.org/10.1016/S0141-0229(97)00264-0]

[28] Kurita K, Tomita K, Tada T, Ishii S, Nishimura S-I, Shimoda K. Squid chitin as a potencial alternative chitin source: deacetylation behavior and characteristic properties. J Polym Sci A Polym Chem 1993; 31(2): 485-91. [http://dx.doi.org/10.1002/pola.1993.080310220]

[29] Acosta N, Jimenez C, Borau V, Heras A. Extraction and characterization of chitin from crustaceans. Biomass Bioenergy 1993; 5(2): 145-53. [http://dx.doi.org/10.1016/0961-9534(93)90096-M]

[30] Lavall RL, Assis OB, Campana-Filho SP. Beta-chitin from the pens of Loligo sp.: extraction and characterization. Bioresour Technol 2007; 98(13): 2465-72. [http://dx.doi.org/10.1016/j.biortech.2006.09.002] [PMID: 17070041]

[31] Paillet M, Dufresne A. Chitin whisker reinforced thermoplastic nanocomposites. Macromolecules 2001; 34(19): 6528-31. [http://dx.doi.org/10.1021/ma002049v]

[32] Almeida EV, Mariano MS, Campana-Filho SP. Nanoparticles and films from ultrasound-processed chitin, chitosan and cellulose. In: Valery Varlamov, Svetlana Bratskaya, Irina Yakovleva, Sevda Senel, Eds. Advances in Chitin Science, vol XIII. 96-101.

[33] Shigemasa Y. Minami Application of chitin and chitosan for biomaterials. Biotechnol Genet Eng Ver 1996; 13: 383-420. [http://dx.doi.org/10.1080/02648725.1996.10647935]

[34] Wysokowski M, Petreanko I, Stelling AL, Stawski D, Jesionowski T, Erlich H. Poriferan chitin as a versatile template for extreme biomimetics. Polymers (Basel) 7: 235-65. [http://dx.doi.org/10.3390/polym7020235]

[35] Ahmeda AB, Mat Tahaa R, Mohajera S, Elnaiem Elaagiba M, Kwon Kimb S. Preparation, properties and biological applications of water soluble chitin Oligosaccharides from marine organisms. Russ J Mar Biol 2012; 38(4): 349-56.

[36] Kumar MN, Muzzarelli RA, Muzzarelli C, Sashiwa H, Domb AJ. Chitosan chemistry and pharmaceutical perspectives. Chem Rev 2004; 104(12): 6017-84. [http://dx.doi.org/10.1021/cr030441b] [PMID: 15584695]

[37] Cho YW, Jang J, Park CR, Ko SW. Preparation and solubility in acid and water of partially deacetylated chitins. Biomacromolecules 2000; 1(4): 609-14. [http://dx.doi.org/10.1021/bm000036j] [PMID: 11710189]

[38] Lamarque G, Viton C, Domard A. Comparative study of the first heterogeneous deacetylation of alpha- and beta-chitins in a multistep process. Biomacromolecules 2004; 5(3): 992-1001. [http://dx.doi.org/10.1021/bm034498j] [PMID: 15132692]

[39] Lamarque G, Viton C, Domard A. Comparative study of the second and third heterogeneous deacetylations of alpha- and beta-chitins in a multistep process. Biomacromolecules 2004; 5(5): 1899-907.
[http://dx.doi.org/10.1021/bm049780k] [PMID: 15360304]

[40] Chang KL, Tsai G, Lee J, Fu WR. Heterogeneous N-deacetylation of chitin in alkaline solution. Carbohydr Polym 1997; 303(3): 327-32.
[http://dx.doi.org/10.1016/S0008-6215(97)00179-1]

[41] Sannan T, Kurita K, Iwakura Y. Studies on chitin, 1. Solubility change by alkaline treatment and film casting. Makromol Chem 1975; 176(4): 1191-5.
[http://dx.doi.org/10.1002/macp.1975.021760426]

[42] Rogovina SZ, Akopova TA, Vikhoreva GA. Investigation of properties of chitosan obtained by solid-phase and suspension methods. J Appl Polym Sci 1998; 70(5): 933.
[http://dx.doi.org/10.1002/(SICI)1097-4628(19981031)70:5<927::AID-APP13>3.0.CO;2-Q]

[43] BELTRAME PL, NAGGI A, TORRI G. Alkaline N-deacetyltion of chitin enhanced by flash treatments. Reaction kinetics and structure modifications. Carbohydr Polym 1990; 12(4): 405-18.
[http://dx.doi.org/10.1016/0144-8617(90)90090-F]

[44] Campana Filho SP, Signini R, Barreto M. Effects of sonication on the reactivity of chitin toward its heterogeneous deacetylation. Int J Polym Mater 2002; 51(8): 695-700.
[http://dx.doi.org/10.1080/714975832]

[45] Delezuk JA, Cardoso MB, Domard A, Campana-Filho SP. Ultrasound-assisted deacetylation of beta-chitin: influence of processing parameters. Polym Int 2011; 60(6): 903-9. [Article].
[http://dx.doi.org/10.1002/pi.3037]

[46] Sahu A, Goswami P, Bora U. Microwave mediated rapid synthesis of chitosan. Journal of Materials Science: Materials in Medicine [Research Support, Non-US Gov't] 2009 Jan; 20(1): 171-5.
[http://dx.doi.org/10.1007/s10856-008-3549-4]

[47] Campana-Filho SP, Signini R. Effects of additives and inert gas bubbling on the deacetylation of chitosan. Int J Polym Mater 2002; 51(8): 701-9.
[http://dx.doi.org/10.1080/714975830]

[48] Lamarque G, Cretenet M, Viton C, Domard A. New route of deacetylation of alpha- and beta-chitins by means of freezepump outthaw cycles. Biomacromolecules 2005; 6(3): 1380-8.
[http://dx.doi.org/10.1021/bm049322b] [PMID: 15877356]

[49] CHAUSSARD G, DOMARD A. Thermodynamic aspects of the heterogeneous deacetylation of b-chitin: reaction mechanisms. Biomacromolecules 2007; 8(6): 1942-50.
[http://dx.doi.org/10.1021/bm070021m] [PMID: 17474777]

[50] Fiamingo A, Delezuk JA, Trombotto S, David L, Campana-Filho SP. Extensively deacetylated high molecular weight chitosan from the multistep ultrasound assisted deacetylation of beta-chitin. Ultrason Sonochem 2016; 32: 79-85.
[http://dx.doi.org/10.1016/j.ultsonch.2016.02.021] [PMID: 27150748]

[51] Brugnerotto J, Lizardi J, Goycoolea FM, Argüelles-Monal W, Desbrières J, Rinaudo M. An infrared investigation in relation with chitin and chitosan characterization. Polymer (Guildf) 2001; 42(8): 3569-80.
[http://dx.doi.org/10.1016/S0032-3861(00)00713-8]

[52] Terbojevich M, Carraro C, Cosani A, Marsano E. Solution studies of the chitin-lithium chloride-N,N-di-methylacetamide system. Carbohydr Res 1988; 180(1): 73-86.
[http://dx.doi.org/10.1016/0008-6215(88)80065-X]

[53] Sashiwa H, Saimoto H, Shigemasa Y, Tokura S. *N*-Acetyl group distribution in partially deacetylated chitins prepared under homogeneous conditions. Carbohydr Res 1993; 242: 167-72.
[http://dx.doi.org/10.1016/0008-6215(93)80031-9] [PMID: 8495437]

[54] Kumirska J, Weinhold MX, Sauvageau JC, Thöming J, Kaczyński Z, Stepnowski P. Determination of the pattern of acetylation of low-molecular-weight chitosan used in biomedical applications. J Pharm Biomed Anal 2009; 50(4): 587-90. [http://dx.doi.org/10.1016/j.jpba.2008.09.048] [PMID: 19013738]

[55] Vårum KM, Anthonsen MW, Grasdalen H, Smidsrød O. Determination of the degree of N-acetylation and the distribution of N-acetyl groups in partially N-deacetylated chitins (chitosans) by high-field n.m.r. spectroscopy. Carbohydr Res 1991; 211(1): 17-23. [http://dx.doi.org/10.1016/0008-6215(91)84142-2] [PMID: 1773428]

[56] Nordtveit RJ, Vårum KM, Smidsrød O. Degradation of fully water-soluble, partially N-acetylated chitosans with lysozyme. Carbohydr Polym 1994; 23(4): 253-60. [http://dx.doi.org/10.1016/0144-8617(94)90187-2]

[57] Aiba S. Studies on chitosan: 4. Lysozymic hydrolysis of partially N-acetylated chitosans. Int J Biol Macromol 1992; 14(4): 225-8. [http://dx.doi.org/10.1016/S0141-8130(05)80032-7] [PMID: 1504044]

[58] No HK, Park NY, Lee SH, Meyers SP. Antibacterial activity of chitosans and chitosan oligomers with different molecular weights. Int J Food Microbiol 2002; 74(1-2): 65-72. [http://dx.doi.org/10.1016/S0168-1605(01)00717-6] [PMID: 11929171]

[59] Jeon YJ, Park PJ, Kim SK. Antimicrobial effect of chitooligosaccharides produced by bioreactor. Carbohydr Polym 2001; 44(1): 71-6. [http://dx.doi.org/10.1016/S0144-8617(00)00200-9]

CHAPTER 3

Chitosan Nanofibers

Eduardo H. Tanabe*, Daliomar L. de O. Júnior, Guilherme L. Dotto and **Daniel A. Bertuol**

Chemical Engineering Department, Federal University of Santa Maria-UFSM, Santa Maria-RS, Brazil

Abstract: Chitosan is a natural polymer produced from the chitin deacetylation. Chitosan has good biocompatibility, biodegradability, antimicrobial activity, wound healing property and antitumor effect. In recent years, chitosan nanofibers have become interesting, due to amazing characteristics such as, large surface area to volume ratio and high porosity with very small pore size. It can be readily produced using different methods. These nanofibers can be promising for many applications including filtration, removal of metals, drug-delivery, wound dressing, cell culture, tissue engineering, cosmetic, gas sensing and catalysis. The chapter will address diverse types of nanofibers produced from chitosan, their derivatives, and blends with several polymers.

Keywords: Analytical characterization, Biomedical applications, Chitosan, Nanofiber, Polymer, Water treatment.

INTRODUCTION

In the field of nanotechnology, polymer nanofibers have become a great interest for various applications such as, filters, skin masks, semi-permeable membranes, clothing and medical materials [1]. Electrospinning is one of the conventional, simple, versatile and efficient methods used to prepare nanofibers [2]. The generated nanofibers have unique characteristics like large surface area to volume ratio, high porosity, good pore interconnectivity and the possibility to incorporate active components on a nanoscale [1].

Recently, nanofibers have been used in drug delivery systems, tissue engineering scaffolds, vascular grafts, filter formation, sensing applications and biological wound dressings [3, 4]. A large number of research groups are dedicated to produce new and improved nanofibers, synthesizing and modifying biocompatible

* **Corresponding author Eduardo H. Tanabe:** Environmental Processes Laboratory (LAPAM), Chemical Engineering Department, Federal University of Santa Maria-UFSM, Santa Maria-RS, Brazil; Tel: +55 (55)3220-8448; E-mail: edutanabe@yahoo.com.br

Guilherme Luiz Dotto, Sergio Paulo Campana-Filho and Luiz Antonio de Almeida Pinto (Eds)

materials, in particular using biological materials such as, chitosan. The natural polysaccharide chitosan has attracted much attention due to its natural origin, biocompatibility, biodegradability, functionalizability, antibacterial activity, non toxic and excellent processability [5, 6].

Chitosan is a linear copolymer of β-(1-4) linked 2-acetamido-2-deoxy-β-D-glucopyranose and 2-amino-2-deoxy-β-D-d-glucopyranose (Fig. **1**). It is obtained by deacetylation (alkaline conditions or enzymatic hydrolysis) of chitin, a polysaccharide widely distributed in nature (*e.g.* crustaceans, insects and certain fungi) [7 - 9]. The deacetylation degree (DD) of chitosan (indication of the number of amino groups along the chains) is calculated as the ratio of D-glucosamine to the sum of D-glucosamine and N-acetyl D-glucosamine. Chitosan DD greatly varies between 60 and 100% while its molecular weight typically ranges from 300 to 1000 kDa, depending on the source and preparation [9].

Chitosan is also widely applied for the removal of heavy metal ions from aqueous solutions, due the presence of amino and hydroxyl groups in its structure. Therefore, it is used for the treatment of wastewaters, purifying them by recovering heavy metals [9]. Chitosan offer advantages, being processed into gels, membranes, beads, microparticles, nanoparticles, scaffolds, sponges and nanofibers [8, 10, 11]. In this way, chitosan and its derivatives can be used in various industrial fields [12].

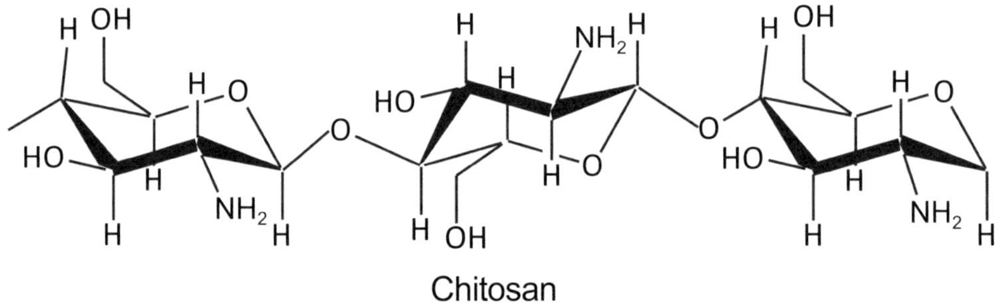

Fig. (1). Chitosan structure [8].

CHITOSAN NANOFIBER PRODUCTION

There are several ways to produce chitosan nanofiber such as, Eletrospinning [13], Forcespinning® [14], Solution Blowing [15], Coaxial Electrospinning [16], Centrifugal Electrospinning [17], Ultrasonic Atomization and Freeze Casting [18]. In this chapter, the different methods of preparation of chitosan nanofibers are reported and their applications in several fields are detailed.

Electrospinning

Electrospinning is a simple and versatile method which produces fibers ranging from the submicron level to several nanometers in diameter in a high voltage electrostatic field [13, 19]. This method is a relatively old approach to fiber production, but is currently one of the most advanced techniques in manufacturing high performance nanofibers [13]. These have been introduced into various technological fields due of their distinct characteristics, such as high surface area to mass ratio (typically 40-100 m^2/g), high porosity, and special chemical and physical properties, which result from their unique structure [13, 20].

The formation of nanofibers through electrospinning is based on the uniaxial stretching of a viscoelastic solution. A typical electrospinning setup consists of three basic components: a capillary tube with a needle, a high voltage supply, and a collector, as shown in Fig. (**2**) [21].

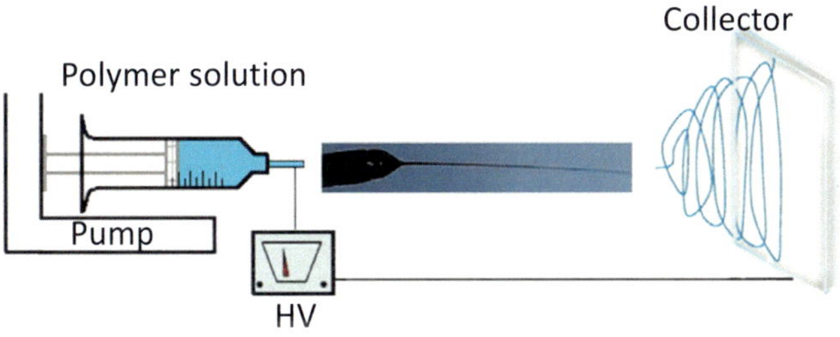

Fig. (2). Schematic diagram of the typical electrospinning setup [21].

In an electrospinning technique, an electric field is applied between the tip of a syringe and a collector plate. Polymer solution is put into a syringe and mounted on a pump, which produces a constant flow rate. When the electric field reaches a certain critical value (which the repulsive electrostatic forces overcome the surface tension), the charged fluid jet is ejected from the tip of the conical protrusion commonly known to as Taylor cone. The jet is continuously lengthened and whipped, resulting in solid nanofibers deposited on the collector with simultaneous rapid evaporation of the solvent [21 - 24].

Important parameters in electrospinning process are solution parameters and process parameters. Solution parameters are determined by solvent and polymer characteristics, including polymer type and concentration, molecular weight, viscosity, conductivity, surface tension and dielectric constant. Process parameters include voltage, nozzle tip-to-collector distance, flow rate of polymer solution, nozzle design and placement, collector composition, geometry, and rotation speed. Ambient parameters include temperature, humidity, and air velocity [21, 25 - 27].

Regarding to the solution parameters, an increase in viscosity or polymer concentration, results in an increase in fiber diameter. However, if the viscosity of polymer solutions becomes too low, the ejected jet may break down into droplets and result in electrospraying [6, 24]. Generally, low surface tension values result in the formation of fibers without beads and, low voltages can be applied in electrospinning [21]. Polymer solutions with low conductivity cannot be electrospun due to the absence of a surface charge on the fluid droplet, which is needed for Taylor cone formation. On the other hand, very high conductivity leads to a depleted tangential electric field along the surface of the fluid droplet, preventing Taylor cone formation [28]. In the dielectric constant, the successful electrospinning occurs when solvents with high dielectric constants are used resulting also in thinner nanofiber formation [21].

In relation to the process parameters, high voltages applied to the solution can increase the electrostatic force, which improve stretching, and whipping of the jet, resulting in the thinner fiber diameter. Voltages between 5 and 40 kV are usually applied in the electrospinning. The longer distance between spinneret and collector can increase the stretching time, which also attribute to the thinner fiber diameter. The temperature and humidity of the experimental environment directly can influence the solvent evaporation of the jet. Some solvents with high volatility, high temperature or low humidity would enhance the solvent evaporation resulting in the porous surface of the fiber [21, 24].

Chitosan has proven challenging to electrospin. Unfortunately, chitosan is difficult to electrospin into a nanofibrous structure owing to its polycationic character in an acidic aqueous solution as a result of the many amino groups in its backbone. Until now, nanofibrous structures were generated successfully by electrospinning chitosan solutions in 90 wt.% aqueous acetic acid or by using a solvent like trifluoroacetic acid (TFA), dichloromethane (DCM) and formic acid. The trace toxic organic solvent or acid in products is harmful, mainly when it is applied to wounded human skin or tissue [20, 29, 30].

Recently, electrospun chitosan composite nanofibrous mats have been fabricated using chitosan and synthetic biodegradable polymers such as, polyvinyl alcohol (PVA), polyethylene oxide (PEO), poly(vinyl pyrrolidine) (PVP), poly(lactic acid) (PLA), polycaprolactone (PCL) and poly(ethylene terpthalate (PET). These composite fiber mats are more advantageous over the electrospinning of pure chitosan, because, the mechanical, biocompatible, antibacterial and other properties of the chitosan nanofibers were drastically enhanced by the addition of PVA, PEO, PLA, PVP, PCL and PET [19, 31]. For improving the biological performance of the chitosan derived nanofibers, researchers have attempted electrospinning of chitosan with natural polymers such as gelatin, collagen, and silk fibroin [31].

Forcespinning®

Forcespinning® has been successfully used as an alternative way to produce a variety of polymer nanofibers. Unlike electrospinning, which produce fibers using electrostatic force, Forcespinning® utilizes centrifugal force to extrude polymer solutions or melts, which allows a high production rate of fine fibers and low cost of fibers production [32]. As reported, the productivity of this method (over 1 g/min per nozzle) at the lab scale is significantly higher than lab scale electrospinning (0.3 g/h) [14 - 33].

Fiber jets are formed by high rotational speeds (0-20,000 rpm) of a spinneret using a nozzle of different diameters. When the centrifugal force and associated hydrostatic pressure exceeds capillary forces (that tend to restrict the flow of fluid in the orifice), a jet of polymer solution is ejected. Reduction of fiber diameter occurs by the inertial drag between the fiber and the atmosphere, while the polymer solution jet is dried (Fig. **3**) [34, 35].

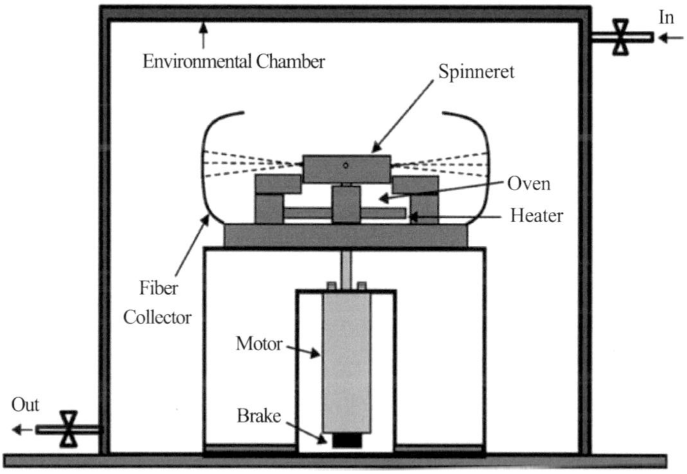

Fig. (3). Forcespinning® equipment [33].

The Forcespinning® equipments with a heater accessory can form fibers from the melt process. In this way, it can eliminate toxic solvents, and enable a wider variety of polymer materials to be spun into fibers (including insoluble polymers) [36]. During nanofiber production, parameters like rotational speed, polymer viscoelasticity, evaporation rate (for solutions), temperature (for melts), collection distance, shape and size of the needles or nozzles affects the fiber morphology and diameter [37, 38].

In the case of solution spinning, the solution concentration and the rotational speed were considered as the main factors [37]. Badrossamay and co-workers [39] concluded that at lower polymer concentrations, thinner and narrower fibers were obtained, while, the rotational speed has a subtle effect on the fiber diameter. For melt spinning, Shanmuganathan and co-workers [40] concluded that spinning temperature had a significant effect on the fiber properties and ultra-fine fibers can be made at higher temperature. Conversely, theoretical studies on the mechanisms of fiber formation were explored simultaneously.

Several materials have been recently produced using the Forcespinning® methods in the nanometer range such as, polyamide, polyethylene oxide, polytetra-fluoroethylene, polylactic acid, polycaprolactone, polypropylene, poly(vinylidene fluoride), poly(3'-hexylthiophene (P3HT)), Teflon®AF, cellulose, indium tin oxide and titanium dioxide nanowires [35, 38].

Solution Blowing

Solution blowing process has been recently reported as an innovative nanofiber spinning process. In this process, polymer solution and pressurized gas are simultaneously ejected. The high-velocity gas flow deforms the solution streams, evaporates the solvent, and solidifies them into fibers, as shown in Fig. (**4**) [15, 41, 42].

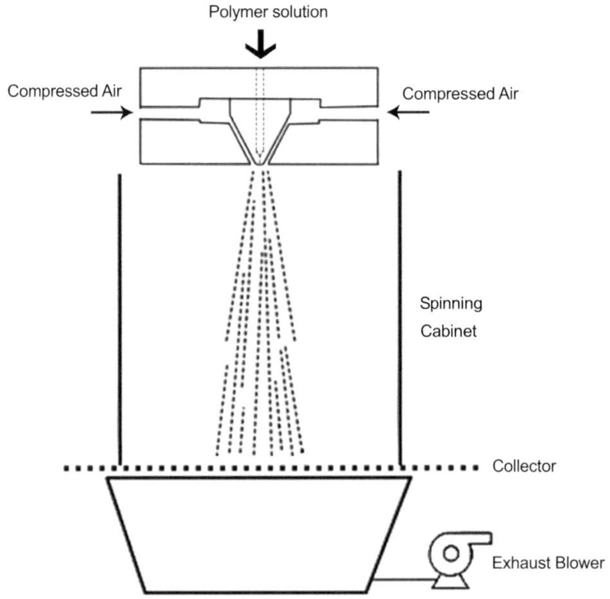

Fig. (4). Solution blowing apparatus [15].

The process is similar to electrospinning but, with different fiber formation driving forces. In solution blowing process, high-voltage equipment is not required. In this method, it is easy to use an assembly with many nozzles without considering of electric field interference. Therefore, solution blowing process is considered a secure and energy saving method for mass production of nanofibers [15].

A wide range of natural and synthetic polymers, such as polyvinylpyrrolidone (PVP), polyacrylonitrile (PAN), polylactic acid (PLA), and cellulose has been solution blown into nanofibers [15].

Coaxial Electrospinning

Coaxial electrospinning is a modern and cost effective technique to produce coaxial structured nanofibers. A typical coaxial setup consists of two concentric capillaries (two different solutions), which core and shell solutions come out and meet at the tip to generate core-sheath nanofibers (Fig. **5**) [16, 31, 43]. The main advantage of this method is the enhancement of a less spinnable material, using a highly spinnable material.

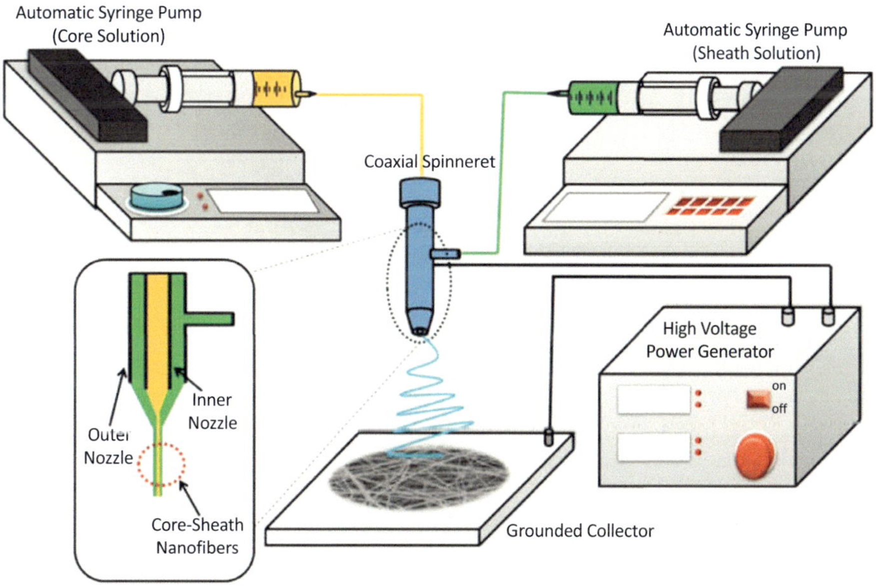

Fig. (5). Schematic equipment of coaxial electrospinning for core-sheath nanofibers [43].

Accordingly, a variety of materials including inorganic and organic compounds have been successfully combined as functional nano-composites for high value-added applications. It has been applied widely in controlling fiber secondary structures, encapsulating drugs or biological agents into fibers, preparing nanofibers from materials that lack filament-forming, and enclosing functional liquids within the fiber matrix [43, 44]. Various drugs and bioactive agents such as antibiotics, DNA, proteins, or growth factors could be incorporated directly into the core, which was protected by the shell layer [45].

It was reported that some polymer solutions that are difficult to process, could be co-electrospun to form ultrafine cores within the shells of other polymer materials [45]. Recently, chitosan nanofibers were produced by coaxial electrospinning. The core-shell nanofibers can be fabricated with highly spinnable synthetic polymers as core or shell templates [31]. Ojha and others [46] have manufactured chitosan nanofibers *via* coaxial electrospinning using polyethylene oxide (PEO) as shell template. The chitosan nanofibers were obtained by removing the PEO by dipping in distilled water. Recently, Pakravan and others [47] produced core-shell nanofibers with chitosan shell and PEO core in aqueous acetic acid by coaxial electrospinning. PEO was removed by dipping the core-shell nanofibers in water for 24 h to obtain hollow chitosan nanofibers.

Centrifugal Electrospinning

In many applications, the deposition of nanofibers in a specific direction is a requirement. For example, in the fabrication of electronic and photonic devices or for tissue engineering applications, aligned and highly ordered architectures are often required [17, 48].

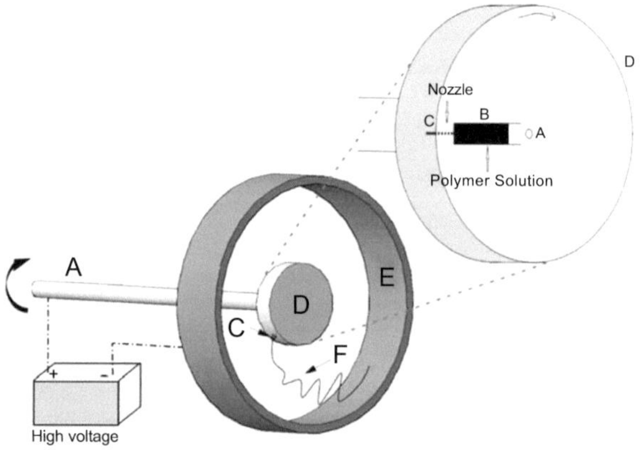

Fig. (6). Schematic representation of centrifugal electrospinning: (**A**) axle of rotation, (**B**) polymer solution container, (**C**) nozzle tip, (**D**) encircling cylinder, (**E**) collector and (**F**) polymeric jet [48].

Recently, a new method to produce aligned nanofibers, called centrifugal electrospinning was introduced. This method combines electrical and centrifugal fields [17, 48, 49]. The centrifugal field ensures additional stretching of the spinning jet, leading to further orientation of the polymeric chains in the nanofibers. Thus, this force can be used to apply high tension on a polymer solution. The centrifugal electrospinning is composed by a reservoir containing the solution, connected to a high voltage power supply through a contactor, nozzle tip, electric motor, cylindrical collector or collector composed of strips, as shown in Fig. (**6**) [48, 49].

Ultrasonic Atomization and Freeze Casting

Freeze casting uses only water as the main solvent. Freeze casting is extensively investigated for simple freeze-drying and jet-rapid freezing, which it has also been used in fiber fabrication. In a typical ice templating process, an aqueous polymer solution or suspension is frozen using a refrigerator, liquid nitrogen, or dry ice. The freezing rate strongly impacts the structures. When the freezing rate exceeds the polymer's Brownian diffusion, the polymer solutes segregate between ice clusters, and upon ice sublimation under vacuum. The polymer solutes form an aligned fiber structure, which is the reverse replica of the ice template. However, when the freezing rate is lower than the polymer's Brownian diffusion, the growing ice crystals steadily drive out the polymer solutes into a macroscopically separated phase and form a random polymer structure. Thus, uniformly oriented polymer fibers are generated by rapid freezing [18, 50]. The schematic illustration and preparation of the ultrasonic atomization-freeze casting of chitosan nanofibers was presented in Fig. (**7**).

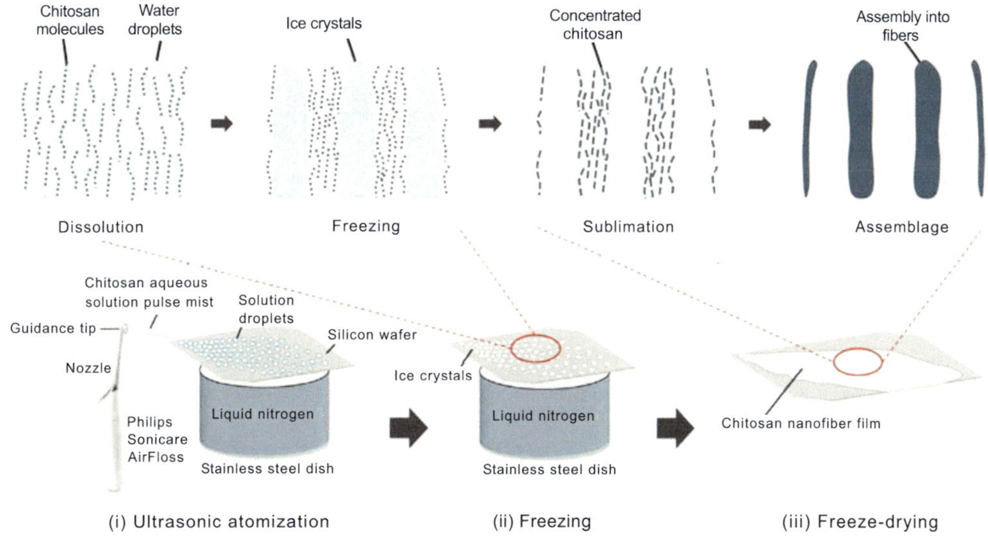

Fig. (7). Schematic illustration of the ultrasonic atomization-freeze casting of chitosan nanofibers [18].

In the first step (i), the nozzle is kept parallel to the silicon wafer surface, cooled using liquid nitrogen, and the aqueous chitosan solution is atomized along the edge. The solution droplets froze rapidly into ice crystals, which grew perpendicular to the cold surface, displaced, and phase separated. In second step (ii), these crystals are removed by lyophilization; and in the last step (iii), the concentrated chitosan assembled into a nanofiber film [18].

CHITOSAN NANOFIBER CHARACTERIZATION

Nanofibers present high surface area, small pore size and the possibility to produce three dimensional structures [51]. Physical, chemical and biological properties of nanomaterials have attracted attention in material science, mechanics, electronics, medicine, plastics and energy. Thus, characterization techniques to allow a better control of morphology, size and dimension of nanomaterials are necessary [52]. To evaluate and develop the manufacturing process, it is necessary to know nanofibers properties. The analysis of these properties will dictate whether the produced material is suitable for a specific application [53, 54]. Examples of characterization techniques that will be discussed in this chapter are described in Table **1**.

Table 1. Examples of nanofiber characterization techniques (adapted) [21].

Technique	Application
SEM	Investigation of nanofiber morphology
TEM	Investigation of nanofiber morphology and internal structure
DRX	Crystallinity and chemical characterization
FTIR	Identification of the compounds structure
DSC	Crystallinity determination
TGA	Changes of the conformational structure
Tensile Test	Mechanical properties of nanofiber

SCANNING ELECTRON MICROSCOPY (SEM)

In Scanning Electron Microscopy (SEM), a high energy electron beam interact with the sample producing secondary electrons that are detected, amplified and displayed in a high-resolution image of the sample surface [55]. SEM micrographs can provide an idea of the material nature and quality, like purity level. The use of software make possible to calculate area, diameter and perimeter of each fiber [54]. Hadipour-Goudarzi and others [2] obtained SEM images of a nanofibrous membrane of chitosan/sericin/poly(vinyl alcohol) with and without silver nitrate ($AgNO_3$), as shown in Fig. (**8**). The authors concluded that the presence of $AgNO_3$ results in the reduction of the mean diameter of the nanofiber.

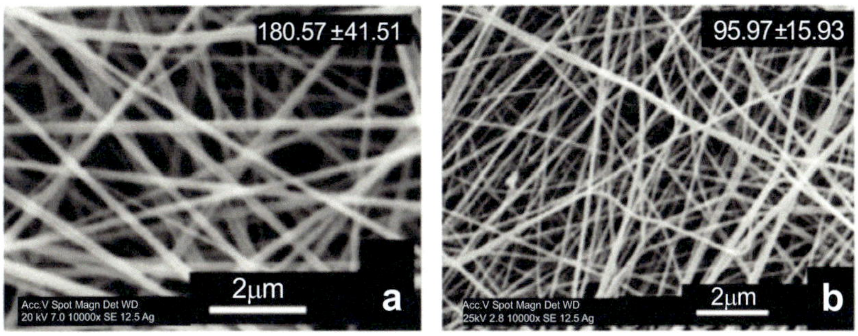

Fig. (8). SEM images of chitosan/sericin/poly(vinyl alcohol) (25:25:50) nanofibrous mats (**a**) without AgNO3, (**b**) with (AgNO3) [2].

Jalaja and co-workers [31] observed the morphology and fiber diameter distribution of a gelatin-chitosan core-shell nanofiber through SEM images (Fig. **9**). The authors concluded that the fibers are mostly randomly oriented with narrow fiber diameter distribution.

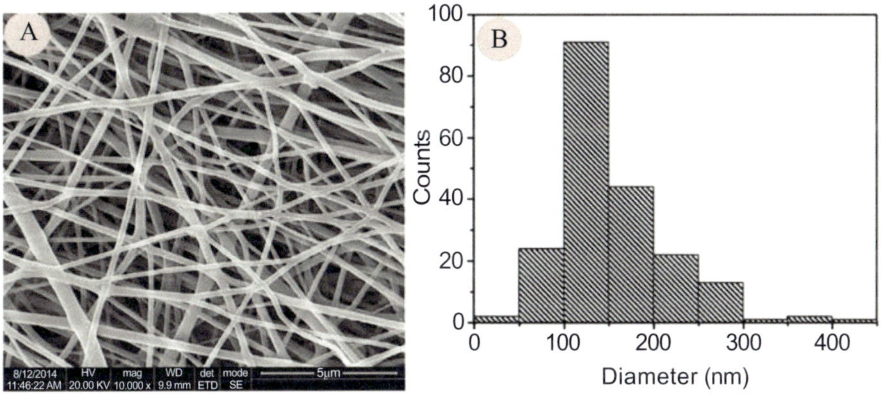

Fig. (9). (**a**) SEM micrograph and (**b**) diameter histogram of the gelatin/chitosan core-shell nanofibers [31].

Transmission Electron Microscopy (TEM)

In this technique, a beam of electrons is transmitted through a sample and forms an image, which is magnified and directed to a fluorescent screen and detected by a sensor.

The volume and density of the sample affect the intensity of the transmitted beam. TEM can be used to characterize fiber diameter, diameter distribution, fiber orientation and morphology [53].

Zhuang and others [56] obtained the TEM image of chitosan/gelatin nanofiber containing silver nanoparticles (AgNPs), as shown in Fig. (**10**). The authors observed that the AgNPs were successfully incorporated into the electrospun nanofibers.

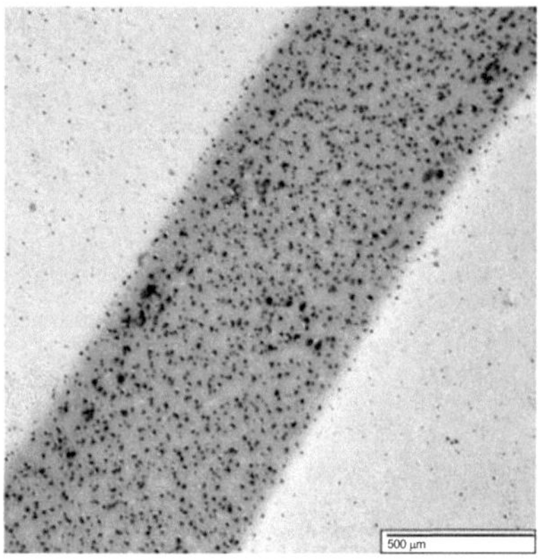

Fig. (10). Section TEM image of an electrospun fiber containing silver nanoparticles [56].

Jalaja and others [31] obtained the TEM micrographs of a single electrospun gelatin-chitosan core-shell nanofiber (Fig. **11**). The authors observed a clear difference in the contrast between the core and shell region.

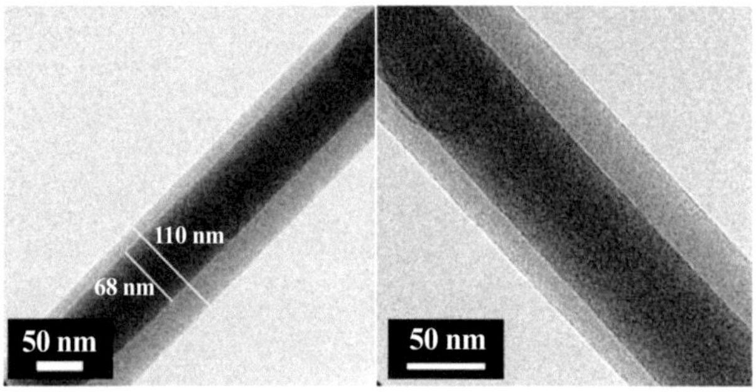

Fig. (11). TEM micrographs of gelatin/chitosan core-shell nanofibers [31].

X-Ray Diffraction (XRD)

XRD is a technique in which a beam of X-rays is aimed at the sample, generating a diffraction pattern. This pattern is a fingerprint of the atomic level periodicity in the polymer sample, used to identify sample's crystalline phases and to measure its structural properties [53, 57]. Zhang and others [1] showed the XRD diffraction patterns of a nanofiber with different weight ratio of nylon-6 to chitosan (Fig. **12**). The authors observed that increasing the chitosan ratio in the fiber, the pure nylon and pure chitosan peaks disappeared, suggesting a strong interaction between chitosan and nylon molecules.

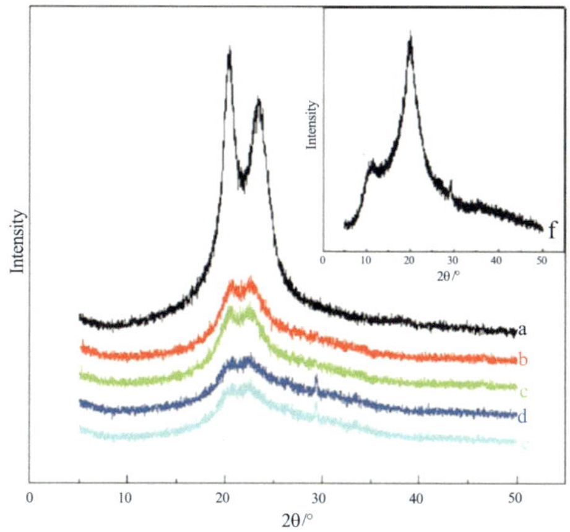

Fig. (12). X-ray diffraction patterns of electrospun complex nanofibers with different weight ratio of nylon-6 to chitosan: (**a**) 100/0; (**b**) 90/10; (**c**) 85/15; (**d**) 80/20; (**e**) 75/25. The insert illustrates (**f**) is the XRD spectra of the chitosan [1].

Nista and co-workers [16] displayed the XRD analyses aiming to monitor a possible change of crystallinity in an alginate-chitosan-poly(ethylene oxide) (SA/CHI/PEO) nanofiber membrane (Fig. **13**). The nanofiber XRD diffraction pattern was compared to the ones of the individual membranes of SA/PEO and CHI/PEO. The intensity of the peaks varied among the samples, showing that SA/PEO membrane was the most crystalline while the CHI/PEO membrane was the least. The ternary mixture of SA/CHI/PEO has an intermediate crystallinity value between chitosan and alginate. This can be attributed to the reorganization of the chain structures of the binary blends.

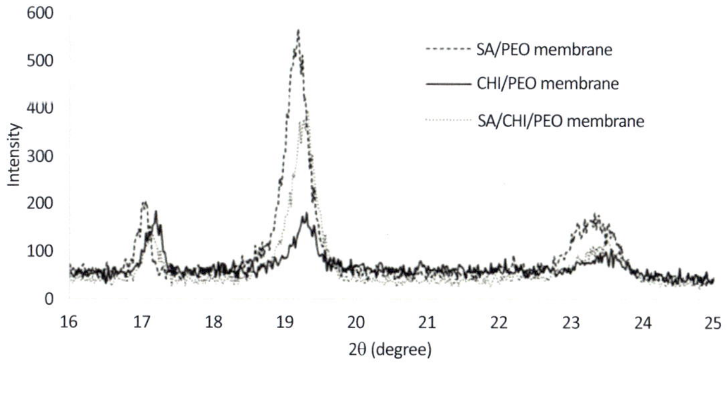

Fig. (13). XRD diffraction pattern of nanofibers obtained from solutions SA/PEO and CHI/PEO and of coaxial nanofibers SA/CHI/PEO [16].

Thermal Analysis

In order to compare nanofibers with their pure polymers, it is important to check thermal characteristics, like glass transition temperature, melting temperature, enthalpy of fusion, degree of crystallinity and thermal degradation. These properties provide evidences of structural changes after the nanofiber production [58].

Differential Scanning Calorimetry (DSC)

In DSC, differences in the amount of heat required to increases the temperature of a sample and a reference, are recorded and measured against the programmed temperature. This difference in heat flow between sample and reference makes DSC able to measure the amount of heat absorbed or released during a physical transformation. Events such as, phase transitions, melting temperature, glass transition temperature, chemical reaction or solvent evaporation can be observed as distinctive patterns in thermal behavior [53, 57].

Zhang and others [1] evaluated intermolecular interactions in the nylon-6/chitosan nanofibers by DSC. The effects of chitosan content are demonstrated in Fig. (**14**). Increasing chitosan contents, a systematic shift of the melting point towards a lower temperature was observed. This endothermic melting indicated a change of crystalline phase.

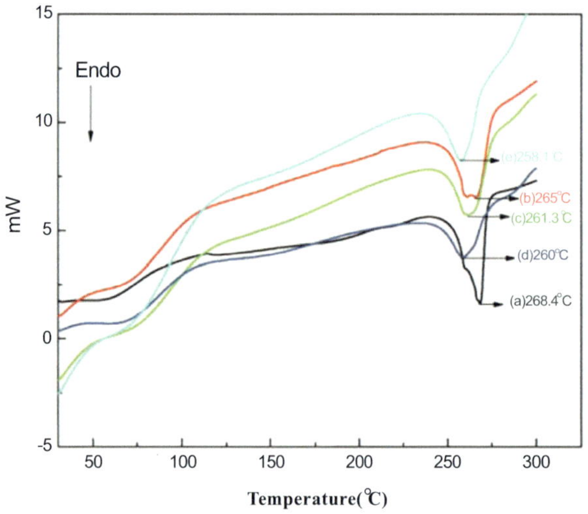

Fig. (14). DSC thermograms of electrospun complex nanofibers with varying content weight ratio of nylon-6 to chitosan: (**a**) 100/0; (**b**) 90/10; (**c**) 85/15; (**d**) 80/20; (**e**) 75/25 [1].

Nista and others [16] obtained the DSC curve of the alginate-chitosan-poly(ethylene oxide) (SA/CHI/PEO) nanofiber membrane, along with the curves of SA/PEO and CHI/PEO (Fig. **15**). The shifting of the endothermic and exothermic peaks indicated the formation of the alginate-chitosan ionic complex.

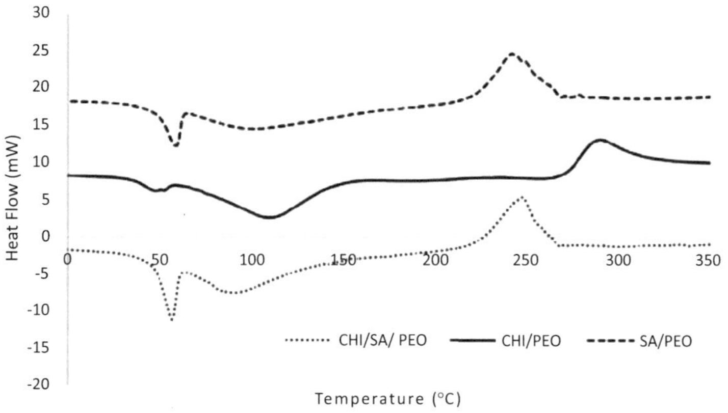

Fig. (15). DSC thermograms for the nanofiber obtained from SA/PEO, CHI/PEO as well as the thermogram for the coaxial nanofibers of CHI/SA/PEO [16].

Thermogravimetric Analysis (TGA)

TGA technique measures changes in sample mass according to the temperature variation by using a thermo balance. A thermo balance combines a suitable electronic microbalance with a controlled atmosphere furnace, a temperature programmer and a computer for control. The mass, time and temperature data can be captured while the sample is simultaneously weighed and heated or cooled in a controlled manner. TGA has been widely used to study changes in conformational structures of polymers during the nanofiber production process [53].

Nista and others [16] obtained the thermogram of the nanofiber membrane from the SA/CHI/PEO membrane, which showed two regions of weight loss (Fig. **16**). Comparing the onset degradation temperatures of the three membranes, it is noticed that the CHI/PEO membrane presented a better thermal stability, followed by the SA/PEO and SA/CHI/PEO membranes. The shift to a lower temperature in the thermal degradation indicates a reorganization of the chitosan and alginate chains due to their complexation.

Jalaja and others [31] obtained the TGA and DTG thermograms of a gelatin-chitosan core-shell nanofiber, as presented in Fig. (**17**). The DTG thermogram of core-shell nanofiber exhibits two decomposition temperatures at 282 °C and 305 °C. This decomposition indicates the presence of both chitosan and gelatin in the resulting composite nanofiber.

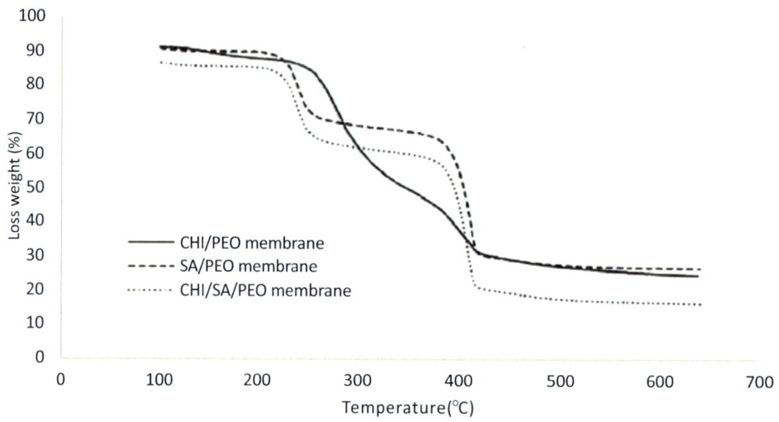

Fig. (16). TGA thermograms for the nanofiber obtained from SA/PEO and CHI/PEO and of coaxial nanofibers SA/CHI/PEO [16].

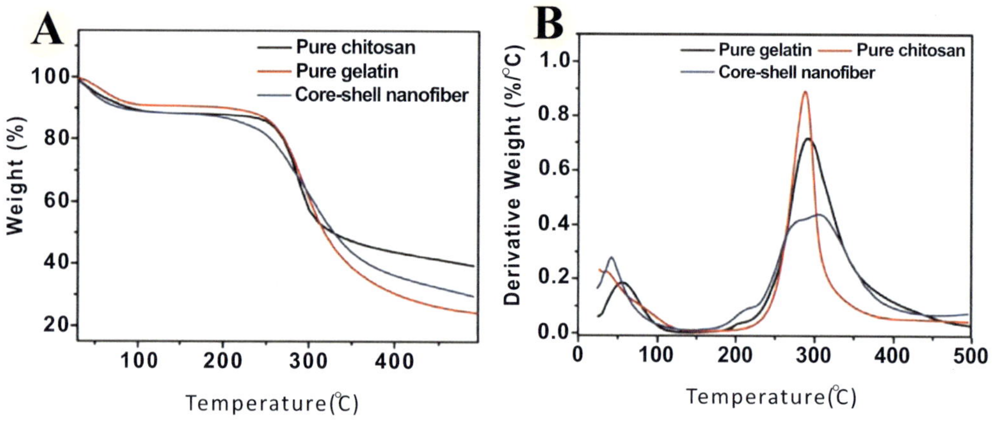

Fig. (17). (**a**) TGA thermograms, (**b**) DTG thermograms of pure gelatin pure chitosan and gelatin/chitosan core shell mats [31].

Fourier Transform Infrared Spectroscopy (FTIR)

In FTIR spectroscopy, the sample is submitted to an infrared beam. This radiation is then absorbed, promoting the excitation of its molecular vibrations. At specific frequencies, the chemical bonds absorb the infrared energy, corresponding to its molecular modes of vibration in the region of the electromagnetic spectrum. The resulting spectrum is characteristic of the molecules present in the sample. Because each specific chemical group has a specific frequency, FTIR is an effective analytical tool for compounds identification [58, 59].

Jalaja and co-workers [31] observed the FTIR spectra of pure chitosan, pure gelatin and gelatin-chitosan core-shell nanofibers (Fig. **18**). All the three spectra exhibit characteristic peaks for both chitosan and gelatin. The spectrum of nanofiber is similar to the pure chitosan spectrum, indicating the presence of chitosan on the surface. The analysis confirms the presence of chitosan and gelatin in the core-shell system.

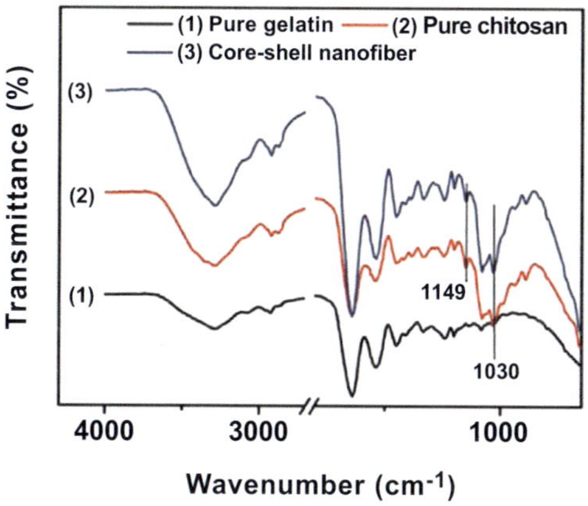

Fig. (18). FTIR spectra of pure gelatin, pure chitosan and gelatin-chitosan core-shell mats [31].

Hadipour-Goudarzi and others [2] studied the chemical structure of chitosan/sericin/poly(vinyl alcohol) (Cs/SS/PVA) nanofiber membrane with FTIR. The FTIR spectra of Cs, SS and PVA powders and Cs/SS/PVA nanofiber are shown in Fig. (**19**). Similar bands can be observed, indicating the presence of all compounds in the final nanofiber structure.

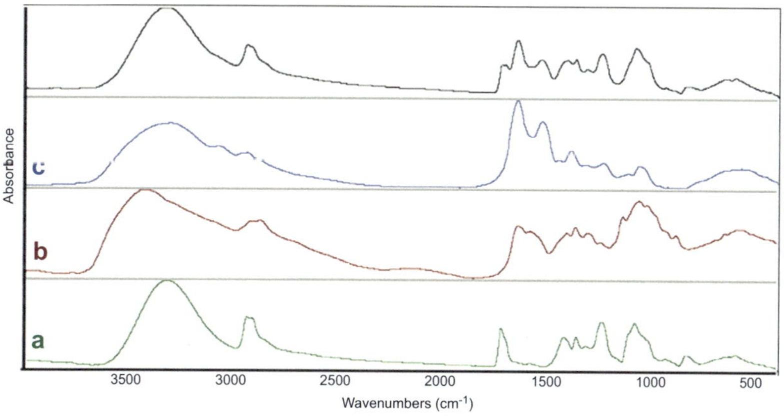

Fig. (19). FTIR spectrums of (**a**) PVA powder, (**b**) Cs powder, (**c**) SS powder and (**d**) Cs/SS/PVA nanofibrous mat [2].

Tensile Test

Mechanical properties of materials describe their responses when loads and displacements are applied [53]. The mechanical properties of nanofibers are influenced by its structure, geometrical arrangement, individual properties and interaction between the fibers. Therefore, it is fundamental determines the tensile deformation of a fiber [60]. The most common technique is tensile property measurement using a universal testing machine [57].

Zhang and others [1] studied the tensile strength and tensile strain for the electrospun nylon-6/chitosan nanofibrous with various chitosan contents (Fig. **20**). By increasing the weight ratio of chitosan, a decrease in the average tensile strength was observed. The addition of chitosan disturbed the original nylon-6 polymer chains and the disturbance was more significant using higher chitosan content.

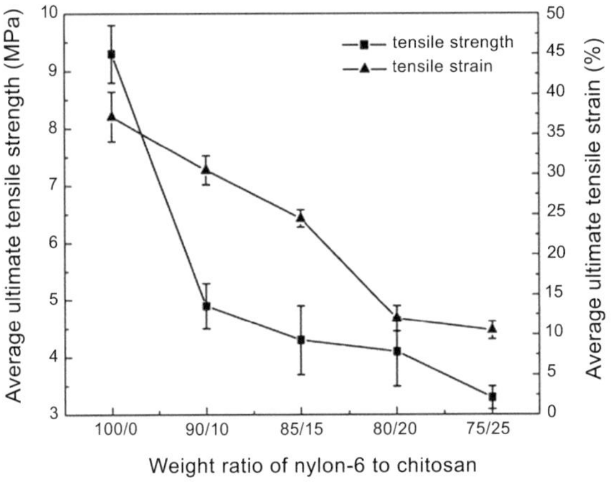

Fig. (20). Tensile properties of electrospun nylon-6/chitosan blend nanofibers with different weight ratio of nylon-6 to chitosan [1].

Liu and others [61] presented a tensile test for poly(vinyl alcohol) (PVA)/chitosan (CS)/graphene oxide (GO) nanofibers, as shown in Fig. (**21**). The changes in mechanical properties of nanofiber with different contents of GO was verified. It was observed that the tensile strengths of the nanofibers were significantly increased using high GO contents. However, at 0.4 wt% GO loading, the tensile strength starts to saturate and then decreases. Thus, GO content within the limits of reasonable concentration may be utilized to improve the nanofiber mechanical properties.

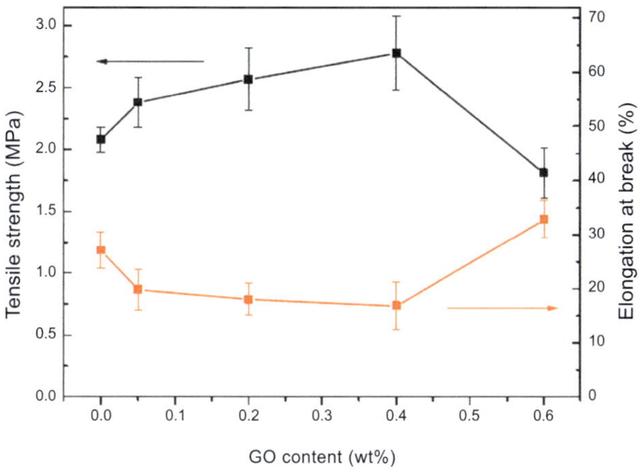

Fig. (21). Tensile stress and elongation at break of nanofibers with different contents of GO [61].

APPLICATIONS OF CHITOSAN NANOFIBERS

Chitosan nanofibers can be used in a wide range of applications, mainly due to the presence of -NH_2 and OH groups. Some of the potential applications of chitosan nanofibers are in the areas of drug delivery, biosensor, food packing, water purification and treatment, tissue engineering and wound dressing. This biopolymer has huge structural possibilities for chemical modifications, which generate novel properties especially for biomedical applications [12].

Water Purification and Treatment

In recent decades, there is a growing demand for high quality water. Drinking water contaminated may cause millions of deaths annually, due mainly to pathogen contamination [62, 63]. Membrane filtration is an accepted technology for the reduction of contaminants from drinking water. Bacteria are easily removed with an ultrafiltration membrane. Antibacterial surfaces made of silver and other metal ions have been produced. Viruses have been removed by nanofiltration membranes using the size-exclusion principle [1]. One material used for filtration applications is chitosan nanofibers. Chitosan has been used as an antimicrobial and antiviral material in the fields of biotechnology, pharmaceutics, wastewater treatment, cosmetics, agriculture, food science, and textiles because of its advantageous biological properties [64]. Chitosan nanofibers have been extensively investigated in the last decades for water purification. This nanofibrous material is described as a suitable natural polymer for removing heavy metals and dyes, due to the presence of large number of -NH_2 and -OH groups [12, 65].

Bai and others [62] discovered that electrospun nanofibers made of chitosan and functionalized with a quaternary amine (HTCC) have the ability to remove pathogens, such as porcine parvovirus (PPV). Graphene was added to improve the virus removal. The hydrophobicity of graphene and the high charge of the HTCC create a system that can bind 95% of PPV. Mi and others [64] also studied chitosan functionalized with a quaternary amine to form quaternized chitosan N-[(2-hydroxyl-3-trimethylammonium) propyl] chitosan (HTCC) to remove pathogens from drinking water. The authors concluded that HTCC has potential to be incorporated into a microfiltration membrane to remove virus. This could create an inexpensive, low pressure filtration membrane for drinking water purification.

Askari and co-workers [63] fabricated a chitosan (CS)/Poly vinyl alcohol (PVA) mat, by electrospinning, to apply in a dynamic process of copper and nickel adsorption. The authors concluded that the structural characteristics of the composite membrane played a significant role in the performance characteristics and efficiency of the nanofibrous chitosan-based filter.

Cooper and others [66] developed chitosan-polycaprolactone (PCL) nanofibrous membranes to utilize in water filtration. The authors observed that chitosan-PCL nanofiber membranes significantly reduced *Staphylococcus aureus*, compared to PCL fibrous membranes. In water permeability and particulate size removal tests, fibrous membranes with 25% chitosan supported the greatest water flux (~7000 L h^{-1} m^{-2}) with 100% removal of 300 nm particulates, maintaining the membrane integrity.

Aliabadi and others [67] prepared polyethylene oxide (PEO)/chitosan nanofibers by the electrospinning technique for the adsorption of nickel (Ni), cadmium (Cd), lead (Pb) and copper (Cu) from aqueous solution. The authors observed that the sorption selectivity of lead, copper, cadmium and nickel onto the membrane was in order of Pb(II)<Cd(II)<Cu(II)<Ni(II). The maximum adsorption capacity of nickel, copper, cadmium and lead ions by the PEO/Chitosan nanofiber membrane followed the order: nickel(II)>copper(II)>cadmium(II)>lead(II). The calculated thermodynamic parameters ($\Delta G°$, $\Delta H°$ and $\Delta S°$) showed that the adsorption of metal ions onto the PEO/chitosan nanofiber membrane was feasible, spontaneous and endothermic. The reusability of nanofiber membrane was determined after five adsorption-desorption cycles and the results showed that PEO/chitosan nanofiber membrane could be utilized extensively in industrial activities.

Razzaz and others [68] produced chitosan nanofibers functionalized by TiO_2 nanoparticles for the removal of heavy metal ions. The authors showed that the chitosan/TiO_2 nanofibers prepared by entrapped method could be reused frequently without significant loss in adsorption performance after five adsorption/desorption cycles. Whereas, lower than 60% of total adsorption in first cycle took place for metal ions sorption using nanofibers prepared by coating method. The sorption studies of metal ions in a binary system revealed that the selectivity of metal sorption using chitosan/TiO_2 nanofibrous adsorbent was in order of Cu (II)>Pb (II).

Wound Dressing

Antibacterial resistance of microorganisms is generally one of the major problems in the field of wound care and management, resulting in complications such as, infection and delayed wound healing [10]. An ideal wound dressing material must be able to absorb the exuded liquid from the wounded area and, should permit the water evaporation at a certain rate and allow no microbial transport [69]. Currently a lot of researches are focused on the developing of new antibacterials, to treat wounds infected with resistant microorganisms. Chitosan nanofibers have been considered as a wound dressing material, due its high efficacy against bacteria and fungi [10, 19, 69].

Abdelgawad and others [70] obtained antimicrobial nanofiber mats from colloidal dispersions of chitosan (CS) based Ag nanoparticles (Ag-NPS) blended with polyvinyl alcohol (PVA), for wound dressing applications. Non-woven mats of PVA/CS blends and PVA/CS-Ag-NPs blends were successfully fabricated by the electrospinning method. The authors concluded that the PVA/CS blends presented good bactericidal activity against the Gram-negative bacteria *E. coli*. However, e-spun mats of PVA/CS-Ag-NPs blends were better. The presence of Ag-NPs in PVA/CS blend solutions enhanced not only the electrospinning performance, but also, the antibacterial ability of the e-spun mats.

Zhao and others [71] fabricated chitosan/sericin composite nanofibers by electrospinning for biomedical applications. The authors observed that the composite nanofibers presented good morphology with diameter between 240 nm and 380 nm. *In vitro* methyl thiazolyl tetrazolium (MTT) assays demonstrated that the chitosan/sericin composite nanofibers were biocompatible and could promote the cell proliferation. Furthermore, the composite nanofibers showed good bactericidal activity against both of Gram-positive and Gram-negative bacteria. Thus, the chitosan/sericin composite nanofibers are promising for wound dressing applications.

Sarhan and Azzazy [72] produced polyvinyl alcohol (P) with honey (H) and chitosan (HP-chitosan nanofiber) for biomedical applications. The authors showed that the new HP-chitosan nanofibers presented pronounced antibacterial activity against *Staphylococcus aureus*, but weak antibacterial activity against *Escherichia coli*. The developed HP-chitosan nanofibers revealed no cytotoxicity effects on cultured fibroblasts. In conclusion, biocompatible, antimicrobial crosslinked honey/polyvinyl alcohol/chitosan nanofibers were developed as effective wound dressing.

Tissue Engineering

Polymeric nanofibers are of great interest in tissue engineering as scaffolding materials to restore, maintain or improve the function of human tissues, lost due to aging, disease, damage, or congenital defects [19, 21]. Tissue engineering can be subdivided into various segments based on the tissue/organ type, like bone, ligament, cartilage, tendon, liver, neural and skin regeneration [11]. There are some basic requirements that have been widely accepted for designing of polymer scaffolds. The scaffold should possess a high porosity with an appropriate pore size distribution, high surface area, biodegradability, appropriate mechanical properties and biocompatibility. Different biocompatible polymers from natural materials (collagen, elastin, laminin, etc.) have been developed [10, 21].

Shalumon and co-workers [73] prepared chitosan/poly(caprolactone) (CS/PCL) nanofibrous scaffold by electrospinning technique, for bone and skin tissue engineering. The presence of CS in CS/PCL scaffold provided a significant improvement in the hydrophilicity. The cyto-compatibility of the CS/PCL scaffold was examined using human osteosarcoma cells (MG63) and found to be non toxic. Cell attachment and proliferation was further confirmed by nuclear staining, using 4,6-diamidino-2-phenylindole (DAPI). All these results indicated that CS/PCL nanofibrous scaffold would be an excellent system for bone and skin tissue engineering.

Tsai and others [74] fabricated chitosan-gelatin-polyvinyl alcohol (PVA) hybrid nanofibrous mats using electrospinning. The authors concluded that the optimum weight ratio of chitosan/gelatin/PVA (2:2:4), produced fiber diameters around 150 nm. The chitosan-gelatin-PVA hybrid nanofibrous mats were further reinforced by crosslinking using glutaraldehyde. The crosslinked mats exhibited better stability and tensile strength (229.6±39.1 N/g). The cyto-compatibility test revealed better proliferation of mesenchymal stem cells on the nanofibrous mats than on the sponge. In this way, the electrospun chitosan-gelatin-PVA hybrid nanofibrous mats showed excellent potential for tissue engineering applications.

Sambudi and others [75] produced chitosan/poly(vinyl alcohol) (PVA) fibers by the incorporation of $CaCO_3$. The authors observed that the tensile test on 4 wt% $CaCO_3$-reinforced fibers, revealed a Young's modulus sixty times higher than the value of the fibers without $CaCO_3$. The biocompatibilities of the $CaCO_3$-loaded fibers were also tested in a 14-day simulated body fluid (SBF) incubation, to induce the formation of apatite minerals and in a 14-day cell proliferation assay with ATDC5 cells. The chitosan/PVA-$CaCO_3$ was found to provide the most suitable environment for cell growth in comparison with chitosan/PVA and chitosan/PVA-apatite. The $CaCO_3$-reinforced chitosan/PVA has potential to be a promising alternative as an artificial cartilage.

Drug Delivery

Nanofibers as drug delivery systems have many advantages due to the variety of polymers used. Depending on the polymer type and polymer-drug interactions, drug release can be immediate, delayed, or modified. When a combination of a water-soluble drug and water-soluble polymer is used, an immediate release is usually observed. However, the delayed release profile can be adjusted by drug incorporation into other nanocarriers like nanoparticles, liposomes, dendrimers, followed by their loading into nanofibers or by using hydrophobic polymers [21]. Recently, chitosan nanofibers have been investigated for controlled drug delivery applications. This material could be useful for the delivery of proteins/peptides, growth factors, anti-inflammatory drugs, antibiotics and imaging agents [7].

Meng and others [24] prepared poly(d,l-lactide-co-glycolide) (PLGA)/chitosan nanofibrous scaffold by electrospinning. The authors used fenbufen (FBF) as a model drug to investigate the optimization of release profile. The results showed that the drug release rate increased with the increase of chitosan content, because the addition of chitosan enhanced the hydrophilicity. Moreover, for the aligned PLGA/chitosan nanofibrous scaffold, the release rate was lower in relation to the randomly oriented PLGA/chitosan nanofibrous scaffold.

Yan and others [76] fabricated gold nanorods (AuNRs) containing polyvinyl alcohol/chitosan (PVA/CS) hybrid nanofibers with dual functions using a simple electrospinning method. The authors concluded that these nanofibers can be used as a carrier for anticancer agent doxorubicin (DOX), and the drug is delivered into the cell nucleus. The AuNRs and DOX incorporated fibers were effective to inhibit the growth and proliferation of ovary cancer cells and they can also be used as the cell imaging agent due to the unique optical properties. The nanofiber matrix combining two functions of cell imaging and drug delivery may be of great application potential in biomedical related areas.

CONCLUDING REMARKS

This chapter discussed various techniques to obtain and characterize nanofibers composed by chitosan, as well as, showed the examples of their applications. Features such as, high surface area, very high porosity and very small pore size allow the use of chitosan nanofibers and their derivatives in various industrial fields. Thus, to evaluate and develop the manufacturing process of chitosan nanofibers, it is necessary to analyze their morphology, crystallinity, chemical composition, change of structure and mechanical properties. Since chitosan is a natural polymer with advantageous biological properties, their nanofibers are promising for various applications. The presence of groups -NH_2 and OH in chitosan nanofibers allows its application in areas such as, biotechnology, wastewater treatment, medicinal and pharmaceutical.

CONFLICT OF INTEREST

The authors confirm that they have no conflict of interest to declare for this publication.

ACKNOWLEDGEMENTS

The authors would like to express their thanks to the CAPES, CNPq, FAPERGS and Federal University of Santa Maria (UFSM) for supporting this work.

REFERENCES

[1] Zhang H, Li S, White CJ, Ning X, Nie H, Zhu L. Studies on electrospun nylon-6/chitosan complex nanofiber interactions. Electrochim Acta 2009; 54(24): 5739-45. [http://dx.doi.org/10.1016/j.electacta.2009.05.021]

[2] Hadipour-Goudarzi E, Montazer M, Latifi M, Aghaji AA. Electrospinning of chitosan/sericin/PVA nanofibers incorporated with *in situ* synthesis of nano silver. Carbohydr Polym 2014; 113: 231-9. [http://dx.doi.org/10.1016/j.carbpol.2014.06.082] [PMID: 25256480]

[3] Elsabee MZ, Naguib HF, Morsi RE. Chitosan based nanofibers. Mater Sci Eng C 2015; 32(7): 1711-26. [review]. [http://dx.doi.org/10.1016/j.msec.2012.05.009]

[4] Lee SJ, Heo DN, Moon J-H, *et al.* Electrospun chitosan nanofibers with controlled levels of silver nanoparticles. Preparation, characterization and antibacterial activity. Carbohydr Polym 2014; 111: 530-7. [http://dx.doi.org/10.1016/j.carbpol.2014.04.026] [PMID: 25037384]

[5] Pillai CK, Paul W, Sharma CP. Chitin and chitosan polymers: Chemistry, solubility and fiber formation. Prog Polym Sci 2009; 34(7): 641-78. [http://dx.doi.org/10.1016/j.progpolymsci.2009.04.001]

[6] Jiang T, Deng M, James R, Nair LS, Laurencin CT. Micro- and nanofabrication of chitosan structures for regenerative engineering. Acta Biomater 2014; 10(4): 1632-45. [http://dx.doi.org/10.1016/j.actbio.2013.07.003] [PMID: 23851172]

[7] Dash M, Chiellini F, Ottenbrite RM, Chiellini E. Chitosan—A versatile semi-synthetic polymer in biomedical applications. Prog Polym Sci 2011; 36(8): 981-1014. [http://dx.doi.org/10.1016/j.progpolymsci.2011.02.001]

[8] Jayakumar R, Prabaharan M, Sudheesh Kumar PT, Nair SV, Tamura H. Biomaterials based on chitin and chitosan in wound dressing applications. Biotechnol Adv 2011; 29(3): 322-37. [http://dx.doi.org/10.1016/j.biotechadv.2011.01.005] [PMID: 21262336]

[9] Croisier F, Jérôme C. Chitosan-based biomaterials for tissue engineering. Eur Polym J 2013; 49(4): 780-92.

[http://dx.doi.org/10.1016/j.eurpolymj.2012.12.009]

[10] Jayakumar R, Menon D, Manzoor K, Nair SV, Tamura H. Biomedical applications of chitin and chitosan based nanomaterials—A short review. Carbohydr Polym 2010; 82(2): 227-32. [http://dx.doi.org/10.1016/j.carbpol.2010.04.074]

[11] Anitha A, Sowmya S, Kumar PT, *et al.* Chitin and chitosan in selected biomedical applications. Prog Polym Sci 2014; 39(9): 1644-67. [http://dx.doi.org/10.1016/j.progpolymsci.2014.02.008]

[12] Shukla SK, Mishra AK, Arotiba OA, Mamba BB. Chitosan-based nanomaterials: a state-of-the-art review. Int J Biol Macromol 2013; 59: 46-58. [http://dx.doi.org/10.1016/j.ijbiomac.2013.04.043] [PMID: 23608103]

[13] Homayoni H, Ravandi SA, Valizadeh M. Electrospinning of chitosan nanofibers: Processing optimization. Carbohydr Polym 2009; 77(3): 656-61. [http://dx.doi.org/10.1016/j.carbpol.2009.02.008]

[14] Xu F, Weng B, Gilkerson R, Materon LA, Lozano K. Development of tannic acid/chitosan/pullulan composite nanofibers from aqueous solution for potential applications as wound dressing. Carbohydr Polym 2015; 115: 16-24. [http://dx.doi.org/10.1016/j.carbpol.2014.08.081] [PMID: 25439862]

[15] Liu R, Xu X, Zhuang X, Cheng B. Solution blowing of chitosan/PVA hydrogel nanofiber mats. Carbohydr Polym 2014; 101: 1116-21. [http://dx.doi.org/10.1016/j.carbpol.2013.10.056] [PMID: 24299882]

[16] Nista SV, Bettini J, Mei LH. Coaxial nanofibers of chitosan-alginate-PEO polycomplex obtained by electrospinning. Carbohydr Polym 2015; 127: 222-8. [http://dx.doi.org/10.1016/j.carbpol.2015.03.063] [PMID: 25965478]

[17] Erickson AE, Edmondson D, Chang F-C, *et al.* High-throughput and high-yield fabrication of uniaxially-aligned chitosan-based nanofibers by centrifugal electrospinning. Carbohydr Polym 2015; 134: 467-74. [http://dx.doi.org/10.1016/j.carbpol.2015.07.097] [PMID: 26428148]

[18] Wang Y, Wakisaka M. Chitosan nanofibers fabricated by combined ultrasonic atomization and freeze casting. Carbohydr Polym 2015; 122: 18-25. [http://dx.doi.org/10.1016/j.carbpol.2014.12.080] [PMID: 25817638]

[19] Jayakumar R, Prabaharan M, Nair SV, Tamura H. Novel chitin and chitosan nanofibers in biomedical applications. Biotechnol Adv 2010; 28(1): 142-50. [http://dx.doi.org/10.1016/j.biotechadv.2009.11.001] [PMID: 19913083]

[20] Pakravan M, Heuzey M-C, Ajji A. A fundamental study of chitosan/PEO electrospinning. Polymer (Guildf) 2011; 52(21): 4813-24. [http://dx.doi.org/10.1016/j.polymer.2011.08.034]

[21] Pelipenko J, Kocbek P, Kristl J. Critical attributes of nanofibers: preparation, drug loading, and tissue regeneration. Int J Pharm 2015; 484(1-2): 57-74. [http://dx.doi.org/10.1016/j.ijpharm.2015.02.043] [PMID: 25701683]

[22] Kriegel C, Kit KM, McClements DJ, Weiss J. Electrospinning of chitosan-poly(ethylene oxide) blend nanofibers in the presence of micellar surfactant solutions. Polymer (Guildf) 2009; 50(1): 189-200. [http://dx.doi.org/10.1016/j.polymer.2008.09.041]

[23] Charernsriwilaiwat N, Opanasopit P, Rojanarata T, Ngawhirunpat T. Lysozyme-loaded, electrospun chitosan-based nanofiber mats for wound healing. Int J Pharm 2012; 427(2): 379-84. [http://dx.doi.org/10.1016/j.ijpharm.2012.02.010] [PMID: 22353400]

[24] Meng ZX, Zheng W, Li L, Zheng YF. Fabrication, characterization and *in vitro* drug release behavior of electrospun PLGA/chitosan nanofibrous scaffold. Mater Chem Phys 2011; 125(3): 606-11. [http://dx.doi.org/10.1016/j.matchemphys.2010.10.010]

[25] Ziani K, Henrist C, Jérôme C, Aqil A, Maté JI, Cloots R. Effect of nonionic surfactant and acidity on chitosan nanofibers with different molecular weights. Carbohydr Polym 2011; 83(2): 470-6. [http://dx.doi.org/10.1016/j.carbpol.2010.08.002]

[26] Aliabadi M, Irani M, Ismaeili J, Najafzadeh S. Design and evaluation of chitosan/hydroxyapatite composite nanofiber membrane for the removal of heavy metal ions from aqueous solution. J Taiwan Inst Chem Eng 2014; 45(2): 518-26. [http://dx.doi.org/10.1016/j.jtice.2013.04.016]

[27] Fouda MM, El-Aassar MR, Al-Deyab SS. Antimicrobial activity of carboxymethyl chitosan/polyethylene oxide nanofibers embedded silver nanoparticles. Carbohydr Polym 2013; 92(2): 1012-7. [http://dx.doi.org/10.1016/j.carbpol.2012.10.047] [PMID: 23399122]

[28] Angammana CJ, Jayaram SH. Analysis of the effects of solution conductivity on electrospinning process and fiber morphology. Industry Applications. IEEE Trans Ind Appl 2011; 47: 1109-17. [http://dx.doi.org/10.1109/TIA.2011.2127431]

[29] Zhou Y, Yang H, Liu X, Mao J, Gu S, Xu W. Electrospinning of carboxyethyl chitosan/poly(vinyl alcohol)/silk fibroin nanoparticles for wound dressings. Int J Biol Macromol 2013; 53: 88-92. [http://dx.doi.org/10.1016/j.ijbiomac.2012.11.013] [PMID: 23164753]

[30] Nam YS, Park WH, Ihm D, Hudson SM. Effect of the degree of deacetylation on the thermal decomposition of chitin and chitosan nanofibers. Carbohydr Polym 2010; 80(1): 291-5. [http://dx.doi.org/10.1016/j.carbpol.2009.11.030]

[31] Jalaja K, Naskar D, Kundu SC, James NR. Potential of electrospun core-shell structured gelatin-chitosan nanofibers for biomedical applications. Carbohydr Polym 2016; 136: 1098-107. [http://dx.doi.org/10.1016/j.carbpol.2015.10.014] [PMID: 26572452]

[32] Padron S, Patlan R, Gutierrez J, Santos N, Eubanks T, Lozano K. Production and characterization of hybrid BEH-PPV/PEO conjugated polymer nanofibers by Forcespinning™. J Appl Polym Sci 2012; 125(5): 3610-6. [http://dx.doi.org/10.1002/app.36420]

[33] McEachin Z, Lozano K. Production and characterization of polycaprolactone nanofibers *via* Forcespinning™ technology. J Appl Polym Sci 2012; 126(2): 473-9. [http://dx.doi.org/10.1002/app.36843]

[34] Weng B, Xu F, Lozano K. Development of hierarchical structured carbon nanotube-nylon nanofiber mats. J Appl Polym Sci 2015; 132(38)

[35] Weng B, Xu F, Lozano K. Mass production of carbon nanotube-reinforced polyacrylonitrile fine composite fibers. J Appl Polym Sci 2014; 131(11)

[36] Zander NE. Formation of melt and solution spun polycaprolactone fibers by centrifugal spinning. J Appl Polym Sci 2015; 132(2)

[37] Xu H, Chen H, Li X, Liu C, Yang B. A comparative study of jet formation in nozzle- and nozzle-less centrifugal spinning systems. J Polym Sci, B, Polym Phys 2014; 52(23): 1547-59. [http://dx.doi.org/10.1002/polb.23596]

[38] Vazquez B, Vasquez H, Lozano K. Preparation and characterization of polyvinylidene fluoride nanofibrous membranes by Forcespinning™. Polym Eng Sci 2012; 52(10): 2260-5. [http://dx.doi.org/10.1002/pen.23169]

[39] Badrossamay MR, McIlwee HA, Goss JA, Parker KK. Nanofiber assembly by rotary jet-spinning. Nano Lett 2010; 10(6): 2257-61. [http://dx.doi.org/10.1021/nl101355x] [PMID: 20491499]

[40] Shanmuganathan K, Fang Y, Chou DY, Sparks S, Hibbert J, Ellison CJ. Solventless high throughput manufacturing of Poly(butylene terephthalate) nanofibers. ACS Macro Lett American Chemical

Society 2012; 1(8): 960-4.
[http://dx.doi.org/10.1021/mz3001995]

[41] Zhuang X, Jia K, Cheng B, Feng X, Shi S, Zhang B. Solution blowing of continuous carbon nanofiber yarn and its electrochemical performance for supercapacitors. Chem Eng J 2014; 237: 308-11.
[http://dx.doi.org/10.1016/j.cej.2013.10.038]

[42] Bonan RF, Bonan PR, Batista AU, *et al. In vitro* antimicrobial activity of solution blow spun poly(lactic acid)/polyvinylpyrrolidone nanofibers loaded with Copaiba (Copaifera sp.) oil. Mater Sci Eng C Mater Biol Appl 2015; 48: 372-7.
[http://dx.doi.org/10.1016/j.msec.2014.12.021] [PMID: 25579936]

[43] Lan T, Shao Z, Wang J, Gu M. Fabrication of hydroxyapatite nanoparticles decorated cellulose triacetate nanofibers for protein adsorption by coaxial electrospinning. Chem Eng J 2015; 260: 818-25.
[http://dx.doi.org/10.1016/j.cej.2014.09.004]

[44] Yu D-G, Yu J-H, Chen L, Williams GR, Wang X. Modified coaxial electrospinning for the preparation of high-quality ketoprofen-loaded cellulose acetate nanofibers. Carbohydr Polym 2012; 90(2): 1016-23.
[http://dx.doi.org/10.1016/j.carbpol.2012.06.036] [PMID: 22840034]

[45] Nguyen TT, Chung OH, Park JS. Coaxial electrospun poly(lactic acid)/chitosan (core/shell) composite nanofibers and their antibacterial activity. Carbohydr Polym 2011; 86(4): 1799-806.
[http://dx.doi.org/10.1016/j.carbpol.2011.07.014]

[46] Ojha SS, Stevens DR, Hoffman TJ, *et al.* Fabrication and characterization of electrospun chitosan nanofibers formed *via* templating with polyethylene oxide. Biomacromolecules 2008; 9(9): 2523-9.
[http://dx.doi.org/10.1021/bm800551q] [PMID: 18702544]

[47] Pakravan M, Heuzey M-C, Ajji A. Core-shell structured PEO-chitosan nanofibers by coaxial electrospinning. Biomacromolecules 2012; 13(2): 412-21.
[http://dx.doi.org/10.1021/bm201444v] [PMID: 22229633]

[48] Dabirian F, Hosseini Ravandi SA, Pishevar AR, Abuzade RA. A comparative study of jet formation and nanofiber alignment in electrospinning and electrocentrifugal spinning systems. J Electrost 2011; 69(6): 540-6.
[http://dx.doi.org/10.1016/j.elstat.2011.07.006]

[49] Kancheva M, Toncheva A, Manolova N, Rashkov I. Advanced centrifugal electrospinning setup. Mater Lett 2014; 136: 150-2.
[http://dx.doi.org/10.1016/j.matlet.2014.08.045]

[50] Lee D, Zhang C, Gao H. Facile production of polypyrrole nanofibers using a freeze-drying method. Macromol Chem Phys 2014; 215(7): 669-74.
[http://dx.doi.org/10.1002/macp.201400039]

[51] Bhardwaj N, Kundu SC. Electrospinning: a fascinating fiber fabrication technique. Biotechnol Adv 2010; 28(3): 325-47.
[http://dx.doi.org/10.1016/j.biotechadv.2010.01.004] [PMID: 20100560]

[52] Joshi M, Bhattacharyya A, Ali SW. Characterization techniques for nanotechnology applications in textiles. Indian J of Fibre & Textile Res 2008; 33: 304-17.

[53] Ko FK. WAN Y. Characterization of nanofibers. Introduction to Nanofiber Materials. 1st ed. United Kingdom: Cambridge University Press 2014; pp. 102-46.
[http://dx.doi.org/10.1017/CBO9781139021333.007]

[54] Regi M. Nanofibers and nanotechnology in textiles nanofibers and nanotechnology in textiles. Elsevier 2007; pp. 113-93.
[http://dx.doi.org/10.1533/9781845693732.2.113]

[55] Goldstein JI, Newbury DE, Echlin P, *et al.* Scanning electron microscopy and X-ray microanalysis A text for biologists, material scientists, and geologists. 2nd ed. New York: Plenum Press 1992; pp. 149-

271.
[http://dx.doi.org/10.1007/978-1-4613-0491-3_4]

[56] Zhuang X, Cheng B, Kang W, Xu X. Electrospun chitosan/gelatin nanofibers containing silver nanoparticles. Carbohydr Polym 2010; 82(2): 524-7.
[http://dx.doi.org/10.1016/j.carbpol.2010.04.085]

[57] Andrady AL. Science and technology of polymer nanofibers hoboken. NJ, USA: John Wiley & Sons, Inc. 2008; pp. 111-52.
[http://dx.doi.org/10.1002/9780470229842.ch5]

[58] Abdel Sayed AS. Electrospinning of polymeric nanofibers and nanocomposite materials: fabrication, physicochemical characterization and medical applications 2011.

[59] Cao G. Nanostructures and nanomaterials: synthesis, properties and applications. London, UK: Imperial College Press 2004; pp. 329-90.
[http://dx.doi.org/10.1142/9781860945960_0008]

[60] Baji A, Mai Y-W, Wong S-C, Abtahi M, Chen P. Electrospinning of polymer nanofibers: Effects on oriented morphology, structures and tensile properties. Compos Sci Technol 2010; 70(5): 703-18.
[http://dx.doi.org/10.1016/j.compscitech.2010.01.010]

[61] Liu Y, Park M, Shin HK, *et al.* Facile preparation and characterization of poly(vinyl alcohol)/chitosan/graphene oxide biocomposite nanofibers. J Ind Eng Chem 2014; 20(6): 4415-20.
[http://dx.doi.org/10.1016/j.jiec.2014.02.009]

[62] Bai B, Mi X, Xiang X, Heiden PA, Heldt CL. Non-enveloped virus reduction with quaternized chitosan nanofibers containing graphene. Carbohydr Res 2013; 380: 137-42.
[http://dx.doi.org/10.1016/j.carres.2013.08.020] [PMID: 24041631]

[63] Askari M, Rezaei B, Mousavi Shoushtari A, Abdouss M. Performance improvements in structural characteristics of chitosan-based nanofibrous composite membrane for using in liquid filtration. J Taiwan Inst Chem Eng 2015; 56: 77-83.
[http://dx.doi.org/10.1016/j.jtice.2015.05.007]

[64] Mi X, Vijayaragavan KS, Heldt CL. Virus adsorption of water-stable quaternized chitosan nanofibers. Carbohydr Res 2014; 387: 24-9.
[http://dx.doi.org/10.1016/j.carres.2014.01.017] [PMID: 24561959]

[65] Li L, Li Y, Cao L, Yang C. Enhanced chromium (VI) adsorption using nanosized chitosan fibers tailored by electrospinning. Carbohydr Polym 2015; 125: 206-13.
[http://dx.doi.org/10.1016/j.carbpol.2015.02.037] [PMID: 25857976]

[66] Cooper A, Oldinski R, Ma H, Bryers JD, Zhang M. Chitosan-based nanofibrous membranes for antibacterial filter applications. Carbohydr Polym 2013; 92(1): 254-9.
[http://dx.doi.org/10.1016/j.carbpol.2012.08.114] [PMID: 23218292]

[67] Aliabadi M, Irani M, Ismaeili J, Piri H, Parnian MJ. Electrospun nanofiber membrane of PEO/Chitosan for the adsorption of nickel, cadmium, lead and copper ions from aqueous solution. Chem Eng J 2013; 220: 237-43.
[http://dx.doi.org/10.1016/j.cej.2013.01.021]

[68] Razzaz A, Ghorban S, Hosayni L, Irani M, Aliabadi M. Chitosan nanofibers functionalized by TiO2 nanoparticles for the removal of heavy metal ions. J Taiwan Inst Chem Eng 2016; 58: 333-43.
[http://dx.doi.org/10.1016/j.jtice.2015.06.003]

[69] Kong M, Chen XG, Xing K, Park HJ. Antimicrobial properties of chitosan and mode of action: a state of the art review. Int J Food Microbiol 2010; 144(1): 51-63.
[http://dx.doi.org/10.1016/j.ijfoodmicro.2010.09.012] [PMID: 20951455]

[70] Abdelgawad AM, Hudson SM, Rojas OJ. Antimicrobial wound dressing nanofiber mats from multicomponent (chitosan/silver-NPs/polyvinyl alcohol) systems. Carbohydr Polym 2014; 100: 166-78.

[http://dx.doi.org/10.1016/j.carbpol.2012.12.043] [PMID: 24188851]

[71] Zhao R, Li X, Sun B, *et al.* Electrospun chitosan/sericin composite nanofibers with antibacterial property as potential wound dressings. Int J Biol Macromol 2014; 68: 92-7. [http://dx.doi.org/10.1016/j.ijbiomac.2014.04.029] [PMID: 24769088]

[72] Sarhan WA, Azzazy HM. High concentration honey chitosan electrospun nanofibers: biocompatibility and antibacterial effects. Carbohydr Polym 2015; 122: 135-43. [http://dx.doi.org/10.1016/j.carbpol.2014.12.051] [PMID: 25817652]

[73] Shalumon KT, Anulekha KH, Chennazhi KP, Tamura H, Nair SV, Jayakumar R. Fabrication of chitosan/poly(caprolactone) nanofibrous scaffold for bone and skin tissue engineering. Int J Biol Macromol 2011; 48(4): 571-6. [http://dx.doi.org/10.1016/j.ijbiomac.2011.01.020] [PMID: 21291908]

[74] Tsai R-Y, Hung S-C, Lai J-Y, Wang D-M, Hsieh H-J. Electrospun chitosan-gelatin-polyvinyl alcohol hybrid nanofibrous mats: Production and characterization. J Taiwan Inst Chem Eng 2014; 45(4): 1975-81. [http://dx.doi.org/10.1016/j.jtice.2013.11.003]

[75] Sambudi NS, Sathyamurthy M, Lee GM, Park S. Bin. Electrospun chitosan/poly(vinyl alcohol) reinforced with CaCO3 nanoparticles with enhanced mechanical properties and biocompatibility for cartilage tissue engineering. Compos Sci Technol 2015; 106: 76-84. [http://dx.doi.org/10.1016/j.compscitech.2014.11.003]

[76] Yan E, Cao M, Wang Y, *et al.* Gold nanorods contained polyvinyl alcohol/chitosan nanofiber matrix for cell imaging and drug delivery. Mater Sci Eng C Mater Biol Appl 2016; 58: 1090-7. [http://dx.doi.org/10.1016/j.msec.2015.09.080] [PMID: 26478408]

CHAPTER 4

Chitosan Nanoparticles and Nanocapsules

Alexandre L. Parize*, **Adalberto Enumo Jr.** and **Natália B. Caon Branco**

Chemistry Department, Federal University of Santa Catarina, Trindade - Florianópolis, Santa Catarina, 88040-900 Brazil

Abstract: Technological advances achieved by biocompatible systems in several areas have attracted increasing interest in the development of nanometric systems based on biopolymers. In this context, this chapter describes the use of chitosan biopolymer as a carrier and base matrix for the development of nanoparticles and nanocapsules. Chitosan is an abundant natural biopolymer in nature, which is renewable, biocompatible, has hemostatic and mucoadhesive characteristics, and is an interesting polymer material in the development of nanometric systems. Most of the preparation methods are based on green routes without the presence of organic solvents and toxic agents. Chitosan based nanoparticulated systems can be easily reproduced through simple and easy routes, and applied in various fields such as: food and nutrition industry, agricultural, pharmaceutical, medical, renewable materials, controlled delivery, nanoencapsulation, environmental remediation, adsorption, among others.

Keywords: Characterization, Incorporation methods, Nanoparticles, Nanocapsules.

INTRODUCTION

Chitosan is a polysaccharide widely found in nature. It is obtained by deacetylation of chitin, which is derived from crustacean exoskeletons and is naturally found in fungi and yeasts [1, 2]. Chitosan has a chemical structure similar to cellulose, presenting amine and acetamide functional groups which can modify the polymer properties [3], and also contains many amine and hydroxyl groups that can effectively bind to negatively charged substances by electrostatic interactions or hydrogen bonding [4].

This biopolymer attracts attention by presenting biological properties and uses in several areas such as biomedical [5, 6], food industry [7], agriculture [8], among others [9, 10]. Chitosan therapeutic properties such as biocompatibility [11], bio-

* **Corresponding author Alexandre L. Parize:** Chemistry Department, Federal University of Santa Catarina, Trindade - Florianópolis, Santa Catarina, 88040-900 Brazil; Tel/Fax: +55 48 37214535; E-mail: alexandre.parize@ufsc.br

Guilherme Luiz Dotto, Sergio Paulo Campana-Filho and Luiz Antonio de Almeida Pinto (Eds)

availability [12], microorganisms growth inhibition [13, 14], pain relief, hemostasis promotion and cell growth assistance are also reported [12, 15], making this biopolymer attractive in medical and pharmaceutical areas [16].

As vector system, chitosan has been attracting attention in recent years in the development of carrier materials, due to its biological and physicochemical properties, leading to the recognition of chitosan as a drug delivery medium and a promising material specifically for the delivery of macromolecules [1, 17, 18]. Chitosan and its derivatives have been widely used for the development of nanoparticles which could increase the drug oral bioavailability, present good mucoadhesive character, enhance the bioavailability and dissolution rates of hydrophobic drugs and enhance the stability of labile drugs, drug bioavailability and controlled drug release owing to the fact that particles in the nanosize ranges are efficient in crossing permeability barriers.

In this sense, chitosan-based delivery systems ranging from microparticles to nanoparticles (NP) and composite films are widely described in the literature [19 - 21]. However, there are some disadvantages of this polysaccharide as drug delivery system, such as poor solubility at physiological pH that ends up impacting the administration [14, 22, 23], causing presystemic metabolism when drugs are orally administered [14], and the ability of chitosan alone to encapsulate water-insoluble drugs is poor, because the loading efficiency is usually rather low [24]. Various chitosan derivatives have been made and presented success, especially when they are in the NPs form [14, 22, 23], in order to transcend these disadvantages.

The examples of nanoparticulated systems obtained by chitosan and its derivatives are presented in Table **1**. This description presents preparation methods, main characteristics of the final materials and their applications. Preparation, characterization methods and applications of chitosan and its derivatives NPs are described later.

Polymeric NPs are commonly described as solid colloidal particles, ranging in size from 10 nm to 1 μm. NPs are used in several areas as system for incorporation and storage of active compounds, and are designed from macromolecular and/or molecular assemblies, in which the active principles such as drugs are dissolved, entrapped, encapsulated, or even adsorbed or attached to the external interface. Nanoparticles can be categorized into two main groups based on their morphology: nanospheres, in general, are solid systems, presenting a homogeneous, matrix structure in which the actives are uniformly dispersed; and nanocapsules, showing a typical core-shell structure present a polymeric membrane an inner core [42, 43].

Table 1. Nanoparticulated systems of chitosan and its derivatives: characteristics and applications.

Biopolymer	Obtaining Method	Functionalizer Agent	System	Characteristics	Application	Ref.
CS and poly(ethylene glycol)-grafted (PEGylated) CS	Ionic gelation	TPP	Anti-β-catenin siRNA-loaded CS NP and anti-β-catenin siRNA-loaded PEGylated CS NP	100-150 nm Encapsulated up to 3 nmol of si-RNA, 48 h after transfection, NPs as effective as Lipofectamine 2000	Cancer treatment	[25]
CS	Ionic gelation	TPP	Extracts of *Gymnema sylvestre* leaves and *Cinnamomum zeylanicum* bark-loaded CS NP	58-120 nm Cytotoxic activity against the SiHa cells	Cancer treatment	[26]
CS	Ionic gelation	Sulphobutylether-β-cyclodextrin	Nag-loaded sulfobutylether-β-CD/CS NP	165-215 nm Higher ophthalmic bioavailability and prolonged residence time of Nag	Drug delivery	[27]
CS	Ionic gelation	Sulphobutylether-β-cyclodextrin	CS, HA and sulphobutyl-ether-β-CD and with or without curcumin	169-338 nm Encapsulation efficiency for lipophilic molecules and stability in water	Cancer treatment	[28]
CS	Emulsion (o/w) and ionic gelation	Essential oils	CS and TPP with essential oil extracted of dried leaves of summer savory (*S. hortensis L.*)	135 - 237 nm; −7.54 to -21.12 mV; Size and zeta potential are dependent of pH of CS and TPP concentration	Antimicrobial, cytotoxicity	[29]
CS	Emulsion (o/w) and ionic gelation	Ascorbyl palmitate (AsP)	CS and TPP	250-930 nm Dependent with the increasing of AsP concentration	Antioxidant	[30]
CS	Emulsion (o/w) and polyelectrolyte complexation	Curcumin	CS and AG	Size and zeta potential are dependent: pH of CS solution and mixing ratio of CS:AG	Antioxidant	[31]

(Table 1) contd.....

Biopolymer	Obtaining Method	Functionalizer Agent	System	Characteristics	Application	Ref.
CS	Oil-in-water emulsion-ionic gelation	Tween 60-TPP	Eugenol-loaded CS NPs	< 100 nm Encapsulation could reduce eugenol loss during thermal process	Food packing	[32]
CS	Oil-in-water emulsion-ionic gelation	Tween 80-CH_2Cl_2	Oregano essential oil-loaded CS NPs	40-80 nm spherical shape Burst effect followed by slow release	Drug delivery	[33]
CS	Ionic gelation	TPP	Chlorogenic acid loaded CS NPs	200-600 nm spherical shape Sustained release	Drug delivery	[34]
CS	Ionic gelation	TPP	Aspirin and probucol in combination loaded CS NPs	100-300 nm Continuous release	Drug delivery	[35]
CS	Ionic gelation	TPP	Vitamin C loaded CS NPs	185 nm, positive charger, spherical shape *In vivo* controlled release until 48h	Drug delivery	[36]
CS and AG	Ionic gelation	TPP	Interaction between CS and AG for loading insulin	170-245 nm +35.7 to +43.4mV Release profile of insulin is dependent of pH	Drug delivery	[37]
N,N-dimethyl CS (DMC) and N,N,N-trimethyl CS (TMC)	Emulsion (o/w) and ionic gelation	Curcumin	N,N-dimethyl CS (DMC) and N,N,N-trimethyl CS (TMC) with TPP, to develop NPs	DMC/TPP > 316 nm and TMC/TPP ~99nm	Cytotoxicity	[38]
CS and CMCS	Spray drying	Idebenone	CS and CMCS contain idebenone and crosslinked with TPP	Characteristics of developed NPs are dependent of polymers and drug ratio	Drug delivery, antioxidant	[39]
N,N,N-trimethyl CS (TMC)	Self-assembly	Ropinirole Hydrochloride	NanoPEC between TMC and DS at different pH conditions	241- 306 nm; −3.9 to −30.4mV Properties are dependents of pH for nasal applications	Drug delivery	[40]

(Table 1) contd.....

Biopolymer	Obtaining Method	Functionalizer Agent	System	Characteristics	Application	Ref.
N-(2-hydroxyl) propyl-3-trimethyl ammonium CS chloride (HTCC)	Ionic gelation and polyelectrolyte complexation	Bovine serum albumin (BSA)	BSA incorporated into HTCC NPs	110-180 nm Incorporation efficiency of BSA is dependent of TPP concentration	Protein encapsulation	[41]

Legend: AG: Arabic gum. AsP: Ascorbyl palmitate. BSA: bovine serum albumin. CD: cyclodextrin. CMCS: carboxymethylchitosan. CS: chitosan. DMC: N,N-dimethyl CS. DS: dextran sulphate. HA: hyaluronic acid. HTCC: N-(2-hydroxyl) propyl-3-trimethyl ammonium CS chloride. Nag: Naringenin. Si-RNA: small interfering RNA. TMC: N,N,N-trimethyl CS. TPP: pentasodium tripolyphospate.

Among the different types of nanotechnology innovations found in nanoparticles and nanocapsules, it is important to highlight that they are both materials with the main purpose to incorporate active substances, have high load capacity and provide controlled release [44 - 46]. The term nanoparticles is commonly used to describe both nanospheres as nanocapsules with differences in their morphology and architecture [47].

The relative interest of using nanoparticulate/nanocapsules rather than microparticulate systems in biotechnology applications is in their potential to increase the absorption rate, improve bioavailability, enable target drug delivery for cancer therapy and intravenous delivery systems and so forth [45]. This systems as drug delivery vehicles enables targeted delivery of drugs into cells rather than non-specific diffusion, improves solubilization and delivery of poorly water-soluble drugs, allows co-delivery of two or more drugs and protects labile drugs from degradation [24].

PREPARATION METHODS OF CHITOSAN NANOPARTICLES AND NANOCAPSULES

The preparation method that decreased the size of chitosan particles to the nanometers length was firstly reported in 1994 by Ohya and collaborators [48]. Chitosan NPs were synthesized by the emulsion and crosslinking method, and made possible the intravenous release of an anticancer drug (5-fluorouracil).

The formulation of nanoparticles involves: size control and distribution, morphology, therapeutic goal of active delivery, biocompatibility of the polymer and compatibility of the physicochemical properties of the incorporated active. The methods used to produce nanoparticles can be divided into three main groups: (i) physicochemical methods, which yield nanoparticles using preformed polymers and inducing their precipitation by emulsification-solvent evaporation, diffusion or reverse salting-out (ii) *in situ* chemical synthesis methods of macromolecules, giving rise for instance to polymerization or interfacial polycondensation reactions (iii) mechanical methods using high-energy devices like high-pressure homogenizers, sonifiers or high-energy wet mills.

The main obtaining methods for chitosan nanoparticles and nanocapsules are ionic gelation; emulsion and crosslinking; polyelectrolyte complexation; self-assembly and drying processes. Some characteristics of each method are presented next, focusing on leading aspects of the processes.

Ionic Gelation

Ionic gelation is probably the most common preparation method of chitosan-based nanoparticles. Chitosan is a cationic polysaccharide able to gel upon contact with multivalent polyanions such as tripolyphosphate (TPP) due to the establishment of electrostatic interactions between the opposite charges of constituents [49 - 51]. In acidic pH, it can spontaneously gel into submicron-sized particles. This process appeared to be very attractive since it allowed to elaborate nanoparticles, with reproducible sizes, in a mild environment (aqueous medium) and prevents the use of organic solvent and high temperatures [52].

Chitosan-tripolyphosphate NPs are prepared by dissolving chitosan in diluted glacial acetic acid solution. The resulting solution is stirred with a magnetic bar at room temperature until complete dissolution. The pH is thus raised to the expected value by addition of NaOH. TPP solution can be prepared in various concentrations by dissolving TPP in ultrapure water. Both solutions are then filtered through a small-pore filter. Next, TPP solution is added dropwise to the chitosan solution under constant stirring until the formation of an opaque suspension, which indicates the formation of particles [52 - 56]. The mean diameter of the NPs is controlled by chitosan/TPP molar ratio [49]. The following parameters determine the mechanisms of the chitosan NPs formation by ionic gelation using TPP, their inner structure and surface topology: degree of acetylation and molecular weight of chitosan, intrinsic viscosity, concentration and $-NH_3^+$/TPP molar ratio [17, 52, 55, 57 - 60]. Other factors may also influence the nanoparticle characteristics, such as preparation temperature or heat treatment can affect the particle size and polydispersity because it seems to reduce significantly the aggregation number of NP. Filtration and centrifugation can isolate the smaller particles from bigger particles [55, 66].

This technique allowed to efficiently trap hydrophobic and hydrophilic small molecules as well as peptides, proteins and genes. The establishment of electrostatic interactions within the matrix components is considered the main mechanism governing the retention of these bioactive compounds [51]. Besides TPP, other polyanions can be used to carry out the cross-linking, such as the case of the active molecule adenosine triphosphate (ATP) [61], fucoidan [62] and sodium metabisulphite ($Na_2S_2O_5$) [63].

Molecules containing catechol as hydrocaffeic acid have shown the ability to increase mucoadhesion of several polymers including chitosan. CS-TPP-hydrocaffeic acid conjugated NPs prepared by ionic gelation showed mucoadhesion to rabbit small intestine 6 times higher than unmodified nanoparticles [64].

Reproducibility issues, such as different physical and biological responses, are often found in different batches of chitosan NPs prepared by ionic gelation. This situation may be due to the poor characterization of chitosan samples, or because of the poor experimental design and inadequate selection of the involved variables on the process [57].

Polyelectrolyte Complex (PEC)

The formation of chitosan nanoparticles *via* PEC is based on the ionic interaction of chitosan with negatively charged polymers. The process occurs when polymers of opposite charges are mixed, and results in self-assembly or a formation of strong reversible electrostatic binding in self-organized systems. These direct interactions between polymeric chains lead to formation of complex poly-electrolytic networks with non-permanent structures, avoiding the use of covalent crosslinking agents. Generally, these systems resulting from polymeric networks are widely used for obtaining hydrogels, which are biocompatible and sensitive to changes on environmental conditions [65, 66].

Polyelectrolyte complexes formation and stability are based on: degree of ionization of each oppositely charged polyelectrolyte, the density of charges present on the structure of each polyelectrolyte, concentration and proportion of polyelectrolytes in the mixture, the mixture order, nature and position of the ionic groups in the polymer chains, the molecular weight of the polyelectrolytes, flexibility of the polymer chain, and also temperature, ionic strength and pH of the reaction medium [65 - 67].

When polyelectrolytes are mixed in a ratio that there is a charge excess (positive or negative), a non-stoichiometric complex is formed, which is usually soluble. In contrast, a stoichiometric polyelectrolytic complex contains equal quantities of each opposite charge, with net charge equals to zero in the resulting complex. In the case of chitosan, various polyelectrolytes complexes can be formed by the interaction of the positively charged surface of chitosan with anionic biopolymers, such as sodium alginate [68], pectin [69], carboxymethylcellulose [70], cashew gum [71, 72], chondroitin sulfate [73, 74], dextran sulfate [40, 75], carrageenan [76], among others.

Drying Processes

This technique is widely used in the pharmaceutical, chemical, materials, cosmetic and food industries. The first patent concerning this technology was in the early 1870s. Drying is the process of water/solvent removal from liquids, solids, or semisolids by evaporation, where the vapor that is produced is removed by vacuum. Hot air, microwave, spray drying, freeze drying, supercritical drying, and natural air drying represent the general methods of drying [77, 78]. Spray drying and supercritical drying are the most commonly used in the preparation of chitosan particulate systems as they are rapid and simple, continuous, reproducible, adapt well with the incorporated active material, single-step thus, scalable without major modifications and have the ability to produce particles of different size and high stability [78 - 81].

Spray drying has been reported for manufacturing of granules and dry powders from drug-excipient mixtures either in a solution or suspension [77]. The expediency of this method extended to include preparation of microparticles from different polymeric materials loaded with protein [82], vaccine antigens, and drugs [6, 77, 83]. Spray drying offers an easy, efficient, one-step, and protein-friendly method for protein-loaded chitosan micro-/nanoparticles [81]. Also, it can be applied to a wide range of compound including heat-sensitive substances, such as drugs, extracts, aromatic oils, pigments, and flavors within different types of carriers resulting in polymeric nanoparticles (NPs) or microparticles (MPs) and nanocomposites [80, 84 - 87].

The basic principle of spray drying relies on using a stream of hot air to dry atomized droplets [77, 88]. The process involves preparation of aqueous chitosan solution that is sprayed through a nozzle into a drying chamber to get the desired particles [88]. The prepared particles are spherical, of smooth surface and narrow size distribution and showed good drug stability and high drug-entrapment efficiency [88]. Other advantages of this method are the possibility of surface cross-linking and coating the spheres during the manufacturing steps, which could enhance stability and control drug release [83].

The size of the prepared particles is influenced by spray flow rate, extent of cross-linking, and inlet air flow rate [89]. Large particles are produced using high spray flow rate from large-sized nozzles and when there is a decrease in the extent of cross-linking, while smaller particle size is obtained at greater airflow rates and not by the inlet air temperature being between 140°C and 180°C [77, 90].

In order to develop particles ranging from 300 nm to 5 μm, nanospray drying was introduced recently by Buchi. This system allows preparation of NPs with effective yield even for small sample amounts [91, 92].

Emulsion System (Oil/Water Emulsification Diffusion Method)

Nanoparticles made by nanoemulsions can be prepared in different ways. Emulsification methods generally require high mechanical energy input to achieve small droplets sizes. However, here the low power method is focused, which is based on the physical behavior of the system, caused by facilitation of oil-water mixture using surfactants [88, 93, 94].

The preparation consists on the dispersion of chitosan in glacial acetic acid solution (water component), then a hydrophobic solvent such as cyclohexane, cyclohexanol, soybean oil or other solvent having similar characteristics (containing or not active substance) is added dropwise to chitosan solution and continuously stirred. The resulting mixture is then titrated with some surfactant, *e.g.* Triton X-100 or Tween, in constant stirring until the mixture becomes transparent or semitransparent, indicating the formation of the nanoparticle by emulsion [30, 95]. Another possibility is mixing the surfactant previously in one of the solutions, saving a step in the process. Generally, the particles formed by the emulsion system have nanocapsule structures [93, 96, 97], high drug loading efficiency and improved bioavailability [49].

Changes in methodology can facilitate the formation of nanocapsules, the use of ultrasonication and increasing reaction temperature cooperate with the process [93, 98]. However, there are several modifications in the preparation of emulsion by the water-in-oil or oil-in-water system, for example, addition of gelling agents as chlorides, TPP or polycations [98, 99]. Although the emulsion technique provides an easy control of particle size, this procedure uses harsh crosslinking agents, and the complete removal of unreacted crosslinking agents could be an issue [49].

Self-assembled Nanoparticles

Self-assembled nanoparticles can be prepared from amphiphilic polymers in aqueous medium. Hydrophobic interactions between parts of the amphiphilic polymer tends to minimize the interfacial energy, allowing the formation of NP [100]. Despite chitosan is not an amphiphilic biopolymer, nanoparticles formed by self-assembly can be made by chitosan structural modification in order to bind a compound as palmitoyl, which forces self-aggregation of chitosan molecules [11, 101]. The hydrophobicity of chitosan has been increased by the introduction of phthaloyl, polyester, fatty acid, cholesterol, among others [11, 102].

Structural changes in derivatives chitosan can make it amphiphilic, resulting in NPs self-assembling and be used as drug delivery vehicles. Quaternary chitosan derivatives were synthesized using (5-pentyl)trimethylammonium and dodecyl aldehyde. The self-aggregates formed in aqueous solution present size distribution which ranges from 250 to 500 nm and exhibits hydrophobic core able to be used as a carrier for hydrophobic substances [103].

CHARACTERIZATION TECHNIQUES FOR CHITOSAN NANO-PARTICLES AND NANOCAPSULES

Size and size distribution are extremely important for nanoparticles characterizations, hence are essential due to the possibility to modify characteristics and morphology of the system, changing the physicochemical and pharmaceutical behaviors of compounds present inside this systems depending on the preparation way. Laser diffraction and Coulter counter can be used to determine particle size; however, the most widely used method is dynamic light scattering (DLS), allowing to evaluate the particle size distribution. Uniformity of particles sizes are obtained by the polydispersity (PDI), where PDI values between 0.10 and 0.25 indicate a narrow size distribution and superior values indicate broad size distribution [96].

Particles containing chitosan or other charged molecules display zeta potential which corresponds to the electric potential presented by the material surface and they are strongly influenced by its composition. Zeta potential are determined using zeta potential measurement, which applies the principle of electrophoresis and reflects the surface electric charges of the particles. An electric field is applied across the dispersed particles in which the particles will migrate to the oppositely charged electrode with a velocity that is proportional to the value of the zeta potential. For dispersion system, high degree of zeta potential, negative or positive, indicates the resistance of the particles to aggregation, and this denotes an apparently stable system. This parameter can be associated with the particles stability, wherein zeta potential values between ± 30 mV promotes low stability which leads to possible particle aggregation [96, 104].

Different microscopy techniques can be employed to confirm the particle size, observe and characterize the morphology and structure of these nanoparticles. Atomic force microscopy (AFM) and scanning electron microscopy (SEM) are some examples of devices used for this purpose, though transmission electron microscopy (TEM) is the most used technique and provides information about structure, porosity and estimate the thickness of the wall in capsules [96].

Chitosan structural modification has been mainly characterized by infrared spectrometry (FTIR) and nuclear magnetic resonance (^{1}H and ^{13}C NMR) techniques. These modifications are typically used to improve the physico-chemical properties of the resulting nanoparticle, increasing interaction with active compounds or providing binding to specific cellular receptors [8, 105].

The specific surface properties of the prepared particulate system could be examined using X-ray photoelectron spectroscopy (XPS). This specific technique provides information about the elements from which the particles originate, chemical bonding, and detailed molecular information with high sensitivity. The former identifies the size-dependent distributions and the chemical states of the elements, while the latter provides detailed surface and near-surface composition analysis, so characterization and composition of the surface chemistry of the particles are illustrated [106, 107].

In drug delivery systems, it is extremely important to determine drug content, encapsulation efficiency, loading efficiency and micro-/nanoparticles yield are other characteristics that should be calculated for the prepared micro-/nanoparticles before therapeutic administration. For this procedure, validated analytical techniques are often used and also employed for simulating the release control in physiological conditions. Among existing techniques used for this purpose, spectrophotometry and high-performance liquid chromatography (HPLC) can be cited [108].

CHITOSAN NANOPARTICLES AND NANOCAPSULES APPLICATIONS

Chitosan-based nanometric systems can be used in several areas and applications, based on desired characteristics of the systems and on the biopolymer chemical properties, such as pKa, degree of deacetylation, molecular weight and chemical modifications. Chitosan presents two types of hydroxyl groups in terms of its structure: a primary hydroxyl on carbon-6 position and a secondary hydroxyl on carbon-3 position, on glucosamine unit. Carbon-2 position is predominantly occupied by-NH_2 groups. Therefore, chitosan structure can be manipulated and modified by the insertion of new functional groups or by crosslinking reactions, providing its derivatives desired characteristics for selected applications.

Chitosan nanoparticles modified with acetophenone *via* Shiff's base condensation were prepared by ionic gelation using tripolyphosphate as gelation agent. Excellent adsorption capacities of different metals (Cu^{2+}, Cd^{2+}, Hg^{2+} and Pb^{2+}) at pH 7.0 were noted. It may be used to remove these divalent metal ions from tap water, sea water and industrial wastewater, with adsorption equilibrium established within 30 minutes of contact time between the absorbent and metal ions [9]. Furthermore, chitosan NPs of 100 nm were also able to remove hexavalent chromium from water, requiring the presence of TPP for effective metal absorption. Cr^{6+} adsorbed on CS-TPP NPs was reduced to Cr^{3+}, the reduced form of chromium is less or nontoxic. Thus, this system can be considered efficient for detoxification and treatment of Cr^{6+} in water [109]. Co^{2+}, Ni^{2+} and Cu^{2+} were removed from water using chitosan-poly(methacrylate) modified NPs prepared by PEC technique and particles size between 100 and 120 nm were obtained [110].

In agriculture, chitosan nanoparticles have been used as herbicidal and pesticidal formulation of some compounds *e.g.* paraquat, clomazone and acetamiprid. Paraquat is a non-selective herbicide widely used for the control of weed, however its aqueous solubility and soil sorption can cause problems of toxicity in humans. This herbicide was encapsulated in chitosan/alginate and CS-TPP NPs prepared by Ca^{2+} complexation [111] and ionic gelation [112], respectively, with diameter around 635 and 300 nm. These systems showed less toxicity than the pure compound, which may reduce adverse impacts on human health and environment [111 - 113]. Clomazone is used in rice fields, and it was encapsulated in chitosan/alginate NPs, with particle size around 200 nm [114]. Acetamiprid pesticide consumption may be reduced mainly in the acid soil by chitosan/alginate NPs due its superior controlled and retarded release compared to commercial formulation. These NPs were prepared using polyelectrolyte complexation method with the particle size of 201 nm and PDI value of 0.39 [8].

Essential oils loaded-chitosan NPs have been prepared effectively using different methods. *In vitro* release profile assays of *Carum copticum* encapsulated in chitosan NPs using TPP and sodium hexametaphosphate separately as cross-linkers, showed initial burst release profile and followed by a sustained release. The NPs displayed average size between 30-80 nm [115]. *Cinnamommum zeylanicum* essential oil was encapsulated in chitosan using water-in-oil technique and tested against *Phytophthora drechsleri* showing size distribution of 100-190 nm [116]. Water-in-oil technique also was used to prepare turmeric (EE% = 71%) and lemongrass (EE% = 77%) oils loaded-chitosan/alginate NPs. Hemo-compatible particles smaller than 300 nm with good stability were obtained, suggesting their use for future biomedical applications [117]. *Zataria multiflora* essential oil was incorporated into chitosan nanoparticles using water-in-oil method, exhibiting average size about 200 nm. The essential oil loaded-chitosan NPs extended the shelf-life of cucumber fruit at storage temperature, offering an edible coating with lowest decay level of cucumbers due to the release of antioxidant compounds by the NPs [118]. Ionic gelation was used to prepare chitosan NPs loaded with cinnamon essential oil with size ranging from 112 to 527 nm, showing good antimicrobial and antioxidant properties for pork during refrigerated storage, keeping the quality of meat and meat products for 15 days [119]. Summer savory (*Satureja hortensis L*) loaded chitosan nanoparticle was prepared by water-in-oil technique followed by ionic gelation. NPs exhibited size between 140-238 nm. The formulations showed strong antibacterial activity against *Staphylococcus aureus, Listeria monocytogenes* and *Escherichia coli* [29].

Polyphenols (PPH) are commonly present in vegetables, fruits, teas and coffees, having great importance for the human diet, due their nutritious value and effects to prevent chronic disease. They have been investigated for their antiallergic, antibacterial, anti-inflammatory, antitumor, antiviral and mainly owing to its strong antioxidant activity [120]. Chitosan-based NPs have been used to encapsulate polyphenols extracted from different sources. Polyphenols from tea extract were successfully encapsulated using chitosan NPs [121] with an average size of 150-350 nm [122]. Other examples are: PPH from strawberry (300-600 nm) [123]; PPH isolated from gallic acid (~ 140 nm) [124] and (~ 300 nm) [125]; catechin extracted from gambier (~ 300 nm) [126, 127]; chlorogenic acid (146 and 255 nm) [34]; quercetin (150-220 nm) [128]; polyphenols contained in red grapes (~300 nm) [129] and cinnamon [119]. By being negative compounds, they interact with amino groups of chitosan and, similarly to TPP, may occur ionic gelation, which can improve the bioavailability of these phytochemicals [123]. Nevertheless, the presence of TPP is essential to ensure the formation of the NP, thus this is the leading method to prepare these NPs. Chitosan NP system may contain sulfobutylether-β-cyclodextrin [121]; quaternary ammonium-chitosan conjugate [129] or self-assembled O-carboxymethyl chitosan-cholic acid conjugates [128]. These phytochemical encapsulated in chitosan were tested regarding its degradation and anticancer activity against Caco-2 colon cells [125]; antibacterial activity (*Escherichia coli* and *Listeria innocua*) [127] and hepatic evaluation (Hep G2 cells) [128].

In food and nutrition industry, chitosan have been used to protect feeds or some nutrient for human diet even as packaging and controlling food pathogens. Nisin loaded chitosan-carrageenan nanocapsules were prepared using PEC method with the smaller particle displaying size around 500 nm, this system was evaluated *in vitro* for antibacterial activity against *Micrococcus luteus*, *Pseudomonas aeruginosa*, and *Salmonella enterica* even as *Enterobactor aerogenes* followed by antibacterial activity in tomato juice [130]. Antidiabetic peptides from Atlantic salmon (*Salmo salar*) protein hydrolysates were developed using chitosan-coated phosphatidylcholine/cholesterol liposomes and exhibiting size of approximately 200 nm [131]. An amphiphilic self-assembled chitosan derivative of N,N-dimethylhexadecyl carboxymethyl chitosan showing an approximate size of 140 nm was synthetized and used to load vitamin D_3, this system produced high encapsulation efficiency and a sustained *in vitro* release [132]. Cinnamon essential oil loaded chitosan NPs (112-527 nm) were impregnated in polyethylene films, exhibiting antimicrobial and antioxidant properties for pork during refrigerated storage while maintaining the pH of the chilled pork [119]. Improvement of postharvest shelf life of fruits as apple has been achieved with chitosan-TPP NPs prepared by ionic gelation with particle size less than 100 nm. Respiration rate (CO_2 production), weight loss, ethylene production and other characteristics also were assessed. Data showed chitosan NPs decreased the natural fruit ripening and extended the quality of apple during storage [7]. As described before, postharvest shelf life has been improved and bioactive components of cucumber were protected by treatment with *Zataria multiflora* essential oil-loaded chitosan NPs (~200 nm) prepared by water-in-oil method [118]. Chitosan-TPP NPs (180 nm) produced *via* ionic gelation also were used for shrimp glazing. Moisture content, weight loss, cutting force, thiobarbituric acid and aerobic counts were analyzed after frozen storage at -21 ºC for 30 days. These experiments revealed higher moisture in glazed than non-glazed, controlling lipid oxidation and reducing aerobic counts in shrimp [133]. Other study demonstrated that chitosan NPs with diameter of 110 nm produced by the same method described before enhanced the quality of silver carp fillets during storage at 4 ºC. Results indicated that CS NPs were effective for preservation of the fish, exhibiting high antimicrobial activity and strong ability to inhibit the spoilage [134].

Small interfering RNA (siRNA) may be a promising tool for gene silencing in undesirable eukaryotic cells engaged in diseases such as cancer, viral infections and other genetic diseases. siRNA is composed of 21-23 base pairs and it is capable of inducing the reduction of target genes [135]. Due to its positive natural characteristic, chitosan-based nanoparticles allow electrostatic interactions with negatively charged siRNA improving its bioavailability. Hence siRNA is a DNA interfering able to achieve high levels of gene silencing without cytotoxicity [136]. Studies in ovarian cancer models administered in mice had resulted in a significant cells transfections and *in vivo* accumulation leading an effective inhibition of proliferation *in vivo*. In this study siRNA/folic acid-poly(ethylene glycol)-chitosan oligosaccharide lactate NPs were prepared by ionic gelation showing average size approximately 200 nm [137]. Several other researches have demonstrated that chitosan-based NPs are useful to encapsulate and delivery siRNA. Small interfering RNA was successfully incorporated by ionic gelation in chitosan-TPP with particle size between 78-200 nm [138, 139]; PEGlated chitosan-TPP (124-294 nm) [136] (100-150 nm) [25] (250-285 nm) [140]; PEGlated chitosan without TPP also can incorporate siRNA efficiently with size between 140-200 nm [135, 141], 235-470 nm [142] or nanocapsules (<100 nm) [143]; PEG grafted carboxymethyl chitosan and calcium phosphate with diameter of 102 nm; urocanic acid-modified chitosan (156-247 nm) prepared by complex coacervation technique [144]; other chitosan NPs with ligands and more complexes grafts were also used to incorporate siRNA [137, 145]. These systems are tested *in vitro* or/and *in vivo* against B16 melanoma cells [136]; Human carcinoma cell line (A549) [135]; lung cancer cell line (H1299) [141, 142, 145]; Raw 264.7 macrophages and colon-26 cells [144]; HepG2 cells [138, 146]; orthotopic U87MG glioblastoma cells [143]; H9c2 cardiomyoblast cells [139];

human colon cancer cell line (HCT-116) [25]; MCF-7 breast cancer [140].

Following the drug delivery line stated by chitosan and modified chitosan NPs, treatment of brain disease such as neurodegenerative, cerebrovascular and cancerous which require that drugs permeate into the brain have been improved for these systems. Once most compounds do not reach brain barrier from bloodstream due to the protection of the brain ensured by regulation of blood-brain barrier (BBB) [147]. To overcome the limitation imposed by the BBB, drugs have been nasally administered because they can be deposited on the respiratory epithelia, be absorbed into the systemic circulation and thus cross the BBB [147, 148]. Chitosan displays mucoadhesive properties which had been extensively studied and enhanced by modifications into chitosan structure [148]. Different systems were utilized to incorporate many drugs such as chitosan coated lipid matrix (115-192 nm) incorporating nile red and dir [148]; buspirone hydrochloride encapsulated in thiolated chitosan NPs prepared by ionic gelation (200-600 nm) [149]; piperine-loaded chitosan with particle size 250 nm [150]; chitosan-TPP nanoparticles containing siRNA (130-300 nm) [151], quetiapine fumarate as an antipsychotic (140-450 nm) [152], rutin [153], vaccine [154]; dopamine-loaded glycol chitosan/sulfobutylether-β-cyclodextrin prepared by ionic gelation with tripolyphosphate exhibiting sizes from 214 nm to more than 1 μm; vaccine encapsulated in O-2-hydroxypropyltrimethyl ammonium chloride chitosan NPs prepared by ionic gelation with TPP [155] or in mannosylated chitosan (60-500 nm) [156]. The intranasal routes using chitosan PNs demonstrated biocompatibility and efficient biodistribution by human bronchial epithelial (16HBE14p) [148] and have been applied in different diseases, for example, general anxiety disorder [149]; neuroprotection in Alzheimer's disease [150]; murine and human glioblastoma multiforme [151]; cerebral ischemia [153]; Newcastle disease [155]; foot and mouth disease [156] and Influenza virus [154].

Chitosan mucoadhesive property can also be useful for drug administration for controlled delivery in the oral [157 - 161], nasal [162 - 166], ocular [51, 167 - 169], vaginal [170], pulmonary [171 - 175], topical and transdermal [176, 177], parental and intravesical pathways [105, 178, 179]. The mechanism responsible for the permeation effect is due to its positive charges. However, a more pronounced cationic feature can be achieved by trimethylation of the primary amino group without improved permeation properties, suggesting a mechanism more complex than supposed [180]. High molecular weight and a high degree of deacetylation enhance epithelial permeability, PEGlation, N-succinylation and thiolation can also improve strongly its mucoadhesive properties. Thiolated chitosan forms disulfide bonds with glycoproteins of the mucus gel layer, and may be raised with mercaptonicotinamide [166, 180].

CONCLUDING REMARKS

Chitosan exhibits cationic character and thus mucoadhesive permeation properties. Nanoparticles and nanocapsules of this polymer can be prepared by several methods and used in different fields, such as water treatment for heavy metal removal; safe pharmaceutical excipient for controlled and targeting drug, vaccine or gene delivery; transport of essential oil and diet nutrients, and in agriculture as herbicidal and pesticide formulations. The potential of these characteristics can be enhanced by modification of the chitosan's structure or in the preparation methods.

In future research on chitosan nanoparticles and nanocapsules, some trends could be identified, such as the assessment of antimicrobial control of chitosan nanoparticles on food-derived cultures, research on the optimization of the nanoencapsulation process of food preservatives using tailor-made functionalized chitosan NPs, enhancement of production yield of chitosan NPs by ionic gelation, as the yield by this method is relatively low, in order to increase the productivity of this preparation method at the industrial scale. Moreover, notable modifications are in chitosan structure for absorption enhancement, which is an important role in future for the pharmaceutical field, since more complex drugs are designed.

CONFLICT OF INTEREST

The authors confirm that they have no conflict of interest to declare for this publication.

ACKNOWLEDGEMENTS

Decleared none.

REFERENCES

[1] Gomathi T, Govindarajan C, Rose H R MH, *et al.* Studies on drug-polymer interaction, *in vitro* release and cytotoxicity from chitosan particles excipient. Int J Pharm 2014; 468(1-2): 214-22. [http://dx.doi.org/10.1016/j.ijpharm.2014.04.026] [PMID: 24742716]

[2] Azzam EM, Eshaq G, Rabie AM, *et al.* Preparation and characterization of chitosan-clay nanocomposites for the removal of Cu(II) from aqueous solution. Int J Biol Macromol 2016; 89: 507-17. [http://dx.doi.org/10.1016/j.ijbiomac.2016.05.004] [PMID: 27151669]

[3] Nithya A, JeevaKumari HL, Rokesh K, Ruckmani K, Jeganathan K, Jothivenkatachalam K. A versatile effect of chitosan-silver nanocomposite for surface plasmonic photocatalytic and antibacterial activity. J Photochem Photobiol B 2015; 153: 412-22. [http://dx.doi.org/10.1016/j.jphotobiol.2015.10.020] [PMID: 26562805]

[4] Chronopoulou L, Nocca G, Castagnola M, *et al.* Chitosan based nanoparticles functionalized with peptidomimetic derivatives for oral drug delivery. N Biotechnol 2016; 33(1): 23-31. [http://dx.doi.org/10.1016/j.nbt.2015.07.005] [PMID: 26257139]

[5] Garcia-Fuentes M, Alonso MJ. Chitosan-based drug nanocarriers: where do we stand? J Control Release 2012; 161(2): 496-504. [http://dx.doi.org/10.1016/j.jconrel.2012.03.017] [PMID: 22480607]

[6] Maestrelli F, Bragagni M, Mura P. Advanced formulations for improving therapies with anti-inflammatory or anaesthetic drugs: A review. J Drug Deliv Sci Technol 2016; 32: 192-205. [http://dx.doi.org/10.1016/j.jddst.2015.09.011]

[7] Sahraei Khosh Gardesh A, Badii F, Hashemi M, Ardakani AY, Maftoonazad N, Gorji AM. Effect of nanochitosan based coating on climacteric behavior and postharvest shelf-life extension of apple cv. Golab Kohanz. LWT -. Food Sci Technol (Campinas) 2016; 70: 33-40.

[8] Kumar S, Chauhan N, Gopal M, Kumar R, Dilbaghi N. Development and evaluation of alginate-chitosan nanocapsules for controlled release of acetamiprid. Int J Biol Macromol 2015; 81: 631-7. [http://dx.doi.org/10.1016/j.ijbiomac.2015.08.062] [PMID: 26321424]

[9] Mahmoud ME, Abou Kana MT, Hendy AA. Synthesis and implementation of nano-chitosan and its acetophenone derivative for enhanced removal of metals. Int J Biol Macromol 2015; 81: 672-80. [http://dx.doi.org/10.1016/j.ijbiomac.2015.08.063] [PMID: 26322815]

[10] Ma X, Zhang B, Cong Q, He X, Gao M, Li G. Organic/inorganic nanocomposites of ZnO/CuO/chitosan with improved properties. Mater Chem Phys 2015; 178: 88-97. [http://dx.doi.org/10.1016/j.matchemphys.2016.04.074]

[11] Yang Y, Wang S, Wang Y, Wang X, Wang Q, Chen M. Advances in self-assembled chitosan nanomaterials for drug delivery. Biotechnol Adv 2014; 32(7): 1301-16. [http://dx.doi.org/10.1016/j.biotechadv.2014.07.007] [PMID: 25109677]

[12] Ngo D-H, Vo T-S, Ngo D-N, Kang K-H, Je J-Y, Pham HN, *et al.* Biological effects of chitosan and its derivatives. Food Hydrocoll 2015; 51: 200-16. [http://dx.doi.org/10.1016/j.foodhyd.2015.05.023]

[13] O'Callaghan KA, Kerry JP. Preparation of low- and medium-molecular weight chitosan nanoparticles and their antimicrobial evaluation against a panel of microorganisms, including cheese-derived cultures. Food Contr 2016; 69: 256-61. [http://dx.doi.org/10.1016/j.foodcont.2016.05.005]

[14] Elgadir MA, Uddin MS, Ferdosh S, Adam A, Chowdhury AJ, Sarker MZ. Impact of chitosan composites and chitosan nanoparticle composites on various drug delivery systems: A review. J Food Drug Anal 2015; 23(4): 619-29. [http://dx.doi.org/10.1016/j.jfda.2014.10.008]

[15] Khor E. The impact of nanotechnology on chitin and chitosan biomaterials research. Chitin: Elsevier 2014; pp. 67-82. [http://dx.doi.org/10.1016/B978-0-08-099939-5.00005-7]

[16] Kumar A, Vimal A, Kumar A. Why Chitosan? From properties to perspective of mucosal drug delivery. Int J Biol Macromol 2016; 91: 615-22. [http://dx.doi.org/10.1016/j.ijbiomac.2016.05.054] [PMID: 27196368]

[17] Bugnicourt L, Alcouffe P, Ladavière C. Elaboration of chitosan nanoparticles: Favorable impact of a mild thermal treatment to obtain finely divided, spherical, and colloidally stable objects. Colloids Surfaces A Physicochem Eng Asp 2014; 457(1): 476-86.

[18] Dodane V, Vilivalam VD. Pharmaceutical applications of chitosan. Pharm Sci Technol Today 1998; 1(6): 246-53. [http://dx.doi.org/10.1016/S1461-5347(98)00059-5]

[19] Fülöp Z, Saokham P, Loftsson T. Sulfobutylether-β-cyclodextrin/chitosan nano- and microparticles and their physicochemical characteristics. Int J Pharm 2014; 472(1-2): 282-7. [http://dx.doi.org/10.1016/j.ijpharm.2014.06.039] [PMID: 24971695]

[20] Kalam MA. Development of chitosan nanoparticles coated with hyaluronic acid for topical ocular delivery of dexamethasone. Int J Biol Macromol 2016; 89: 127-36. [http://dx.doi.org/10.1016/j.ijbiomac.2016.04.070] [PMID: 27126165]

[21] Caro N, Medina E, Díaz-Dosque M, López L, Abugoch L, Tapia C. Novel active packaging based on films of chitosan and chitosan/quinoa protein printed with chitosan-tripolyphosphate-thymol nanoparticles *via* thermal ink-jet printing. Food Hydrocoll 2016; 52: 520-32. [http://dx.doi.org/10.1016/j.foodhyd.2015.07.028]

[22] Pasanphan W, Rattanawongwiboon T, Choofong S, Güven O, Katti KK. Irradiated chitosan nanoparticle as a water-based antioxidant and reducing agent for a green synthesis of gold nanoplatforms. Radiat Phys Chem 2015; 106: 360-70. [http://dx.doi.org/10.1016/j.radphyschem.2014.08.023]

[23] Choi C, Nam JP, Nah JW. Application of chitosan and chitosan derivatives as biomaterials. J Ind Eng Chem 2015; 33: 1-10. [http://dx.doi.org/10.1016/j.jiec.2015.10.028]

[24] Chen Y-Z, Huang Y-K, Chen Y, Ye Y-J, Lou K-Y, Gao F. Novel nanoparticles composed of chitosan and β-cyclodextrin derivatives as potential insoluble drug carrier. Vol. 26. Chin Chem Lett 2015. [http://dx.doi.org/10.1016/j.cclet.2015.05.044]

[25] Rudzinski WE, Palacios A, Ahmed A, Lane MA, Aminabhavi TM. Targeted delivery of small interfering RNA to colon cancer cells using chitosan and PEGylated chitosan nanoparticles. Carbohydr Polym 2016; 147: 323-32. [http://dx.doi.org/10.1016/j.carbpol.2016.04.041] [PMID: 27178938]

[26] Sujima Anbu A, Velmurugan P, Lee J-H, Oh B-T, Venkatachalam P. Biomolecule-loaded chitosan nanoparticles induce apoptosis and molecular changes in cancer cell line (SiHa). Int J Biol Macromol 2016; 88: 18-26. [http://dx.doi.org/10.1016/j.ijbiomac.2016.03.042] [PMID: 27016087]

[27] Zhang P, Liu X, Hu W, Bai Y, Zhang L. Preparation and evaluation of naringenin-loaded sulfobutylether-β-cyclodextrin/chitosan nanoparticles for ocular drug delivery. Carbohydr Polym 2016; 149: 224-30. [http://dx.doi.org/10.1016/j.carbpol.2016.04.115] [PMID: 27261746]

[28] Abruzzo A, Zuccheri G, Belluti F, *et al.* Chitosan nanoparticles for lipophilic anticancer drug delivery: Development, characterization and *in vitro* studies on HT29 cancer cells. Colloids Surf B Biointerfaces 2016; 145: 362-72. [http://dx.doi.org/10.1016/j.colsurfb.2016.05.023] [PMID: 27214786]

[29] Feyzioglu GC, Tornuk F. Development of chitosan nanoparticles loaded with summer savory (Satureja hortensis L.) essential oil for antimicrobial and antioxidant delivery applications. LWT -. Food Sci Technol (Campinas) 2016; 70: 104-10.

[30] Yoksan R, Jirawutthiwongchai J, Arpo K. Encapsulation of ascorbyl palmitate in chitosan nanoparticles by oil-in-water emulsion and ionic gelation processes. Colloids Surf B Biointerfaces 2010; 76(1): 292-7. [http://dx.doi.org/10.1016/j.colsurfb.2009.11.007] [PMID: 20004558]

[31] Tan C, Xie J, Zhang X, Cai J, Xia S. Polysaccharide-based nanoparticles by chitosan and gum arabic polyelectrolyte complexation as carriers for curcumin. Food Hydrocoll 2016; 57: 236-45. [http://dx.doi.org/10.1016/j.foodhyd.2016.01.021]

[32] Woranuch S, Yoksan R. Eugenol-loaded chitosan nanoparticles: I. Thermal stability improvement of eugenol through encapsulation. Carbohydr Polym 2013; 96(2): 578-85. [http://dx.doi.org/10.1016/j.carbpol.2012.08.117] [PMID: 23768603]

[33] Hosseini SF, Zandi M, Rezaei M, Farahmandghavi F. Two-step method for encapsulation of oregano essential oil in chitosan nanoparticles: preparation, characterization and *in vitro* release study.

Carbohydr Polym 2013; 95(1): 50-6.
[http://dx.doi.org/10.1016/j.carbpol.2013.02.031] [PMID: 23618238]

[34] Nallamuthu I, Devi A, Khanum F. Chlorogenic acid loaded chitosan nanoparticles with sustained release property, retained antioxidant activity and enhanced bioavailability. Asian J Pharm Sci 2014; 10(3): 203-11.
[http://dx.doi.org/10.1016/j.ajps.2014.09.005]

[35] Ajun W, Yan S, Li G, Huili L. Preparation of aspirin and probucol in combination loaded chitosan nanoparticles and *in vitro* release study. Carbohydr Polym 2009; 75(4): 566-74.
[http://dx.doi.org/10.1016/j.carbpol.2008.08.019]

[36] Alishahi A, Mirvaghefi A, Tehrani MR, Farahmand H, Koshio S, Dorkoosh FA, *et al.* Chitosan nanoparticle to carry vitamin C through the gastrointestinal tract and induce the non-specific immunity system of rainbow trout (Oncorhynchus mykiss). Carbohydr Polym 2011; 86(1): 142-6.
[http://dx.doi.org/10.1016/j.carbpol.2011.04.028]

[37] Avadi MR, Sadeghi AM, Mohammadpour N, *et al.* Preparation and characterization of insulin nanoparticles using chitosan and Arabic gum with ionic gelation method. Nanomedicine Nanotechnology. Biol Med 2010; 6(1): 58-63.

[38] Facchi SP, Scariot DB, Bueno PV, *et al.* Preparation and cytotoxicity of N-modified chitosan nanoparticles applied in curcumin delivery. Int J Biol Macromol 2016; 87: 237-45.
[http://dx.doi.org/10.1016/j.ijbiomac.2016.02.063] [PMID: 26930578]

[39] Amorim C de M, Couto AG, Netz DJ, de Freitas RA, Bresolin TM. Antioxidant idebenone-loaded nanoparticles based on chitosan and N-carboxymethylchitosan. Nanomedicine Nanotechnology. Biol Med 2010; 6(6): 745-52.

[40] Kulkarni AD, Vanjari YH, Sancheti KH, *et al.* New nasal nanocomplex self-assembled from charged biomacromolecules: N,N,N-Trimethyl chitosan and dextran sulfate. Int J Biol Macromol 2016; 88: 476-90.
[http://dx.doi.org/10.1016/j.ijbiomac.2016.03.045] [PMID: 27017981]

[41] Xu Y, Du Y, Huang R, Gao L. Preparation and modification of N-(2-hydroxyl) propyl-3-trimethyl ammonium chitosan chloride nanoparticle as a protein carrier. Biomaterials 2003; 24(27): 5015-22.
[http://dx.doi.org/10.1016/S0142-9612(03)00408-3] [PMID: 14559015]

[42] Thassu D, Pathak Y, Deleers M. Nanoparticulate Drug-Delivery Systems: An Overview. Nanoparticulate Drug Delivery Systems. CRC Press 2007; pp. 1-31.
[http://dx.doi.org/10.1201/9781420008449.ch1]

[43] Rao CN, Müller A, Cheetham AK. Nanomaterials - An Introduction. In: The Chemistry of Nanomaterials. Weinheim, FRG: Wiley-VCH Verlag GmbH & Co. KGaA 2005; pp. 1-11.

[44] Lu S, Xu L, Kang ET, Mahendran R, Chiong E, Neoh KG. Co-delivery of peptide-modified cisplatin and doxorubicin *via* mucoadhesive nanocapsules for potential synergistic intravesical chemotherapy of non-muscle-invasive bladder cancer. Eur J Pharm Sci 2016; 84: 103-15.
[http://dx.doi.org/10.1016/j.ejps.2016.01.013] [PMID: 26780592]

[45] Claudio P, Reatul K, Brigitte E, Geraldine P. Drug-delivery nanocarriers to cross the blood-brain barrier. Nanobiomaterials in Drug Delivery. Elsevier 2016; pp. 333-70.
[http://dx.doi.org/10.1016/B978-0-323-42866-8.00010-1]

[46] Kapuścińska A, Olejnik A, Nowak I. Nanocapsules as carriers of active substances. Nanobiomaterials in Galenic Formulations and Cosmetics. Elsevier 2016; pp. 175-99.
[http://dx.doi.org/10.1016/B978-0-323-42868-2.00008-5]

[47] Piñón-Segundo E, Mendoza-Muñoz N, Quintanar-Guerrero D. Nanoparticles as dental drug-delivery systems. Nanobiomaterials in Clinical Dentistry. Elsevier 2012; pp. 475-95.

[48] Ohya Y, Shiratani M, Kobayashi H, Ouchi T. Release behavior of 5-Fluorouracil from chitosan-gel nanospheres immobilizing 5-Fluorouracil coated with polysaccharides and their cell specific

cytotoxicity. J Macromol Sci Part A Pure Appl Chem 1994; 31(5): 629-42. [http://dx.doi.org/10.1080/10601329409349743]

[49] Kashyap PL, Xiang X, Heiden P. Chitosan nanoparticle based delivery systems for sustainable agriculture. Int J Biol Macromol 2015; 77: 36-51. [http://dx.doi.org/10.1016/j.ijbiomac.2015.02.039] [PMID: 25748851]

[50] Calvo P, Remunan-Lopez C. Novel hydrophilic chitosan-polyethylene oxide nanoparticles as protein carriers. J Appl Polym Sci 1997; 63(1): 125-32. [http://dx.doi.org/10.1002/(SICI)1097-4628(19970103)63:1<125::AID-APP13>3.0.CO;2-4]

[51] de la Fuente M, Raviña M, Paolicelli P, Sanchez A, Seijo B, Alonso MJ. Chitosan-based nanostructures: a delivery platform for ocular therapeutics. Adv Drug Deliv Rev 2010; 62(1): 100-17. [http://dx.doi.org/10.1016/j.addr.2009.11.026] [PMID: 19958805]

[52] Hashad RA, Ishak RA, Fahmy S, Mansour S, Geneidi AS. Chitosan-tripolyphosphate nanoparticles: Optimization of formulation parameters for improving process yield at a novel pH using artificial neural networks. Int J Biol Macromol 2016; 86: 50-8. [http://dx.doi.org/10.1016/j.ijbiomac.2016.01.042] [PMID: 26783636]

[53] Konecsni K, Low NH, Nickerson MT. Chitosan-tripolyphosphate submicron particles as the carrier of entrapped rutin. Food Chem 2012; 134(4): 1775-9. [http://dx.doi.org/10.1016/j.foodchem.2012.03.070] [PMID: 23442620]

[54] Shah BR, Li Y, Jin W, An Y, He L, Li Z, *et al.* Preparation and optimization of Pickering emulsion stabilized by chitosan-tripolyphosphate nanoparticles for curcumin encapsulation. Food Hydrocoll 2016; 52: 369-77. [http://dx.doi.org/10.1016/j.foodhyd.2015.07.015]

[55] Antoniou J, Liu F, Majeed H, Qi J, Yokoyama W, Zhong F. Physicochemical and morphological properties of size-controlled chitosan-tripolyphosphate nanoparticles. Colloids Surfaces A Physicochem Eng Asp 2015; 465: 137-46. [http://dx.doi.org/10.1016/j.colsurfa.2014.10.040]

[56] Sipoli CC, Radaic A, Santana N, de Jesus MB, de la Torre LG. Chitosan nanoparticles produced with the gradual temperature decrease technique for sustained gene delivery. Biochem Eng J 2015; 103: 114-21. [http://dx.doi.org/10.1016/j.bej.2015.06.017]

[57] Kleine-Brueggeney H, Zorzi GK, Fecker T, El Gueddari NE, Moerschbacher BM, Goycoolea FM. A rational approach towards the design of chitosan-based nanoparticles obtained by ionotropic gelation. Colloids Surf B Biointerfaces 2015; 135: 99-108. [http://dx.doi.org/10.1016/j.colsurfb.2015.07.016] [PMID: 26241921]

[58] de Pinho Neves AL, Milioli CC, Müller L, Riella HG, Kuhnen NC, Stulzer HK. Factorial design as tool in chitosan nanoparticles development by ionic gelation technique. Colloids Surfaces A Physicochem Eng Asp 2014; 445: 34-9. [http://dx.doi.org/10.1016/j.colsurfa.2013.12.058]

[59] Abdel-Hafez SM, Hathout RM, Sammour OA. Towards better modeling of chitosan nanoparticles production: screening different factors and comparing two experimental designs. Int J Biol Macromol 2014; 64: 334-40. [http://dx.doi.org/10.1016/j.ijbiomac.2013.11.041] [PMID: 24355618]

[60] Sipoli CC, Santana N, Shimojo AA, Azzoni A, de la Torre LG. Scalable production of highly concentrated chitosan/TPP nanoparticles in different pHs and evaluation of the *in vitro* transfection efficiency. Biochem Eng J 2015; 94: 65-73. [http://dx.doi.org/10.1016/j.bej.2014.11.008]

[61] Giacalone G, Hillaireau H, Capiau P, Chacun H, Reynaud F, Fattal E. Stabilization and cellular delivery of chitosan-polyphosphate nanoparticles by incorporation of iron. J Control Release 2014; 194: 211-9.

[http://dx.doi.org/10.1016/j.jconrel.2014.08.022] [PMID: 25192940]

[62] Huang YC, Li RY, Chen JY, Chen JK. Biphasic release of gentamicin from chitosan/fucoidan nanoparticles for pulmonary delivery. Carbohydr Polym 2016; 138: 114-22. [http://dx.doi.org/10.1016/j.carbpol.2015.11.072] [PMID: 26794744]

[63] Bento D, Staats HF, Gonçalves T, Borges O. Development of a novel adjuvanted nasal vaccine: C48/80 associated with chitosan nanoparticles as a path to enhance mucosal immunity. Eur J Pharm Biopharm 2015; 93: 149-64. [http://dx.doi.org/10.1016/j.ejpb.2015.03.024] [PMID: 25818119]

[64] Soliman GM, Zhang YL, Merle G, Cerruti M, Barralet J. Hydrocaffeic acid-chitosan nanoparticles with enhanced stability, mucoadhesion and permeation properties. Eur J Pharm Biopharm 2014; 88(3): 1026-37. [http://dx.doi.org/10.1016/j.ejpb.2014.09.003] [PMID: 25281213]

[65] Luo Y, Wang Q. Recent development of chitosan-based polyelectrolyte complexes with natural polysaccharides for drug delivery. Int J Biol Macromol 2014; 64: 353-67. [http://dx.doi.org/10.1016/j.ijbiomac.2013.12.017] [PMID: 24360899]

[66] Liu Z, Jiao Y, Wang Y, Zhou C, Zhang Z. Polysaccharides-based nanoparticles as drug delivery systems. Adv Drug Deliv Rev 2008; 60(15): 1650-62. [http://dx.doi.org/10.1016/j.addr.2008.09.001] [PMID: 18848591]

[67] Wang J, Ni C, Zhang Y, *et al.* Preparation and pH controlled release of polyelectrolyte complex of poly(L-malic acid-co-D,L-lactic acid) and chitosan. Colloids Surf B Biointerfaces 2014; 115: 275-9. [http://dx.doi.org/10.1016/j.colsurfb.2013.12.018] [PMID: 24384143]

[68] Bernela M, Kaur P, Chopra M, Thakur R. Synthesis, characterization of nisin loaded alginate-chitosa--pluronic composite nanoparticles and evaluation against microbes. LWT - Food Sci Technol 2014; 59(2P1): 1093-9. [http://dx.doi.org/10.1016/j.lwt.2014.05.061]

[69] Rampino A, Borgogna M, Bellich B, Blasi P, Virgilio F, Cesàro A. Chitosan-pectin hybrid nanoparticles prepared by coating and blending techniques. Eur J Pharm Sci 2016; 84: 37-45. [http://dx.doi.org/10.1016/j.ejps.2016.01.004] [PMID: 26772898]

[70] Zhang L, Jin Y, Liu H, Du Y. Structure and control release of chitosan/carboxymethyl cellulose microcapsules. J Appl Polym Sci 2001; 82(3): 584-92. [http://dx.doi.org/10.1002/app.1886]

[71] Silva DA, MacIel JS, Feitosa JP, Paula HC, De Paula RC. Polysaccharide-based nanoparticles formation by polyelectrolyte complexation of carboxymethylated cashew gum and chitosan. J Mater Sci 2010; 45(20): 5605-10. [http://dx.doi.org/10.1007/s10853-010-4625-y]

[72] Abreu FO, Oliveira EF, Paula HC, de Paula RC. Chitosan/cashew gum nanogels for essential oil encapsulation. Carbohydr Polym 2012; 89(4): 1277-82. [http://dx.doi.org/10.1016/j.carbpol.2012.04.048] [PMID: 24750942]

[73] Jardim KV, Joanitti GA, Azevedo RB, Parize AL. Physico-chemical characterization and cytotoxicity evaluation of curcumin loaded in chitosan/chondroitin sulfate nanoparticles. Mater Sci Eng C Mater Biol Appl 2015; 56: 294-304. [http://dx.doi.org/10.1016/j.msec.2015.06.036] [PMID: 26249593]

[74] Yeh MK, Cheng KM, Hu CS, Huang YC, Young JJ. Novel protein-loaded chondroitin sulfate-chitosan nanoparticles: preparation and characterization. Acta Biomater 2011; 7(10): 3804-12. [http://dx.doi.org/10.1016/j.actbio.2011.06.026] [PMID: 21742066]

[75] Shu S, Zhang X, Wu Z, Wang Z, Li C. Delivery of protein drugs using nanoparticles self-assembled from dextran sulfate and quaternized chitosan. J Control Release 2011; 152 (Suppl. 1): e170-2. [http://dx.doi.org/10.1016/j.jconrel.2011.08.069] [PMID: 22195830]

[76] Rodrigues S, da Costa AM, Grenha A. Chitosan/carrageenan nanoparticles: effect of cross-linking with tripolyphosphate and charge ratios. Carbohydr Polym 2012; 89(1): 282-9. [http://dx.doi.org/10.1016/j.carbpol.2012.03.010] [PMID: 24750635]

[77] Ray S, Raychaudhuri U, Chakraborty R. An overview of encapsulation of active compounds used in food products by drying technology. Food Biosci 2016; 13: 76-83. [http://dx.doi.org/10.1016/j.fbio.2015.12.009]

[78] Abdelwahed W, Degobert G, Stainmesse S, Fessi H. Freeze-drying of nanoparticles: formulation, process and storage considerations. Adv Drug Deliv Rev 2006; 58(15): 1688-713. [http://dx.doi.org/10.1016/j.addr.2006.09.017] [PMID: 17118485]

[79] Muzzarelli RA. Biomedical exploitation of chitin and chitosan *via* mechano-chemical disassembly, electrospinning, dissolution in imidazolium ionic liquids, and supercritical drying. Mar Drugs 2011; 9(9): 1510-33. [http://dx.doi.org/10.3390/md9091510] [PMID: 22131955]

[80] Yu H, Tran T-T, Teo J, Hadinoto K. Dry powder aerosols of curcumin-chitosan nanoparticle complex prepared by spray freeze drying and their antimicrobial efficacy against common respiratory bacterial pathogens. Colloids Surfaces A Physicochem Eng Asp 2016; 504: 34-42. [http://dx.doi.org/10.1016/j.colsurfa.2016.05.053]

[81] Ruphuy G, Saralegi A, Lopes JC, Dias MM, Barreiro MF. Spray drying as a viable process to produce nano-hydroxyapatite/chitosan (n-HAp/CS) hybrid microparticles mimicking bone composition. Adv Powder Technol 2015; 27(2): 575-83. [http://dx.doi.org/10.1016/j.apt.2016.02.010]

[82] Muheem A, Shakeel F, Jahangir MA, *et al.* A review on the strategies for oral delivery of proteins and peptides and their clinical perspectives. Saudi Pharm J 2016; 24(4): 413-28. [http://dx.doi.org/10.1016/j.jsps.2014.06.004] [PMID: 27330372]

[83] Popat A, Karmakar S, Jambhrunkar S, Xu C, Yu C. Curcumin-cyclodextrin encapsulated chitosan nanoconjugates with enhanced solubility and cell cytotoxicity. Colloids Surf B Biointerfaces 2014; 117: 520-7. [http://dx.doi.org/10.1016/j.colsurfb.2014.03.005] [PMID: 24698148]

[84] Pourshahab PS, Gilani K, Moazeni E, Eslahi H, Fazeli MR, Jamalifar H. Preparation and characterization of spray dried inhalable powders containing chitosan nanoparticles for pulmonary delivery of isoniazid. J Microencapsul 2011; 28(7): 605-13. [http://dx.doi.org/10.3109/02652048.2011.599437] [PMID: 21793647]

[85] Gamboa A, Araujo V, Caro N, Gotteland M, Abugoch L, Tapia C. Spray freeze-drying as an alternative to the ionic gelation method to produce chitosan and alginate nano-particles targeted to the colon. J Pharm Sci 2015; 104(12): 4373-85. [http://dx.doi.org/10.1002/jps.24617] [PMID: 26305273]

[86] Estevinho BN, Rocha F, Santos L, Alves A. Microencapsulation with chitosan by spray drying for industry applications - A review. Trends Food Sci Technol 2013; 31(2): 138-55. [http://dx.doi.org/10.1016/j.tifs.2013.04.001]

[87] Harris R, Lecumberri E, Mateos-Aparicio I, Mengíbar M, Heras A. Chitosan nanoparticles and microspheres for the encapsulation of natural antioxidants extracted from Ilex paraguariensis. Carbohydr Polym 2011; 84(2): 803-6. [http://dx.doi.org/10.1016/j.carbpol.2010.07.003]

[88] Agnihotri SA, Mallikarjuna NN, Aminabhavi TM. Recent advances on chitosan-based micro- and nanoparticles in drug delivery. J Control Release 2004; 100(1): 5-28. [http://dx.doi.org/10.1016/j.jconrel.2004.08.010] [PMID: 15491807]

[89] Madureira AR, Pereira A, Pintado M. Current state on the development of nanoparticles for use against bacterial gastrointestinal pathogens. Focus on chitosan nanoparticles loaded with phenolic

compounds. Carbohydr Polym 2015; 130: 429-39.
[http://dx.doi.org/10.1016/j.carbpol.2015.05.030] [PMID: 26076644]

[90] Romić MD, Klarić MŠ, Lovrić J, *et al.* Melatonin-loaded chitosan/Pluronic® F127 microspheres as *in situ* forming hydrogel: An innovative antimicrobial wound dressing. Eur J Pharm Biopharm 2016; 107: 67-79.
[http://dx.doi.org/10.1016/j.ejpb.2016.06.013] [PMID: 27329001]

[91] Sosnik A, Seremeta KP. Advantages and challenges of the spray-drying technology for the production of pure drug particles and drug-loaded polymeric carriers. Adv Colloid Interface Sci 2015; 223: 40-54.
[http://dx.doi.org/10.1016/j.cis.2015.05.003] [PMID: 26043877]

[92] Oliveira AM, Guimarães KL, Cerize NN, Tunussi AS, Poço JG. Nano spray drying as an innovative technology for encapsulating hydrophilic active pharmaceutical ingredients (API). J Nanomed Nanotechnol 2013; 4(6): 1-6.
[http://dx.doi.org/10.4172/2157-7439.1000186]

[93] Carneiro SP, Cruz RM, dos Santos OD. Evaluation of short cycles of ultrasound application in nanoemulsions to obtain nanocapsules. Ultrason Sonochem 2015; 27: 536-42.
[http://dx.doi.org/10.1016/j.ultsonch.2015.04.002] [PMID: 25899309]

[94] Matougui N, Boge L, Groo AC, *et al.* Lipid-based nanoformulations for peptide delivery. Int J Pharm 2016; 502(1-2): 80-97.
[http://dx.doi.org/10.1016/j.ijpharm.2016.02.019] [PMID: 26899976]

[95] Jain A, Thakur K, Sharma G, Kush P, Jain UK. Fabrication, characterization and cytotoxicity studies of ionically cross-linked docetaxel loaded chitosan nanoparticles. Carbohydr Polym 2016; 137: 65-74.
[http://dx.doi.org/10.1016/j.carbpol.2015.10.012] [PMID: 26686106]

[96] Pereira dos Santos P, Flôres SH, de Oliveira Rios A, Chisté RC. Biodegradable polymers as wall materials to the synthesis of bioactive compound nanocapsules. Trends Food Sci Technol 2016; 53: 23-33.
[http://dx.doi.org/10.1016/j.tifs.2016.05.005]

[97] Esquerdo VM, Dotto GL, Pinto LA. Preparation of nanoemulsions containing unsaturated fatty acid concentrate-chitosan capsules. J Colloid Interface Sci 2015; 445: 137-42.
[http://dx.doi.org/10.1016/j.jcis.2014.12.094] [PMID: 25617613]

[98] Jia Z, Dumont M-J, Orsat V. Encapsulation of phenolic compounds present in plants using protein matrices. Food Biosci 2016; 15: 87-104.
[http://dx.doi.org/10.1016/j.fbio.2016.05.007]

[99] Panwar R, Pemmaraju SC, Sharma AK, Pruthi V. Efficacy of ferulic acid encapsulated chitosan nanoparticles against Candida albicans biofilm. Microb Pathog 2016; 95: 21-31.
[http://dx.doi.org/10.1016/j.micpath.2016.02.007] [PMID: 26930164]

[100] Vyas R, Gupta N, Nimesh S. Chitosan nanoparticles for efficient and targeted delivery of anticancer drugs. Nanobiomaterials in Cancer Therapy. 7th ed. Elsevier 2016; pp. 281-306.
[http://dx.doi.org/10.1016/B978-0-323-42863-7.00009-8]

[101] Balan V, Redinciuc V, Tudorachi N, Verestiuc L. Biotinylated N-palmitoyl chitosan for design of drug loaded self-assembled nanocarriers. Eur Polym J 2016; 81: 284-94.
[http://dx.doi.org/10.1016/j.eurpolymj.2016.06.014]

[102] Sarvaiya J, Agrawal YK. Chitosan as a suitable nanocarrier material for anti-Alzheimer drug delivery. Int J Biol Macromol 2015; 72: 454-65.
[http://dx.doi.org/10.1016/j.ijbiomac.2014.08.052] [PMID: 25199867]

[103] de Oliveira Pedro R, Schmitt CC, Neumann MG. Syntheses and characterization of amphiphilic quaternary ammonium chitosan derivatives. Carbohydr Polym 2016; 147: 97-103.
[http://dx.doi.org/10.1016/j.carbpol.2016.03.083] [PMID: 27178913]

[104] Arpicco S, Battaglia L, Brusa P, *et al.* Recent studies on the delivery of hydrophilic drugs in nanoparticulate systems. J Drug Deliv Sci Technol 2016; 32: 298-312. [http://dx.doi.org/10.1016/j.jddst.2015.09.004]

[105] Huo M, Zhang Y, Zhou J, *et al.* Synthesis and characterization of low-toxic amphiphilic chitosan derivatives and their application as micelle carrier for antitumor drug. Int J Pharm 2010; 394(1-2): 162-73. [http://dx.doi.org/10.1016/j.ijpharm.2010.05.001] [PMID: 20457237]

[106] Wang L, Sun Y, Yang X. Fabrication and characterization of ZnxCd1-xS nanoparticles in chitosan alginate nanocomposite films. Ceram Int 2014; 40(3): 4869-73. [http://dx.doi.org/10.1016/j.ceramint.2013.09.036]

[107] Di Gioia S, Trapani A, Mandracchia D, *et al.* Intranasal delivery of dopamine to the striatum using glycol chitosan/sulfobutylether-β-cyclodextrin based nanoparticles. Eur J Pharm Biopharm 2015; 94(1): 180-93. [http://dx.doi.org/10.1016/j.ejpb.2015.05.019] [PMID: 26032293]

[108] Kumar SP, Birundha K, Kaveri K, Devi KT. Antioxidant studies of chitosan nanoparticles containing naringenin and their cytotoxicity effects in lung cancer cells. Int J Biol Macromol 2015; 78: 87-95. [http://dx.doi.org/10.1016/j.ijbiomac.2015.03.045] [PMID: 25840152]

[109] Dima JB, Sequeiros C, Zaritzky NE. Hexavalent chromium removal in contaminated water using reticulated chitosan micro/nanoparticles from seafood processing wastes. Chemosphere 2015; 141: 100-11. [http://dx.doi.org/10.1016/j.chemosphere.2015.06.030] [PMID: 26151484]

[110] Shaker MA. Adsorption of Co(II), Ni(II) and Cu(II) ions onto chitosan-modified poly(methacrylate) nanoparticles: Dynamics, equilibrium and thermodynamics studies. J Taiwan Inst Chem Eng 2015; 57: 111-22. [http://dx.doi.org/10.1016/j.jtice.2015.05.027]

[111] Silva M dos S, Cocenza DS, Grillo R, *et al.* Paraquat-loaded alginate/chitosan nanoparticles: preparation, characterization and soil sorption studies. J Hazard Mater 2011; 190(1-3): 366-74. [http://dx.doi.org/10.1016/j.jhazmat.2011.03.057] [PMID: 21493003]

[112] Grillo R, Pereira AE, Nishisaka CS, *et al.* Chitosan/tripolyphosphate nanoparticles loaded with paraquat herbicide: an environmentally safer alternative for weed control. J Hazard Mater 2014; 278: 163-71. [http://dx.doi.org/10.1016/j.jhazmat.2014.05.079] [PMID: 24968252]

[113] Grillo R, Clemente Z, de Oliveira JL, *et al.* Chitosan nanoparticles loaded the herbicide paraquat: the influence of the aquatic humic substances on the colloidal stability and toxicity. J Hazard Mater 2015; 286: 562-72. [http://dx.doi.org/10.1016/j.jhazmat.2014.12.021] [PMID: 25636059]

[114] de Oliveira CR, Fraceto LF, Rizzi GM, *et al.* Hepatic effects of the clomazone herbicide in both its free form and associated with chitosan-alginate nanoparticles in bullfrog tadpoles. Chemosphere 2016; 149: 304-13. [http://dx.doi.org/10.1016/j.chemosphere.2016.01.076] [PMID: 26874058]

[115] Esmaeili A, Asgari A. *In vitro* release and biological activities of Carum copticum essential oil (CEO) loaded chitosan nanoparticles. Int J Biol Macromol 2015; 81: 283-90. [http://dx.doi.org/10.1016/j.ijbiomac.2015.08.010] [PMID: 26257380]

[116] Mohammadi A, Hashemi M, Hosseini SM. Chitosan nanoparticles loaded with Cinnamomum zeylanicum essential oil enhance the shelf life of cucumber during cold storage. Postharvest Biol Technol 2015; 110: 203-13. [http://dx.doi.org/10.1016/j.postharvbio.2015.08.019]

[117] Natrajan D, Srinivasan S, Sundar K, Ravindran A. Formulation of essential oil-loaded chitosan-alginate nanocapsules. J Food Drug Anal 2015; 23(3): 560-8. [http://dx.doi.org/10.1016/j.jfda.2015.01.001]

[118] Mohammadi A, Hashemi M, Hosseini SM. Postharvest treatment of nanochitosan-based coating loaded with Zataria multiflora essential oil improves antioxidant activity and extends shelf-life of cucumber. Innov Food Sci Emerg Technol 2016; 33: 580-8. [http://dx.doi.org/10.1016/j.ifset.2015.10.015]

[119] Hu J, Wang X, Xiao Z, Bi W. Effect of chitosan nanoparticles loaded with cinnamon essential oil on the quality of chilled pork. LWT -. Food Sci Technol (Campinas) 2015; 63(1): 519-26.

[120] Hu Q, Luo Y. Polyphenol-chitosan conjugates: Synthesis, characterization, and applications. Carbohydr Polym 2016; 151: 624-39. [http://dx.doi.org/10.1016/j.carbpol.2016.05.109] [PMID: 27474608]

[121] Liu F, Antoniou J, Li Y, *et al.* Chitosan/sulfobutylether-β-cyclodextrin nanoparticles as a potential approach for tea polyphenol encapsulation. Food Hydrocoll 2016; 57: 291-300. [http://dx.doi.org/10.1016/j.foodhyd.2016.01.024]

[122] Madureira AR, Pereira A, Pintado M. Chitosan nanoparticles loaded with 2,5-dihydroxybenzoic acid and protocatechuic acid: Properties and digestion. J Food Eng 2016; 174: 8-14. [http://dx.doi.org/10.1016/j.jfoodeng.2015.11.007]

[123] Pulicharla R, Marques C, Das RK, Rouissi T, Brar SK. Encapsulation and release studies of strawberry polyphenols in biodegradable chitosan nanoformulation. Int J Biol Macromol 2016; 88: 171-8. [http://dx.doi.org/10.1016/j.ijbiomac.2016.03.036] [PMID: 27005769]

[124] Lamarra J, Rivero S, Pinotti A. Design of chitosan-based nanoparticles functionalized with gallic acid. Mater Sci Eng C Mater Biol Appl 2016; 67: 717-26. [http://dx.doi.org/10.1016/j.msec.2016.05.072] [PMID: 27287172]

[125] Hu B, Wang Y, Xie M, Hu G, Ma F, Zeng X. Polymer nanoparticles composed with gallic acid grafted chitosan and bioactive peptides combined antioxidant, anticancer activities and improved delivery property for labile polyphenols. J Funct Foods 2015; 15: 593-603. [http://dx.doi.org/10.1016/j.jff.2015.04.009]

[126] Kailaku SI, Mulyawanti I, Alamsyah AN. Formulation of nanoencapsulated catechin with chitosan as encapsulation material. Procedia Chem 2014; 9: 235-41. [http://dx.doi.org/10.1016/j.proche.2014.05.028]

[127] Zhang H, Jung J, Zhao Y. Preparation, characterization and evaluation of antibacterial activity of catechins and catechins-Zn complex loaded β-chitosan nanoparticles of different particle sizes. Carbohydr Polym 2016; 137: 82-91. [http://dx.doi.org/10.1016/j.carbpol.2015.10.036] [PMID: 26686108]

[128] Du H, Liu M, Yang X, Zhai G. The role of glycyrrhetinic acid modification on preparation and evaluation of quercetin-loaded chitosan-based self-aggregates. J Colloid Interface Sci 2015; 460: 87-96. [http://dx.doi.org/10.1016/j.jcis.2015.08.049] [PMID: 26319324]

[129] Fabiano A, Mattii L, Braca A, Felice F, Di Stefano R, Zambito Y. Nanoparticles based on quaternary ammonium-chitosan conjugate: A vehicle for oral administration of antioxidants contained in red grapes. J Drug Deliv Sci Technol 2016; 32: 291-7. [http://dx.doi.org/10.1016/j.jddst.2015.06.014]

[130] Chopra M, Kaur P, Bernela M, Thakur R. Surfactant assisted nisin loaded chitosan-carageenan nanocapsule synthesis for controlling food pathogens. Food Contr 2014; 37(1): 158-64. [http://dx.doi.org/10.1016/j.foodcont.2013.09.024]

[131] Li Z, Paulson AT, Gill TA. Encapsulation of bioactive salmon protein hydrolysates with chitosan-coated liposomes. J Funct Foods 2015; 19: 733-43.

[http://dx.doi.org/10.1016/j.jff.2015.09.058]

[132] Li W, Peng H, Ning F, *et al.* Amphiphilic chitosan derivative-based core-shell micelles: synthesis, characterisation and properties for sustained release of Vitamin D3. Food Chem 2014; 152: 307-15. [http://dx.doi.org/10.1016/j.foodchem.2013.11.147] [PMID: 24444942]

[133] Solval KM, Espinoza Rodezno LA, Moncada M, Bankston JD, Sathivel S. Evaluation of chitosan nanoparticles as a glazing material for cryogenically frozen shrimp. LWT -. Food Sci Technol (Campinas) 2014; 57(1): 172-80.

[134] Ramezani Z, Zarei M, Raminnejad N. Comparing the effectiveness of chitosan and nanochitosan coatings on the quality of refrigerated silver carp fillets. Food Contr 2015; 51: 43-8. [http://dx.doi.org/10.1016/j.foodcont.2014.11.015]

[135] Guzman-Villanueva D, El-Sherbiny IM, Vlassov AV, Herrera-Ruiz D, Smyth HD. Enhanced cellular uptake and gene silencing activity of siRNA molecules mediated by chitosan-derivative nanocomplexes. Int J Pharm 2014; 473(1-2): 579-90. [http://dx.doi.org/10.1016/j.ijpharm.2014.07.026] [PMID: 25063077]

[136] Ragelle H, Riva R, Vandermeulen G, *et al.* Chitosan nanoparticles for siRNA delivery: optimizing formulation to increase stability and efficiency. J Control Release 2014; 176(1): 54-63. [http://dx.doi.org/10.1016/j.jconrel.2013.12.026] [PMID: 24389132]

[137] Li TS, Yawata T, Honke K. Efficient siRNA delivery and tumor accumulation mediated by ionically cross-linked folic acid-poly(ethylene glycol)-chitosan oligosaccharide lactate nanoparticles: for the potential targeted ovarian cancer gene therapy. Eur J Pharm Sci 2014; 52(1): 48-61. [http://dx.doi.org/10.1016/j.ejps.2013.10.011] [PMID: 24178005]

[138] Farid MM, Hathout RM, Fawzy M, Abou-Aisha K. Silencing of the scavenger receptor (Class B - Type 1) gene using siRNA-loaded chitosan nanoparticles in a HepG2 cell model. Colloids Surf B Biointerfaces 2014; 123: 930-7. [http://dx.doi.org/10.1016/j.colsurfb.2014.10.045] [PMID: 25466457]

[139] Venkatesan B, Tumala A, Subramanian V, Vellaichamy E. Data on synthesis and characterization of chitosan nanoparticles for *in vivo* delivery of siRNA-Npr3: Targeting NPR-C expression in the heart. Data Brief 2016; 8: 441-7. [http://dx.doi.org/10.1016/j.dib.2016.05.074] [PMID: 27366782]

[140] Şalva E, Turan SO, Kabasakal L, *et al.* Investigation of the therapeutic efficacy of codelivery of psiRNA-vascular endothelial growth factor and pIL-4 into chitosan nanoparticles in the breast tumor model. J Pharm Sci 2014; 103(3): 785-95. [http://dx.doi.org/10.1002/jps.23815] [PMID: 24357345]

[141] Guţoaia A, Schuster L, Margutti S, *et al.* Fine-tuned PEGylation of chitosan to maintain optimal siRNA-nanoplex bioactivity. Carbohydr Polym 2016; 143: 25-34. [http://dx.doi.org/10.1016/j.carbpol.2016.01.010] [PMID: 27083340]

[142] Corbet C, Ragelle H, Pourcelle V, *et al.* Delivery of siRNA targeting tumor metabolism using non-covalent PEGylated chitosan nanoparticles: Identification of an optimal combination of ligand structure, linker and grafting method. J Control Release 2016; 223: 53-63. [http://dx.doi.org/10.1016/j.jconrel.2015.12.020] [PMID: 26699426]

[143] Danhier F, Messaoudi K, Lemaire L, Benoit J-P, Lagarce F. Combined anti-Galectin-1 and anti-EGFR siRNA-loaded chitosan-lipid nanocapsules decrease temozolomide resistance in glioblastoma: *in vivo* evaluation. Int J Pharm 2015; 481(1-2): 154-61. [http://dx.doi.org/10.1016/j.ijpharm.2015.01.051] [PMID: 25644286]

[144] Xiao B, Ma P, Viennois E, Merlin D. Urocanic acid-modified chitosan nanoparticles can confer anti-inflammatory effect by delivering CD98 siRNA to macrophages. Colloids Surf B Biointerfaces 2016; 143: 186-93. [http://dx.doi.org/10.1016/j.colsurfb.2016.03.035] [PMID: 27011348]

[145] Ragelle H, Colombo S, Pourcelle V, *et al.* Intracellular siRNA delivery dynamics of integrin-targeted, PEGylated chitosan-poly(ethylene imine) hybrid nanoparticles: A mechanistic insight. J Control Release 2015; 211(1): 1-9.
[http://dx.doi.org/10.1016/j.jconrel.2015.05.274] [PMID: 25989603]

[146] Xie Y, Qiao H, Su Z, Chen M, Ping Q, Sun M. PEGylated carboxymethyl chitosan/calcium phosphate hybrid anionic nanoparticles mediated hTERT siRNA delivery for anticancer therapy. Biomaterials 2014; 35(27): 7978-91.
[http://dx.doi.org/10.1016/j.biomaterials.2014.05.068] [PMID: 24939077]

[147] Rassu G, Soddu E, Cossu M, Gavini E, Giunchedi P, Dalpiaz A. Particulate formulations based on chitosan for nose-to-brain delivery of drugs. A review. J Drug Deliv Sci Technol 2016; 32: 77-87.
[http://dx.doi.org/10.1016/j.jddst.2015.05.002]

[148] Gartziandia O, Herran E, Pedraz JL, Carro E, Igartua M, Hernandez RM. Chitosan coated nanostructured lipid carriers for brain delivery of proteins by intranasal administration. Colloids Surf B Biointerfaces 2015; 134: 304-13.
[http://dx.doi.org/10.1016/j.colsurfb.2015.06.054] [PMID: 26209963]

[149] Bari NK, Fazil M, Hassan MQ, *et al.* Brain delivery of buspirone hydrochloride chitosan nanoparticles for the treatment of general anxiety disorder. Int J Biol Macromol 2015; 81: 49-59.
[http://dx.doi.org/10.1016/j.ijbiomac.2015.07.041] [PMID: 26210037]

[150] Elnaggar YS, Etman SM, Abdelmonsif DA, Abdallah OY. Intranasal piperine-loaded chitosan nanoparticles as brain-targeted therapy in Alzheimer's disease: optimization, biological efficacy, and potential toxicity. J Pharm Sci 2015; 104(10): 3544-56.
[http://dx.doi.org/10.1002/jps.24557] [PMID: 26147711]

[151] Van Woensel M, Wauthoz N, Rosière R, *et al.* Development of siRNA-loaded chitosan nanoparticles targeting Galectin-1 for the treatment of glioblastoma multiforme *via* intranasal administration. J Control Release 2016; 227: 71-81.
[http://dx.doi.org/10.1016/j.jconrel.2016.02.032] [PMID: 26902800]

[152] Shah B, Khunt D, Misra M, Padh H. Application of Box-Behnken design for optimization and development of quetiapine fumarate loaded chitosan nanoparticles for brain delivery *via* intranasal route. Int J Biol Macromol 2016; 89: 206-18.
[http://dx.doi.org/10.1016/j.ijbiomac.2016.04.076] [PMID: 27130654]

[153] Ahmad N, Ahmad R, Naqvi AA, *et al.* Rutin-encapsulated chitosan nanoparticles targeted to the brain in the treatment of Cerebral Ischemia. Int J Biol Macromol 2016; 91: 640-55.
[http://dx.doi.org/10.1016/j.ijbiomac.2016.06.001] [PMID: 27264648]

[154] Dehghan S, Tafaghodi M, Bolourieh T, *et al.* Rabbit nasal immunization against influenza by dry-powder form of chitosan nanospheres encapsulated with influenza whole virus and adjuvants. Int J Pharm 2014; 475(1-2): 1-8.
[http://dx.doi.org/10.1016/j.ijpharm.2014.08.032] [PMID: 25148732]

[155] Dai C, Kang H, Yang W, *et al.* O-2-hydroxypropyltrimethyl ammonium chloride chitosan nanoparticles for the delivery of live Newcastle disease vaccine. Carbohydr Polym 2015; 130: 280-9.
[http://dx.doi.org/10.1016/j.carbpol.2015.05.008] [PMID: 26076628]

[156] Nanda RK, Hajam IA, Edao BM, *et al.* Immunological evaluation of mannosylated chitosan nanoparticles based foot and mouth disease virus DNA vaccine, pVAC FMDV VP1-OmpA in guinea pigs. Biologicals 2014; 42(3): 153-9.
[http://dx.doi.org/10.1016/j.biologicals.2014.01.002] [PMID: 24656961]

[157] Elzatahry AA, Mohy Eldin MS. Preparation and characterization of metronidazole-loaded chitosan nanoparticles for drug delivery application. Polym Adv Technol 2008; 19: 1787-91.
[http://dx.doi.org/10.1002/pat.1195]

[158] Bowman K, Leong KW. Chitosan nanoparticles for oral drug and gene delivery. Int J Nanomedicine 2006; 1(2): 117-28.
[http://dx.doi.org/10.2147/nano.2006.1.2.117] [PMID: 17722528]

[159] Liu L, Zhou C, Xia X, Liu Y. Self-assembled lecithin/chitosan nanoparticles for oral insulin delivery: preparation and functional evaluation. Int J Nanomedicine 2016; 11: 761-9.
[http://dx.doi.org/10.2147/IJN.S96146] [PMID: 26966360]

[160] Mukhopadhyay P, Mishra R, Rana D, Kundu PP. Strategies for effective oral insulin delivery with modified chitosan nanoparticles: A review. Prog Polym Sci 2012; 37(11): 1457-75.
[http://dx.doi.org/10.1016/j.progpolymsci.2012.04.004]

[161] Chen MC, Mi FL, Liao ZX, *et al.* Recent advances in chitosan-based nanoparticles for oral delivery of macromolecules. Adv Drug Deliv Rev 2013; 65(6): 865-79.
[http://dx.doi.org/10.1016/j.addr.2012.10.010] [PMID: 23159541]

[162] Vidgren MT, Kublik H. Nasal delivery systems and their effect on deposition and absorption. Adv Drug Deliv Rev 1998; 29(1-2): 157-77.
[http://dx.doi.org/10.1016/S0169-409X(97)00067-7] [PMID: 10837586]

[163] Baltzley S, Mohammad A, Malkawi AH, Al-Ghananeem AM. Intranasal drug delivery of olanzapine-loaded chitosan nanoparticles. AAPS PharmSciTech 2014; 15(6): 1598-602.
[http://dx.doi.org/10.1208/s12249-014-0189-5] [PMID: 25142821]

[164] Amidi M, Romeijn SG, Borchard G, Junginger HE, Hennink WE, Jiskoot W. Preparation and characterization of protein-loaded N-trimethyl chitosan nanoparticles as nasal delivery system. J Control Release 2006; 111(1-2): 107-16.
[http://dx.doi.org/10.1016/j.jconrel.2005.11.014] [PMID: 16380189]

[165] Costantino HR, Illum L, Brandt G, Johnson PH, Quay SC. Intranasal delivery: physicochemical and therapeutic aspects. Int J Pharm 2007; 337(1-2): 1-24.
[http://dx.doi.org/10.1016/j.ijpharm.2007.03.025] [PMID: 17475423]

[166] Casettari L, Illum L. Chitosan in nasal delivery systems for therapeutic drugs. J Control Release 2014; 190: 189-200.
[http://dx.doi.org/10.1016/j.jconrel.2014.05.003] [PMID: 24818769]

[167] Diebold Y, Jarrín M, Sáez V, *et al.* Ocular drug delivery by liposome-chitosan nanoparticle complexes (LCS-NP). Biomaterials 2007; 28(8): 1553-64.
[http://dx.doi.org/10.1016/j.biomaterials.2006.11.028] [PMID: 17169422]

[168] De Campos AM, Sánchez A, Alonso MJ. Chitosan nanoparticles: a new vehicle for the improvement of the delivery of drugs to the ocular surface. Application to cyclosporin A. Int J Pharm 2001; 224(1-2): 159-68.
[http://dx.doi.org/10.1016/S0378-5173(01)00760-8] [PMID: 11472825]

[169] Puras G, Zarate J, Aceves M, *et al.* Low molecular weight oligochitosans for non-viral retinal gene therapy. Eur J Pharm Biopharm 2013; 83(2): 131-40. [Internet].
[http://dx.doi.org/10.1016/j.ejpb.2012.09.010] [PMID: 23059418]

[170] Meng J, Sturgis TF, Youan BB. Engineering tenofovir loaded chitosan nanoparticles to maximize microbicide mucoadhesion. Eur J Pharm Sci 2011; 44(1-2): 57-67.
[http://dx.doi.org/10.1016/j.ejps.2011.06.007] [PMID: 21704704]

[171] Jafarinejad S, Gilani K, Moazeni E, Ghazi-Khansari M, Najafabadi AR, Mohajel N. Development of chitosan-based nanoparticles for pulmonary delivery of itraconazole as dry powder formulation. Powder Technol 2012; 222: 65-70.
[http://dx.doi.org/10.1016/j.powtec.2012.01.045]

[172] Sung JC, Padilla DJ, Garcia-Contreras L, *et al.* Formulation and pharmacokinetics of self-assembled rifampicin nanoparticle systems for pulmonary delivery. Pharm Res 2009; 26(8): 1847-55.
[http://dx.doi.org/10.1007/s11095-009-9894-2] [PMID: 19407933]

[173] Grenha A, Remuñán-López C, Carvalho EL, Seijo B. Microspheres containing lipid/chitosan nanoparticles complexes for pulmonary delivery of therapeutic proteins. Eur J Pharm Biopharm 2008; 69(1): 83-93.
[http://dx.doi.org/10.1016/j.ejpb.2007.10.017] [PMID: 18166446]

[174] Choi M, Cho M, Han BS, *et al.* Chitosan nanoparticles show rapid extrapulmonary tissue distribution and excretion with mild pulmonary inflammation to mice. Toxicol Lett 2010; 199(2): 144-52.
[http://dx.doi.org/10.1016/j.toxlet.2010.08.016] [PMID: 20816729]

[175] Rodrigues S, Cordeiro C, Seijo B, Remuñán-López C, Grenha A. Hybrid nanosystems based on natural polymers as protein carriers for respiratory delivery: Stability and toxicological evaluation. Carbohydr Polym 2015; 123: 369-80.
[http://dx.doi.org/10.1016/j.carbpol.2015.01.048] [PMID: 25843870]

[176] Kim DG, Jeong YI, Choi C, *et al.* Retinol-encapsulated low molecular water-soluble chitosan nanoparticles. Int J Pharm 2006; 319(1-2): 130-8.
[http://dx.doi.org/10.1016/j.ijpharm.2006.03.040] [PMID: 16713152]

[177] Al-Kassas R, Wen J, Cheng AE, Kim AM, Liu SS, Yu J. Transdermal delivery of propranolol hydrochloride through chitosan nanoparticles dispersed in mucoadhesive gel. Carbohydr Polym 2016; 153: 176-86.
[http://dx.doi.org/10.1016/j.carbpol.2016.06.096] [PMID: 27561485]

[178] Nagpal K, Singh SK, Mishra DN. Chitosan nanoparticles: a promising system in novel drug delivery. Chem Pharm Bull (Tokyo) 2010; 58(11): 1423-30.
[http://dx.doi.org/10.1248/cpb.58.1423] [PMID: 21048331]

[179] Huang L, Guo S. Nanoparticles escaping RES and endosome: Challenges for siRNA delivery for cancer therapy. J Nanomater 2011; 2011

[180] Bernkop-Schnürch A, Dünnhaupt S. Chitosan-based drug delivery systems. Eur J Pharm Biopharm 2012; 81(3): 463-9.
[http://dx.doi.org/10.1016/j.ejpb.2012.04.007] [PMID: 22561955]

Frontiers in Biomaterials, 2017, *Vol. 3*, 107-132

CHAPTER 5

Chemically Modified Chitosan Derivatives

Adriana Pavinatto, Anderson Fiamingo, Andrea de Lacerda Bukzem, Daniella de Souza e Silva, Danilo Martins dos Santos, Tonimar Arrighi Domiciano Senra, William Marcondes Facchinatto and **Sergio Paulo Campana-Filho**[*]

Instituto de Química de São Carlos – Universidade de São Paulo, Av. Trabalhador são-carlense, 400 – CEP, 13566-590 São Carlos/SP, Brazil.

Abstract: Some chemical modifications aiming to introduce a new functionality, to improve and/or develop a given property of chitosan are addressed. Given the great number of possible reactions, focus is put only on some representative chitosan derivatives, exploring examples of those exhibiting anionic, cationic, amphiphilic and hydrophilic character, as well as on crosslinked derivatives. The main synthetic routes used to produce such chitosan derivatives are discussed as well as their structural and physicochemical characterization and some examples of applications in different fields. The effects of the characteristics of the parent chitosan, the *locus* of chemical modification, the nature and the content of substituent's on the properties of the chitosan derivatives are discussed. Some applications of chitosan derivatives are discussed as well.

Keywords: Amphiphilic derivatives, Anionic derivatives, Chitosan, Chemical modifications, Cationic derivatives, Genipin-crosslinked *N,O*-carboxymehtyl chitosan, *N,O*-carboxymehtylchitosan, *N*-(2-hydroxyl)propyl-3-trimethylchitosan, *N,N,N*-trimethylchitosan, *N,O*-methoxypoly(ethyleneglycol)-*g*-chitosan.

INTRODUCTION

Chitosan exhibits several interesting physical and chemical properties along with biological activities but some drawbacks, such as the effect of pH on its polycationic character, solubility and mucoadhessiveness, limit its applications in drug delivery systems. Additionally, when chitosan is used to develop bandages, biomaterials for implants and scaffolds for biomedical applications, one must be aware that adequate mechanical properties will not develop unless cross-linking is

[*] **Corresponding author Sergio Paulo Campana-Filho:** Instituto de Química de São Carlos – Universidade de São Paulo, Av. Trabalhador são-carlense, 400 – CEP, 13566-590 São Carlos/SP, Brazil; Tel: +55-1633739929; Fax: +55-1633739952; E-mail: scampana@iqsc.usp.br

Guilherme Luiz Dotto, Sergio Paulo Campana-Filho and Luiz Antonio de Almeida Pinto (Eds)

carried out frequently by using cross-linker agents, or if stable polymer networks are formed *via* physical interactions, including hydrogen bonds, hydrophobic and electrostatic interactions. Thus, to overcome some of the inherent limitations of chitosan, several chemical modifications have been proposed to improve a given physical chemical property and/or to enhance specific functionalities and biological activities. Owing to the presence and nucleophilicity of hydroxyl and amino groups, these are the preferential sites for carrying out chemical modifications on chitosan. Although chemical modifications can be carried out at the acetamido groups pertaining to GlcNAc units of chitosan, they are rarely performed owing to low reactivity as compared to hydroxyl and amino groups and due to the effect on chitosan biodegradability as, for instance, lysozyme is active to hydrolyze chitosan only at the chain regions containing at least some diads GlcNAc / GlcNAc. Additionally, unless if preserving the degree of polymerization of chitosan is not important or if the production of chitooligosaccharides is aimed, strongly acid reaction media should be avoided as chitosan is susceptible to depolymerization *via* acid hydrolysis. The preparation, properties and applications of some chitosan derivatives, focusing on the examples of anionic, cationic, hydrophilic, amphiphilic and crosslinked chitosan derivatives, are addressed in the following.

CARBOXYMETHYLCHITOSAN

Carboxymethylchitosan (CMCh), an anionic derivative of chitosan which can be prepared *via* reaction at amino and hydroxyl groups of chitosan, is soluble in an ample range of pH, including neutral and alkaline media, as compared to the parent chitosan which is soluble only in moderately acidic media [1 - 3]. Depending on the reaction conditions, mainly reaction temperature, and on the acylating reagent, namely monochloroacetic acid or glyoxylic acid, the carboxymethylation of chitosan results in *O*-, *N*-, *N,O*- or *N,N*-carboxymethylchitosan [4, 5]. When monochloroacetic acid reacts with chitosan at room temperature in concentrated aqueous NaOH/isopronanol, *O*-carboxymethylation predominates as the strong alkaline medium that promotes the polymer swelling and enhances the nucleophilicity of hydroxyl groups [6 - 8]. However, it is worth mentioning that if chitosan carboxymethylation is carried out at elevated temperatures, *N,O*-carboxymethylchitosan (*N,O*-CMCh) will be produced [9]. When glyoxylic acid is employed as the alkylating agent, a soluble aldimine is firstly produced, which is then reduced by reaction with sodium cyanoborohydride, resulting in *N*-carboxymethylchitosan (*N*-CMCh). According to Muzzarelli *et al.* [10], the production of soluble *N*-CMCh is associated with an equimolar ratio of glyoxylic acid and chitosan amino groups, since the excess of glyoxylic acid leads to insoluble products, while to produce *N,N*-dicarboxymethylchitosan (*N,N*-CMCh), chitosan may react with glyoxylic acid in the presence of water, glacial acetic acid and sodium borohydride [11 - 13]. Additionally, the carboxymethylation of chitosan can be carried out under microwave irradiation to increase the reaction rate and to avoid the use of organic solvents [14].

The structural modifications introduced by carrying out carboxymethylation can be observed by comparing the FTIR and ^{1}H NMR spectra of chitosan and CMCh. The FTIR spectrum of CMCh displays intense bands at 1600 cm^{-1} and 1411 cm^{-1} which are attributed to the symmetric and asymmetric axial deformations of -COO$^-$ group, respectively, confirming the introduction of carboxymethyl groups on chitosan chains [15, 16]. The ^{1}H NMR spectrum of *N,O*-CMCh shows the occurrence of *N*-carboxymethylation as the signals at 3.1 - 3.4 ppm are assigned to mono and disubstitution at amino groups [17], while *O*-substitution is evidenced by the presence of signals at 4.5 - 4.0 ppm due to the hydrogens of carboxymethyl substituents inserted at 6-O and 3-O of the glycopyranose ring [15, 18]. The main characteristics of CMCh are its average degree of substitution ($\overline{DS}$), average molecular weight and dispersity (Đ). The precise determination of $\overline{DS}$ is usually carried out by using ^{1}H NMR [18] and ^{13}C NMR [19] spectroscopy [18, 19] while weight average molecular weight ($\overline{Mw}$) and dispersity of CMCh can be determined by size exclusion chromatography coupled to multi-angle static light scattering detector (SEC-MALS).

Compared to chitosan, the solubility of CMCh in aqueous solution is remarkably improved due to the introduction of carboxymethyl groups along the polymer chains, however the water-solubility of CMCh calls for $\overline{DS} > 0.4$ [20]. In addition, CMCh presents better moisture retention, membrane forming, flocculating, chelating and sorption properties as compared to chitosan [21]. However, carboxymethylchitosan is thermally unstable and adopt a less ordered arrangement as compared to chitosan as demonstrated by thermogravimetric and X-ray diffraction analyses, respectively [6, 16].

Carboxymethylchitosan can affect cell functioning, antioxidant and antiapoptotic activities, while its antimicrobial activity is affected by $\overline{DS}$, the distribution of carboxymethyl groups and polymer average molecular weight. Thus, *N,O*-CMCh exhibits less effective antibacterial activity as compared to chitosan, which is attributed to the decrease in the content of amino groups as a consequence of *N*-carboxymethylation [22].

Due to its improved water-solubility [6, 23], gel-forming capacity [24, 25], biocompatibility [26], biodegradability, low immunogenicity [27], antimicrobial [28 - 31] and antioxidant activities [32, 33], carboxymethylchitosan presents a high potential for applications in Tissue Engineering and Biomedicine. The use of carboxymethylchitosan to prevent the occurrence of pericardial adhesions was first described by Krause *et al.* [34]. Further studies showed that the use of CMCh resulted in significant decrease of postoperative adhesions in rabbit or large animal models without undesirable side effects [35 - 37], owing to the inability of fibroblasts to adhere to pericardial surface covered by CMCh [38], indicating that the anionic derivative of chitosan may act as a biophysical barrier. More recently, CMCh has been used as a carrier for various active biological compounds aiming its application in drug delivery [23]. Lopes *et al.* [36] shown that CMCh and keratinocyte growth factor (KGF) develop a synergistic action to significantly reduce the severity of postoperative pericardial adhesions in large animals. Zhu *et al.* [39] showed that CMCh can help to stabilize a large amount of anticancer drug camptothecin in water *via* aggregation, the camptothecin concentration being much higher (≈ 6x) in the presence of CMCh as compared to pure water. Similar results were reported by Anitha *et al.* [40] who used CMCh nanoparticles to efficiently encapsulate curcumin which were cytotoxic to cancer cells but not to healthy cells, the *in vitro* drug release being faster in the presence of lysozyme. Due to its high biocompatibility and film-forming capacity, CMCh can be used to produce biosensors, electronic devices able to translate a biochemical interaction occurring in the sensor surface into a response directly related to the presence and concentration of biological molecules [23]. Thus, Xu *et al.* [41] reported on the preparation of CMCh/gold nanocomposites with immobilized enzyme, the resulting biosensor exhibiting high sensitivity and stability.

CATIONIC DERIVATIVES

Cationic derivatives of chitosan belong to a class of polymers that have enough positive charges along its macromolecular backbones to impart polyelectrolyte character to them. These derivatives are synthesized to improve the physical chemical properties of chitosan, such as water-solubility, and biological activities as well, including antimicrobial activity [42], mucoadhessiveness [43], higroscopicity and moisture retention [44], contributing to enlarge the possibilities of applications of chitosan. Indeed, the presence of permanent positive charges along the chains of cationic derivatives of chitosan confers to them water-solubility in a wider range of pH as compared to the parent chitosan. There are several cationic derivatives of chitosan but focus will be given mainly on *N,N,N*-trimethylchitosan chloride (TMCh) and *N*-(2-hydroxyl)propyl-3-trimethyl ammonium chitosan chloride (QCh).

Trimethylchitosan

The synthesis of *N,N,N*-trimethylchitosan (TMCh) by the reaction between chitosan and CH_3I in the presence of aqueous NaOH and using *N*-methyl-2-pyrrolidone as reaction medium was firstly reported by Domard *et. al.* [45] who also reported on the occurrence of severe depolymerization of chitosan when submitted to such reaction conditions. Thus, some modifications were introduced by Le Dung *et. al.* [46], Sieval *et. al.* [47] and Snyman *et. al.* [48], such as the change of the molar ratio chitosan:CH_3I and of the number of reaction steps, aiming to prevent the occurrence of depolymerization and to attain a higher degree of quaternization ($\overline{DQ}$), and improved reaction regiosselectivity. Studies using differents types of base (NaOH and dimethylamino-pyridine) and varying the number of reaction steps were carried out by Hamman and Kotzé [49] and Polnok *et.al* [50]. Runarsson *et. al.* [51] used DMF/ H_2O mixture (50:50 v/v) as reaction medium and performed the reaction in the absence of sodium iodide to produce cationic derivative of chitosan exhibiting high degree of substitution ($\overline{DS}$ = 81 – 88%) by carrying out a "one pot" reaction [52]. In an attempt to overcome the disadvantages of the use of methyl iodide, an expensive and highly volatile reagent which is claimed to be carcinogenic, Britto and Assis [53] proposed the use of dimethylsulfate as the alkylating agent to produce TMCh. TMCh can also be prepared after a two steps procedure in which chitosan first reacts with acetaldehyde to form a Schiff's base that is then reduced by treating it with $NaCNBH_3$, preventing chain depolymerization, *N*-deacetylation and *O*-methylation as well [54]. The occurrence of *O*-methylation has also being prevented by Verheul *et.al* [55]. who introduced some modifications in the two step procedure proposed by Muzzarelli and Tanfani [54].

The comparison of the FTIR of chitosan and TMCh spectra highlights the occurrence of *N*-methylation as in the spectrum of the latter polymer the following typical bands are observed [56]: a) at 1475 cm^{-1}, due to asymmetric angular deformation of C-H bonds of methyl groups ; (b) at 1630 – 1660 cm^{-1}, assigned to quaternary ammonium salt; c) at 1415 – 1430 cm^{-1}, attributed to $N\text{-}CH_3$. Additionally, the band at 1577 cm^{-1}, which is attributed to angular deformation of N-H bond of amino groups occurs in both spectra, but it is much weaker in the spectrum of chitosan as compared to TMCh due to the occurrence of *N*-methylation in low extension. The main signals observed in the ^1H-NMR spectrum of TMCh are those at 3,4 ppm and 2,5 ppm, both due to methyl hydrogens, denoting the presence of $-N^+(CH_3)_3$ and $-N(CH_3)_2$ moieties, respectively, evidencing the occurrence of *N*-methylation, while the signals at 3,36 ppm – 356 ppm were taken as evidence of *O*-methylation [56, 57]. The average degree of quaternization ($\overline{DQ}$) of TMCh has been determined by using ^1H-NMR spectroscopy and conductometric titration [56, 57] and ^{13}C-NMR spectroscopy [47]. Solid-state CP-MAS ^{13}C NMR spectroscopy has also been used to characterize TMCh, the characteristic signals being observed at 56,4 ppm

($-N^+(CH_3)_3$), 48,1 ppm ($-N(CH_3)_2$) and 68,5 ppm ($-OCH_3$) [58].

Due to the presence of quaternized amino groups ($-N^+(CH_3)_3$), TMCh exhibits improved water-solubility in an ample range of acidity (pH 1-9) as compared to parent chitosan even at low degree of quaternization (DQ $\geq$ 10%) [59]. A thermogravimetric study of chitosan and TMCh carried out in nitrogen atmosphere also showed that the presence of quaternized amino groups ($-N^+(CH_3)_3$) affect the thermal stability of the polymers, as the latter is less stable [60].

The biological activities of TMCh, mainly mucoadhesiviness [61, 62], absorption enhancement [63] and low cytotoxicity [64], have been explored in a number of applications, such as for permeation enhancement [65, 66]; in colonic delivery [67]; to develop nano- and microparticles to the delivery of drug [68 - 70] and gene [71]. Additionally, the antibacterial properties of TMCh [72] allowed its application as an edible film able to extend the shelf-life in the food industry [73].

N-(2-hydroxyl)propyl-3-trimethylammonium Chitosan Chloride

The most commonly used method to produce *N*-(2-hydroxyl)propyl-3-trimethyl ammonium chitosan chloride (QCh) is the reaction of chitosan with glycidyltrimethylammonium chloride (GTMAC) in alkaline, neutral or acid medium at relatively high temperature (> 70 °C) for long reaction time (4-18 h) [74 - 77]. This reaction occurs by a S_N2 mechanism and its regioselectivity is governed by the pH of medium: in neutral and alkaline medium, *O*-substitution occurs to an appreciable extent [44, 74]. In contrast, when such a synthesis is carried out in acid medium, highly substituted QCh samples are produced and *N*-substitution predominates. Another disadvantage of the use of neutral or alkaline media is the occurrence of the hydrolysis of GTMAC to 2,3-dihydroxypropyltrimethylammonium chloride, negatively affecting the reaction yield [44, 74, 77]. Molar ratio chitosan/GTMAC, reaction time and temperature also affect the reaction yield as well as important properties of QCh, such as its average degree of quaternization and intrinsic viscosity [44, 74, 77]. Several studies reported on the increase of $\overline{DQ}$ with increasing molar ratio GTMAC/chitosan, reaction time and temperature [74, 77, 78], which is attributed to the fact that higher excess of GTMAC and longer reaction times favor a more complete *N*-substitution on the chitosan chains. Additionally, increasing the reaction temperature results in more reactive species having enough energy to overcome the barrier corresponding to the activation energy, resulting in faster reaction. Santos, Bukzem and Campana-Filho [78] found that carrying out the reaction for long time at high temperature and by using high concentration of glycidyltrimethylammonium chloride improves the reaction yield but the resulting

QCh presents lower intrinsic viscosity as compared to the parent chitosan. Moreover, such study have shown that carrying out the reaction under microwave irradiation resulted is much faster and complete reaction as compared to conventional heating.

The structural modifications resulting from the reaction of chitosan with GTMAC are evidenced by comparing the FTIR and ^{1}H NMR spectra of chitosan and QCh. The infrared spectrum of QCh displays a new band centered at 1483 cm^{-1}, which is attributed to the C–H bending of $-N^+(CH_3)_3$ group [77]. Additionally, the band corresponding to the primary amine observed at 1600 cm^{-1} in the spectrum of chitosan is less intense in the spectra of QCh and it is shifted to lower wavenumber while that band observed at 1652 cm^{-1} is more intense in the spectra of QCh [76]. Compared to chitosan, the ^{1}H NMR spectrum of QCh shows additional signals at 3.2 ppm and 3.3 ppm corresponding to the methyl hydrogens of $-N^+(CH_3)_3$ and methylene hydrogens of $-NHCH_2$, respectively, which are due to the introduction of the substituent on the chitosan chains [77, 78]. The physical chemical properties, biological activities and applications of QCh are strongly dependent on its average degree of quaternization ($\overline{DQ}$), which expresses the average content of quaternized amino groups ($-N^+(CH_3)_3$), and average molecular weight. The most widely used method to determine $\overline{DQ}$ consists in dosing the polymer counter-ions (Cl^-) by titration of the QCh aqueous solution with standardized aqueous $AgNO_3$ [77]. The average degree of quaternization ($\overline{DQ}$) can be determined by acquiring the QCh ^{1}H NMR spectrum at 85 °C in solution of HCl/D_2O 1% (v/v), Thus, the average degree of quaternization ($\overline{DQ}$) of QCh is calculated from the intensities of the signals due to the anomeric hydrogen bonded to *N*-substituted GlcN units (≈ 5.0 ppm) and due to the anomeric hydrogen bonded to unsubstituted GlcN units (≈ 4.8 ppm) [78, 79]. The molecular weight of derivatives is determined by size-exclusion chromatography analyzes using multi-angle laser light scattering as detectors (SEC-MALLS).

When QCh presents $\overline{DQ} > 40\%$, it displays water-solubility in acid, neutral and alkaline medium ($2 < pH < 12$) [78]. On the other hand, the crystallinity and thermal stability of this derivative are lower as compared to those of the parent chitosan and both characteristics are progressively lower the higher the $\overline{DQ}$ as revealed by X-ray diffraction and thermogravimetric analyses, respectively [76, 78].

The unique properties of QCh, such as biocompatibility, biodegradability, biological activities and low toxicity have been widely explored in studies aiming to develop applications in Biomedicine and Pharmacy [44, 74, 76]. Thus, QCh has been used in various formulations such as hydrogel [80], wound dressing [81], nanoparticle for the controlled release of drugs [82], genes [76] and proteins [83]. Additionally, QCh has been used in a number of applications as nanofiltration membranes [84], fiber additive in textile industry [85], adsorbent for the removal of dyes [86] and catalyst for the synthesis of cyclic carbonates [87].

AMPHIPHILIC DERIVATIVES

Generally, chemical modifications in the chitosan backbone aim to improve a given polymer property, for instance to improve its solubility which is limited to aqueous acidic media. In this context, acylation reactions have been used to insert hydrophobic groups capable to affect the solubility of chitosan [88 - 93]. Owing to the presence of hydroxyl and amino groups in chitosan, the reaction conditions must be controlled if regioselectivity is aimed, the use of protection/deprotection synthetic strategy being frequently recommended [94, 95]. The acylation of chitosan is frequently carried out by coupling it to carboxylic acids *via* carbodiimine, anhydrides and acyl chlorides. When the synthesis of *N*-acylchitosan is aimed, coupling carboxylic acids with chitosan by using carbodiimide and 1-ethyl-3-(3 dimethylaminopropyl)carbodiimine (EDC) is carried out. If the reaction proceeds in organic solvents, *N,N'*-dicyclohexyl-carbodiimide (DCC) replaces carbodiimide, while *N*-hydroxisuccimide (NHS) or water-soluble sulfo-NHS can replace EDC to improve the reaction efficiency [90, 91]. The acylation reactions *via* anhydrides lead to the formation of *N*- or *O*-acylated chitosan derivatives. In fact, *N*-acylation occurs preferentially given that the amino group is a stronger nucleophile as compared to hydroxyl group but to avoid its occurrence, *O*-acyl chitosan is prepared by using *N*-phtaloyl protection, the protecting group being removed at the end of the synthesis by reaction with hydrazine, such a strategy imparting high regioselectivity to the reaction and being easily carried out in mild conditions [94]. Even though the reactions using *N*-phthaloyl protection are effective to prepare *O*-acyl chitosan, this approach involves several steps, including protection and deprotection, the overall procedure being long and calling for the use of organic solvents. A much simpler process proposes the use of methanesulfonic acid, a strong acid that serves as reaction medium and protects the amino groups of chitosan by promoting its full protonation, and acyl chloride as acylating agent [95]. Among the *O*-acyl derivatives of chitosan, those presenting amphiphilic character are very interesting as they can dissolve in organic solvents and display surface activity properties [96]. Depending on the average degree of substitution ($\overline{DS}$) as well as on the chain length of the acyl substituent, the chitosan amphiphilic derivative will display different physical chemical properties as compared to the parent polymer, such as solubility in organic solvents as chloroform, dichloromethane and tetrahydrofuran [97]. Additionally, such amphiphilic derivatives of chitosan may decrease the surface tension of the liquid surfaces and solutions, allowing the formation of well-organized thin films at the interface [90, 95] and in solid substrates [98]. Spherical micelles can be formed in aqueous solution of amphiphilic derivatives of chitosan when the self-aggregation of hydrophobic moieties is favored as the polymer concentration attains a critical value, namely the critical micellar concentration (CMC) [99]. The formation of micelles from amphiphilic derivatives of chitosan is a thermodynamically favorable spontaneous process ($\Delta G < 0$) which occurs *via* self-assembly [91]. Additionally, amphiphilic derivatives of chitosan can be used to form vesicles [100] and liposomes [101].

As previously stated, the content and the length of the hydrophobic substituent strongly affects the properties of amphiphilic derivatives of chitosan. The average degree of substitution ($\overline{DS}$) quantitatively expresses the average content of hydrophobic substituent in the chains of the amphiphilic derivative of chitosan. Frequently $\overline{DS}$ is determined by using ^{1}H NMR [102] or a solid state ^{13}C NMR and spectroscopy [103], but conductometric titration is used as well [101]. Additionally, the structural characterization of amphiphilic derivatives of chitosan can also be done by using FTIR spectroscopy [103]. The measurements of surface tension and solution conductivity as function of polymer concentration as well as fluorescence spectroscopy have been used to determine the CMC of amphiphilic derivatives of chitosan [96].

Among several applications, amphiphilic derivatives of chitosan are mainly applied in biomedical and agricultural fields owing to its properties, such as antifungal activity [104], as prospective carrier for the controlled release of agrochemicals [105] and drug delivery. Indeed, owing to its biocompatibility, biodegradability and low toxicity, amphiphilic derivatives of chitosan are proposed for the development of drug delivery systems as polymeric micelles (PMs) and vesicles [106]. Thus, Woraphatphadung *et al.* [99] synthesized three different amphiphilic chitosan derivatives by inserting naphthyl, octyl and benzyl groups in the chitosan backbone (*N*- and *O*-substitution occurred regardless of the substituent) which were used to produce polymeric micelles (PMs) to encapsulate and deliver hydrophobic drugs, especially meloxicam. Huo *et. al.* [107] produced amphiphilic chitosan by inserting alkyl chains (*N*-octyl) as hydrophobic moiety and glycol as hydrophilic substituent, the nano-sized micelles of such derivative being used to encapsulate paclitaxel (PTX), an antitumor drug. Peng *et. al.* [101] synthesized a novel amphiphilic chitosan derivative, namely *N,N*-dimethylhexadecylcarboxymethyl chitosan (DCMCh) which was combined with cholesterol to form liposomes used to encapsulate salidroside, an antidepressant drug. The authors stated that the novel chitosan-based liposome system showed better encapsulation efficiency and slower sustained release rate when compared to liposome formed by phosphatidylcholine / cholesterol [92].

PEG-BASED HYDROPHILIC DERIVATIVE

Currently, polyethylene glycol (PEG) is the main synthetic hydrophilic and non-ionic polymer used in carriers, such as gels, micelles, nano and microparticles, to perform bioactive drug administration. Indeed, PEG-based carriers exhibit high water-solubility and the steric hindrance, attributed to the "conformational cloud" of hydrophilic segments, is claimed to prevent the recognition of the immune system by suppressing opsonization due to non-specific interactions of blood components, to reduce the susceptibility to proteolytic enzymes and antibodies, and the toxicity of bioactive. Accordingly, most of the liposomal, micellar and conjugated formulations commercially available as well as those that have reached advanced clinical testing, contain PEG as a constituent, approved by the FDA in formulations aiming nasal, rectal, parenteral and topical administration [107, 108]. Thus, grafting PEG chains to chitosan backbone, also known as pegylation reaction, has been proposed to improve chitosan water-solubility and to reduce its immunogenicity, antigenicity and susceptibility to enzymatic degradation [109]. Additionally, chitosan pegylation is claimed to improve mucoadhesiviness due to the penetration of hydrophilic segments through mucosal glycoproteins [110], and to reduce cytotoxicity as compared to parent chitosan [111 - 113]. Different synthetic routes can be used to graft PEG chains by coupling them to -OH and $-NH_2$ groups present in chitosan. In most cases, the precursor is α-monomethoxy,ω-hydroxy polyethyleneglycol (mPEG), in which only one terminal (-OH) is available to coupling reactions with chitosan [114]. However, activation of mPEG is needed to favor the production of mPEG-*g*-chitosan, the most common being the activation of PEG terminal as aldehyde [115 - 117], carboxylic acid [118 - 120], *N*-hydroxysuccinimide [113, 121], iodide [111, 122], carbonate [123], among others [109, 124]. Due to the higher nucleophilicity of amino groups as compared to hydroxyl groups, chitosan pegylation occurs predominantly at the former sites, *i.e.* *N*-pegylation predominates [116]. The first study on the synthesis of mPEG-*g*-chitosan proposed the use of aldehyde PEG activation to allow coupling to amino groups *via* Schiff's base formation [115]. However, the mPEG-aldehyde instability resulted in PEG-*g*-chitosan exhibiting very low average degree of substitution ($\overline{DS}$ = 5-6%), while using anhydrous DMSO resulted in improved reaction efficiency [117]. The activation as mPEG-NHS is the most used strategy to produce mPEG-*g*-chitosan, resulting in variable average degree of substitution ($5\% < \overline{DS} < 20\%$) by using *N,N'*-diciclohexylcarbodiimide (DCC) in DMSO as the coupling agent [121]. However, the average degree of substitution of mPEG-*g*-chitosan ranged in a narrower range ($1\% < \overline{DS} < 7\%$) when a water-soluble coupling agent, namely *N*-(3- dimethylaminopropyl)-*N'*-ethylcarbodiimide hydrochloride (EDC.HCl), was used [113]. Synthetic routes aiming to carry out chitosan *O*-pegylation, although less explored, can be interesting as it preserves most the -NH_2 groups, which remain available for interaction with cell membranes and mucosal surfaces, affecting charge density and absorption efficiency when used in drug delivery systems [111, 122, 125]. The first study on the *O*-pegylation of chitosan proposed the etherification with PEG through CH_3I, using Ag_2O as catalyst [122]. However, the resulting products showed satisfactory solubility only at high average degree of substitution, and catalyst traces remained in the polymer matrix [111]. Alternatively, the previous reaction of amino groups of chitosan with phthalic anhydride to protect them from pegylation was proposed, the deprotection occurring at the last stage by reaction with hydrazine [126]. However, it was observed that the protection/deprotection process was not complete, phthaloyl groups (20%) remaining in the final product [114], which was partially degraded [125]. Despite the occurrence of *N,O*-substitution due to low efficiency of protection with *N*-phthaloyl groups [118], mPEG-*g*-chitosan prepared from high molecular weight chitosan was used to produce micellar carriers [127, 128]. Later, this route was explored, excluding the amino protection step, to produce water-soluble ($2 < pH < 10$) PEG-*g*-chitosan exhibiting $\overline{DS} \approx$ 42% [120]. A recent study reported on the use of this last synthetic route to pegylate chitosan samples exhibiting variable average degree of deacetylation ($62.0\% < \overline{DD} < 96.0\%$) and viscosity average molecular weight (80×10^3 g mol^{-1} $< \overline{Mv} < 508 \times 10^3$ g mol^{-1}) by using commercial mPEG (5,000 g mol^{-1}) [129]. Regardless of the parent chitosan, the resulting PEG-*g*-chitosan samples exhibited very similar average degrees of substitution ($\overline{DS} \approx 40\%$) but very lower intrinsic viscosities as compared to the parent chitosans. Thus, it was proposed that further grafting of PEG chains in chitosan did not occur, even in the case of extensively deacetylated chitosan ($\overline{DD} \approx 5\%$), due to the steric hindrance of PEG chains. The dramatic reduction of hydrodynamic volume of PEG-*g*-chitosan as compared to the parent chitosans is attributed to the high hydrophilicity of the PEG chains. Thus, given the high content of such hydrophilic and long PEG chains, their interactions with water molecules are highly favored, with the exposition of such polymer segments to the aqueous medium, forcing the chitosan backbone to adopt a coiled conformation and provoking the contraction of PEG-*g*-chitosan chain. However, it was observed that the rheological behavior of the physically crosslinked gelled solution of PEG-*g*-chitosan was directly affected by the characteristics of the parent chitosan. Therefore, it was observed that a critical hydrophilicity/hydrophobicity ratio was attained when mPEG was grafted to chitosan exhibiting intermediate values of $\overline{DD}$ and $\overline{Mv}$; the corresponding hydrogel exhibiting a thermal-induced sol-gel transition.

CROSSLINKED DERIVATIVES

Due to excellent biocompatibility, chitosan and its derivatives have been widely used in biomedical applications. However, the development of stable chitosan-based systems, such as nanoparticles, microspheres, self-sustained films, membranes and hydrogels that exhibit suitable physical chemical characteristics and appropriate mechanical properties is necessary in several biomedical applications. Thus, such chitosan-based systems may be stabilized by covalent crosslinking, a type of chemical modification, which aims to link together the polymer chains imparting insolubility and stability, and improving mechanical properties. Cross-linker reagents are molecules containing at least two reactive groups able to react with the functional groups of different polymeric chains resulting in the formation of a polymer network maintained by covalent bridges [130]. Thus, in the covalent crosslinking of chitosan, the reactive sites are its amino and hydroxyl groups [131]. Several cross-linker agents have been proposed such as formaldehyde [132 - 134], glutaraldehyde [133 - 135], epichlorohydrin [136, 137], 1-ethyl-3(3-dimethylaminopropyl) carbodiimide hydrochloride [138 - 140], poly(*N*-vinyl-2-pyrrolidone) [141], oxidized starch [142], oxidized dextran [143], oxidized sugars [144] and genipin [140, 145, 146]. The most common cross-linkers agents used to crosslinking chitosan are formaldehyde (Fig. **1a**), glyoxal [147] and glutaraldehyde (Fig. **1b**). In these cases, the crosslinking reaction occurs *via* nucleophilic addition of the amino groups of chitosan to the carbonyl of the cross-linker agent, resulting in imine bonds [130]. Mi *et al.* [148] used different concentrations of glutaraldehyde to crosslink chitosan membranes and it was observed that increasing the concentration of cross-linking agent led to the production of extensively crosslinked membranes. Clasen *et al.* [149] also used glutaraldehyde to crosslink chitosan membranes, which presented improved mechanical properties with increasing degree of crosslinking while its permeability was significantly reduced. Liu, Y. *et al.* [150] studied the crosslinking *O*-carboxymethylchitosan with glutaraldehyde to develop a promising carrier to be used in the controlled drug release. It is important to highlight that crosslinking chitosan by using dialdehydes as cross-linker agents can result in materials exhibiting low swelling capacity and permeability, mainly when high degree of crosslinking is attained, due to the consumption of most amino groups of chitosan [148, 151]. Aiming to preserve the amino groups, others crosslinking agents have been proposed such as epichlorohydrin, a chlorinated epoxide that is reactive toward hydroxyl groups of chitosan (Fig. **1c**). Huang, R. *et. al.* [136] reported on the crosslinking of a membrane based on a cationic derivative of chitosan, the resulting crosslinked material presented high swelling capacity and water permeability. Chen, A. *et al.* [152] reported on the high capacity of epichlorohydrin-crosslinked chitosan to adsorb Cu(II), Zn(II) and Pb(II) ions from aqueous solution. Owing to the high cytotoxicity and low biocompatibility of most synthetic crosslinking agents, natural cross-linkers should be preferred to develop materials for applications in Tissue Engineering. In this sense, genipin, a natural compound isolated from the fruits of *Gardenia jasminoides* Ellis [153], which exhibits very low cytotoxicity as compared to glutaraldehyde [154, 155], has been proposed as a cross-linker agent to produce biocompatible, noncytotoxic polymer-based scaffolds potentially useful in Tissue Engineering and for the development of new drug delivery systems [141, 156, 157].

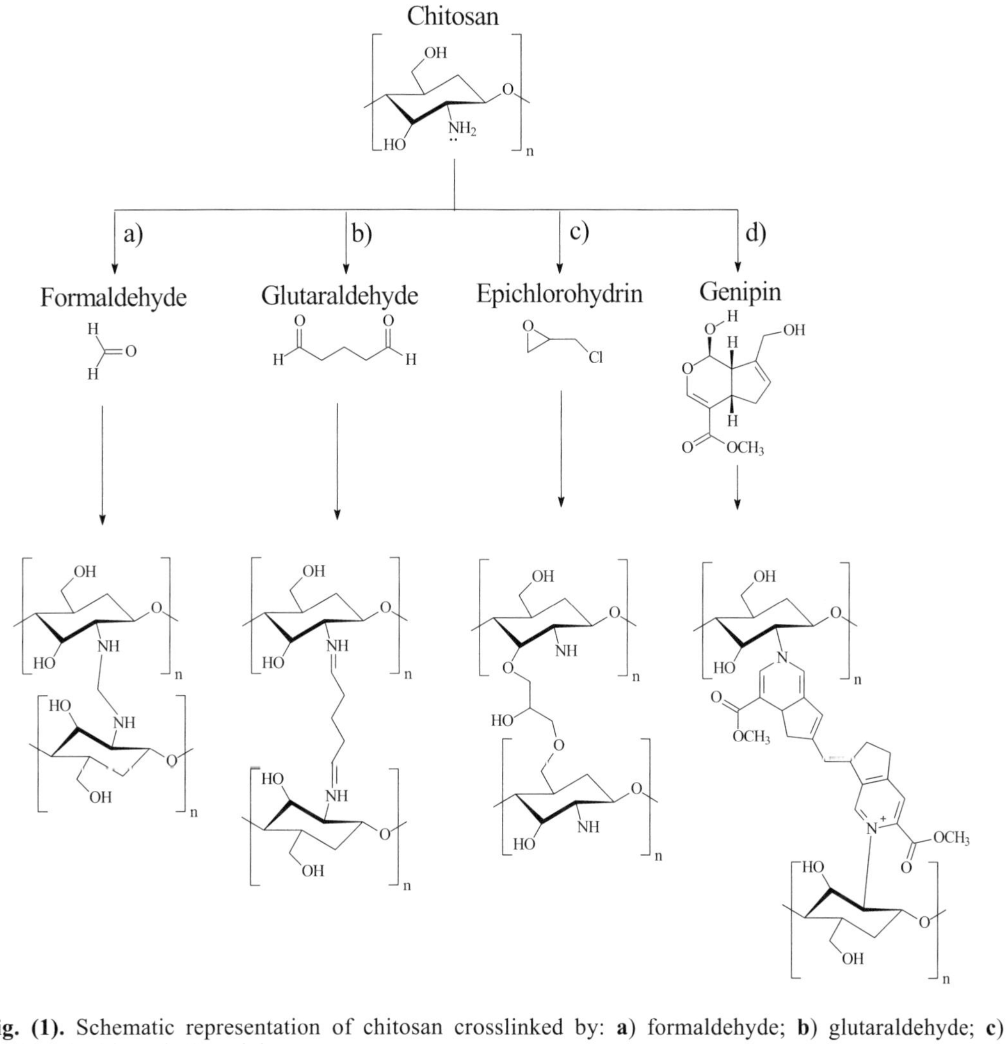

Fig. (1). Schematic representation of chitosan crosslinked by: **a**) formaldehyde; **b**) glutaraldehyde; **c**) epichlorohydrin and; **d**) genipin.

Additionally, Sung, H. *et al.* [158] observed that the genipin-crosslinked bioadhesive was 5,000 times less cytotoxic as compared to the glutaraldehyde-crosslinked bioadhesive. Fiamingo, A. and Campana-Filho, S.P [159]. reported on the production of genipin-crosslinked porous carboxymethylchitosan membranes, showing that the morphology, physical chemical and mechanical properties of the resulting material are strongly affected by the degree of crosslinking, as already reported in the literature [148, 160], but they also depend on the characteristics of the parent carboxymethylchitosan. Thus, when high molecular weight carboxymethylchitosan is used, even lightly crosslinking membrane is stable due to the establishment of extensive chain entanglement, and it also exhibits superior mechanical properties and susceptibility to lysozyme degradation.

CONCLUDING REMARKS

Some chemical modifications of chitosan aimed to introduce new functionalities, to improve a given property of the polymer and to introduce new properties were discussed with focus on the synthetic routes used to produce the chitosan derivatives as well as on its structural and physicochemical characterization, and some examples of applications in different fields. Such an approach highlighted the diversity of chemical modifications one can carry out on amino and hydroxyl groups of chitosan to confer, for instance, a dominant physical chemical characteristic, such as water-solubility or amphiphilic character, to be explored in a given application. The effects of the characteristics of the parent chitosan, the *locus* of chemical modification, the nature and the content of substituents on the properties of the chitosan derivatives have been discussed too, revealing the close associations among them.

CONFLICT OF INTEREST

The authors confirm that they have no conflict of interest to declare for this publication.

ACKNOWLEDGEMENTS

The authors are grateful to the Brazilian agencies CAPES, CNPq and FAPESP for financial support.

REFERENCES

[1] Dumont VC, Mansur AA, Carvalho SM, Medeiros Borsagli FG, Pereira MM, Mansur HS. Chitosan and carboxymethyl-chitosan capping ligands: Effects on the nucleation and growth of hydroxyapatite nanoparticles for producing biocomposite membranes. Mater Sci Eng C Mater Biol Appl 2016; 59: 265-77.
[http://dx.doi.org/10.1016/j.msec.2015.10.018] [PMID: 26652373]

[2] Baran T, Mentes A. Cu (II) and Pd (II) complexes of water soluble O-carboxymethyl chitosan Schiff bases : Synthesis , characterization. Int J Biol Macromol 2015; 79: 542-54. [http://dx.doi.org/10.1016/j.ijbiomac.2015.05.021] [PMID: 26021275]

[3] Mohamed NA, Mohamed RR, Seoudi RS. Synthesis and characterization of some novel antimicrobial thiosemicarbazone O-carboxymethyl chitosan derivatives. Int J Biol Macromol 2014; 63: 163-9. [http://dx.doi.org/10.1016/j.ijbiomac.2013.10.044] [PMID: 24211430]

[4] Abreu FR, Campana-Filho SP. Preparation and characterization of carboxymethylchitosan. Polímeros 2005; 15(2): 79-83. [http://dx.doi.org/10.1590/S0104-14282005000200004]

[5] An NT, Thien DT, Dong NT, Dung P, Le . Water-soluble N-carboxymethylchitosan derivatives: Preparation, characteristics and its application. Carbohydr Polym 2009; 75(3): 489-97.

[6] Abreu FR, Campana-Filho SP. Characteristics and properties of carboxymethylchitosan. Carbohydr Polym 2009; 75(2): 214-21. [http://dx.doi.org/10.1016/j.carbpol.2008.06.009]

[7] Zhu A, Jin W, Yuan L, Yang G, Yu H, Wu H. O-Carboxymethylchitosan-based novel gatifloxacin delivery system. Carbohydr Polym 2007; 68(4): 693-700. [http://dx.doi.org/10.1016/j.carbpol.2006.08.007]

[8] Li GY, Zhong M, Zhou ZD, Zhong YD, Ding P, Huang Y. Formulation optimization of chelerythrine loaded O-carboxymethylchitosan microspheres using response surface methodology. Int J Biol Macromol 2011; 49(5): 970-8. [http://dx.doi.org/10.1016/j.ijbiomac.2011.08.019] [PMID: 21889529]

[9] Tokura S, Nishi N, Tsutsumi A, Somorin O. Studies on chitin VII. Some properties of water soluble chitin derivatives. Polym J 1983; 15(6): 485-9. [http://dx.doi.org/10.1295/polymj.15.485]

[10] Muzzarelli RA, Ilari P, Petrarulo M. Solubility and structure of N-carboxymethylchitosan. Int J Biol Macromol 1994; 16(4): 177-80. [http://dx.doi.org/10.1016/0141-8130(94)90048-5] [PMID: 7848964]

[11] An NT, Dung PL, Thien DT, Dong NT, Nhi TT. An improved method for synthesizing N,N′-dicarboxymethylchitosan. Carbohydr Polym 2008; 73(2): 261-4. [http://dx.doi.org/10.1016/j.carbpol.2007.11.025]

[12] Muzzarelli RA, Ramos V, Stanic V, Mattioli-Belmonte M, Tosi G, Giardino R. Osteogenesis promoted by calcium phosphate N,N-dicarboxymethyl chitosan. Carbohydr Polym 1998; 36(4): 267-76. [http://dx.doi.org/10.1016/S0144-8617(98)00008-3]

[13] le Dung P, Milas M, Rinaudo M, Desbrières J. Water soluble derivatives obtained by controlled chemical modifications of chitosan. Carbohydr Polym 1994, 24(3): 209-14. [http://dx.doi.org/10.1016/0144-8617(94)90132-5]

[14] Ge HC, Luo DK. Preparation of carboxymethyl chitosan in aqueous solution under microwave irradiation. Carbohydr Res 2005; 340(7): 1351-6. [http://dx.doi.org/10.1016/j.carres.2005.02.025] [PMID: 15854605]

[15] Chen X-G, Park H-J. Chemical characteristics of O-carboxymethyl chitosans related to the preparation conditions. Carbohydr Polym 2003; 53(4): 355-9. [http://dx.doi.org/10.1016/S0144-8617(03)00051-1]

[16] Bukzem AL, Signini R, Dos Santos DM, Lião LM, Ascheri DP. Optimization of carboxymethyl chitosan synthesis using response surface methodology and desirability function. Int J Biol Macromol 2016; 85: 615-24. [http://dx.doi.org/10.1016/j.ijbiomac.2016.01.017] [PMID: 26778157]

[17] Muzzarelli RA, Ilari P, Petrarulo M. Solubility and structure of N-carboxymethylchitosan. Int J Biol Macromol 1994; 16(4): 177-80. [Internet]. [http://dx.doi.org/10.1016/0141-8130(94)90048-5] [PMID: 7848964]

[18] Hjerde RJ, Vårum KM, Grasdalen H, Tokura S, Smidsrød O. Chemical composition of O-(carboxymethyl)-chitins in relation to lysozyme degradation rates. Carbohydr Polym 1997; 34(3): 131-9. [http://dx.doi.org/10.1016/S0144-8617(97)00113-6]

[19] Rinaudo M, Le Dung P, Gey C, Milas M. Substituent distribution on O,N-carboxymethylchitosans by 1H and 13C n.m.r. Int J Biol Macromol 1992; 14(3): 122-8. [http://dx.doi.org/10.1016/S0141-8130(05)80001-7] [PMID: 1390443]

[20] Chen L, Du Y, Zeng X. Relationships between the molecular structure and moisture-absorption and moisture-retention abilities of carboxymethyl chitosan. II. Effect of degree of deacetylation and carboxymethylation. Carbohydr Res 2003; 338(4): 333-40. [http://dx.doi.org/10.1016/S0008-6215(02)00462-7] [PMID: 12559731]

[21] Mourya V, Inamdar NN, Tiwari A. Carboxymethyl chitosan and its applications. Adv Mater Lett , 2010 May 1; [[cited 2013 Nov 8]];1(1): 11-33. Available from: http://www.amlett.com/journal_Jun2010_Vol1_Issu1/article/11-33.pdf

[22] Fei Liu X, Lin Guan Y, Zhi Yang D, Li Z, De Yao K. Antibacterial action of chitosan and carboxymethylated chitosan. J Appl Polym Sci. John Wiley & Sons. Inc 2001; 79(7): 1324-35.

[23] Upadhyaya L, Singh J, Agarwal V, Tewari RP. Biomedical applications of carboxymethyl chitosans. Carbohydr Polym 2013; 91(1): 452-66. [http://dx.doi.org/10.1016/j.carbpol.2012.07.076] [PMID: 23044156]

[24] Chen SC, Wu YC, Mi FL, Lin YH, Yu LC, Sung HW. A novel pH-sensitive hydrogel composed of N,O-carboxymethyl chitosan and alginate cross-linked by genipin for protein drug delivery. J Control Release 2004; 96(2): 285-300. [http://dx.doi.org/10.1016/j.jconrel.2004.02.002] [PMID: 15081219]

[25] Lin YH, Liang HF, Chung CK, Chen MC, Sung HW. Physically crosslinked alginate/N,O-carboxymethyl chitosan hydrogels with calcium for oral delivery of protein drugs. Biomaterials 2005; 26(14): 2105-13. [http://dx.doi.org/10.1016/j.biomaterials.2004.06.011] [PMID: 15576185]

[26] Chen XG, Wang Z, Liu WS, Park HJ. The effect of carboxymethyl-chitosan on proliferation and collagen secretion of normal and keloid skin fibroblasts. Biomaterials 2002; 23(23): 4609-14. [http://dx.doi.org/10.1016/S0142-9612(02)00207-7] [PMID: 12322982]

[27] Fu D, Han B, Dong W, Yang Z, Lv Y, Liu W. Effects of carboxymethyl chitosan on the blood system of rats. Biochem Biophys Res Commun 2011; 408(1): 110-4. [http://dx.doi.org/10.1016/j.bbrc.2011.03.130] [PMID: 21463605]

[28] Liu X, Wang Q, Zhuang X, Wu B, Yang F, Zeng A. Study on antibacterial activity of O-carboxymethyl chitosan sodium salt and spinnability of O-carboxymethyl chitosan sodium salt/cellulose polyblends in N-methylmorpholine-N-oxide system. Carbohydr Polym 2012; 89(1): 104-10. [http://dx.doi.org/10.1016/j.carbpol.2012.02.057] [PMID: 24750610]

[29] Liu XF, Guan YL, Yang DZ, Li Z, De Yao K. Antibacterial action of chitosan and carboxymethylated chitosan. J Appl Polym Sci 2001; 79(7): 1324-35. [http://dx.doi.org/10.1002/1097-4628(20010214)79:7<1324::AID-APP210>3.0.CO;2-L]

[30] Kim KW, Thomas RL, Lee C, Park HJ. Antimicrobial activity of native chitosan, degraded chitosan, and O-carboxymethylated chitosan. J Food Prot 2003; 66(8): 1495-8. [http://dx.doi.org/10.4315/0362-028X-66.8.1495] [PMID: 12929845]

[31] Rabea EI, Badawy ME, Stevens CV, Smagghe G, Steurbaut W. Chitosan as antimicrobial agent: applications and mode of action. Biomacromolecules 2003; 4(6): 1457-65. [http://dx.doi.org/10.1021/bm034130m] [PMID: 14606868]

[32] Zhao D, Huang J, Hu S, Mao J, Mei L. Biochemical activities of N,O-carboxymethyl chitosan from squid cartilage. Carbohydr Polym 2011; 85(4): 832-7. [http://dx.doi.org/10.1016/j.carbpol.2011.04.007]

[33] Sun T, Zhou D, Xie J, Mao F. Preparation of chitosan oligomers and their antioxidant activity. Eur Food Res Technol 2007; 225(3-4): 451-6. [http://dx.doi.org/10.1007/s00217-006-0439-1]

[34] Krause TJ, Zazanis GA, McKinnon RD. Prevention of postoperative adhesions with the chitin derivative N-O-carboxymethylchitosan. Wound Repair Regen 1996; 4(1): 53-7. [http://dx.doi.org/10.1046/j.1524-475X.1996.40110.x] [PMID: 17129348]

[35] Krause TJ, Zazanis G, Malatesta P, Solina A. Prevention of pericardial adhesions with N-O carboxymethylchitosan in the rabbit model. J Invest Surg 2001; 14(2): 93-7. [http://dx.doi.org/10.1080/08941930152024219] [PMID: 11396625]

[36] Lopes JB, Dallan LA, Moreira LF, *et al.* Synergism between keratinocyte growth factor and carboxymethyl chitosan reduces pericardial adhesions. Ann Thorac Surg 2010; 90(2): 566-72. [http://dx.doi.org/10.1016/j.athoracsur.2010.03.086] [PMID: 20667351]

[37] Costain DJ, Kennedy R, Ciona C, McAlister VC, Lee TD. Prevention of postsurgical adhesions with N,O-carboxymethyl chitosan: examination of the most efficacious preparation and the effect of N,O-carboxymethyl chitosan on postsurgical healing. Surgery 1997; 121(3): 314-9. [http://dx.doi.org/10.1016/S0039-6060(97)90360-3] [PMID: 9068673]

[38] Zhou J, Liwski RS, Elson C, Lee TD. Reduction in postsurgical adhesion formation after cardiac surgery in a rabbit model using N,O-carboxymethyl chitosan to block cell adherence. J Thorac Cardiovasc Surg 2008; 135(4): 777-83. [http://dx.doi.org/10.1016/j.jtcvs.2007.09.033] [PMID: 18374756]

[39] Zhu AP, Liu JH, Ye WH. Effective loading and controlled release of camptothecin by O-carboxymethylchitosan aggregates. Carbohydr Polym 2006; 63(1): 89-96. [http://dx.doi.org/10.1016/j.carbpol.2005.08.006]

[40] Anitha A, Maya S, Deepa N, Chennazhi KP, Nair SV, Tamura H, *et al.* Efficient water soluble O-carboxymethyl chitosan nanocarrier for the delivery of curcumin to cancer cells. Carbohydr Polym 2011; 83(2): 452-61. [http://dx.doi.org/10.1016/j.carbpol.2010.08.008]

[41] Xu Q, Mao C, Liu N-N, Zhu J-J, Sheng J. Direct electrochemistry of horseradish peroxidase based on biocompatible carboxymethyl chitosan-gold nanoparticle nanocomposite. Biosens Bioelectron 2006; 22(5): 768-73. [http://dx.doi.org/10.1016/j.bios.2006.02.010] [PMID: 16600589]

[42] Rabea EI, Badawy ME, Stevens CV, Smagghe G, Steurbaut W. Chitosan as antimicrobial agent: applications and mode of action. Biomacromolecules 2003; 4(6): 1457-65. [http://dx.doi.org/10.1021/bm034130m] [PMID: 14606868]

[43] Sharma TA. *In vitro* evaluation of N-(2-hydroxy) propyl-3-trimethyl ammonium chitosan for oral insulin delivery. Carbohydr Polym 2011; 84: 103-9. [http://dx.doi.org/10.1016/j.carbpol.2010.10.070]

[44] Prado HJ, Matulewicz MC. Cationization of polysaccharides: A path to greener derivatives with many industrial applications. Eur Polym J 2014; 52: 53-75. [http://dx.doi.org/10.1016/j.eurpolymj.2013.12.011]

[45] Domard A, Rinaudo M, Terrassin C. New method for the quaternization of chitosan. Int J Biol Macromol 1986; 8: 105-7.

[http://dx.doi.org/10.1016/0141-8130(86)90007-3]

[46] Le Dung P, Milas M, Rinaudo M, Desbrières J. Water soluble derivatives obtained by controlled chemical modifications of chitosan. Carbohydr Polym 1994; 24: 209-14. [http://dx.doi.org/10.1016/0144-8617(94)90132-5]

[47] Sieval AB, Thanou M, Kotzé AF, Verhoef JC, Brussee J, Junginger HE. Preparation and NMR characterization of highly substituted N-trimethyl chitosan chloride. Carbohydr Polym 1998; 36: 157-65. [http://dx.doi.org/10.1016/S0144-8617(98)00009-5]

[48] Snyman D, Hamman JH, Kotze JS, Rollings JE, Kotzé AF. The relationship between the absolute molecular weight and the degree of quaternization of N -trimethyl chitosan chloride. Carbohydr Polym 2002; 50: 145-50. [http://dx.doi.org/10.1016/S0144-8617(02)00008-5]

[49] Hamman JH, Kotzé AF. Effect of the type of base and number of reaction steps on the degree of quaternization and molecular weight of N-trimethyl chitosan chloride. Drug Dev Ind Pharm 2001; 27(5): 373-80. [http://dx.doi.org/10.1081/DDC-100104312] [PMID: 11448044]

[50] Polnok A, Borchard G, Verhoef JC, Sarisuta N, Junginger HE. Influence of methylation process on the degree of quaternization of N-trimethyl chitosan chloride. Eur J Pharm Biopharm 2004; 57(1): 77-83. [http://dx.doi.org/10.1016/S0939-6411(03)00151-6] [PMID: 14729082]

[51] Runarsson ÖV, Holappa J, Nevalainen T, *et al.* Antibacterial activity of methylated chitosan and chitooligomer derivatives: Synthesis and structure activity relationships. Eur Polym J 2007; 43: 2660-71. [http://dx.doi.org/10.1016/j.eurpolymj.2007.03.046]

[52] Runarsson ÖV, Holappa J, Jónsdóttir S, Steinsson H, Másson M. N-selective "one pot" synthesis of highly N-substituted trimethyl chitosan (TMC). Carbohydr Polym 2008; 74: 740-4. [http://dx.doi.org/10.1016/j.carbpol.2008.03.008]

[53] de Britto D, Assis OB. A novel method for obtaining a quaternary salt of chitosan. Carbohydr Polym 2007; 69: 305-10. [http://dx.doi.org/10.1016/j.carbpol.2006.10.007]

[54] Muzzarelli RA, Tanfani F. The N-permethylation of chitosan and the preparation of N-trimethyl chitosan iodide. Carbohydr Polym 1985; 5: 297-307. [http://dx.doi.org/10.1016/0144-8617(85)90037-2]

[55] Verheul RJ, Amidi M, van der Wal S, van Riet E, Jiskoot W, Hennink WE. Synthesis, characterization and *in vitro* biological properties of O-methyl free N,N,N-trimethylated chitosan. Biomaterials 2008; 29(27): 3642-9. [http://dx.doi.org/10.1016/j.biomaterials.2008.05.026] [PMID: 18556059]

[56] Curti E, de Britto D, Campana-Filho SP. Methylation of Chitosan with Iodomethane: Effect of Reaction Conditions on Chemoselectivity and Degree of Substitution. Macromol Biosci 2003; 3: 571-6. [http://dx.doi.org/10.1002/mabi.200300030]

[57] Curti E, Campana-Filho SP. Viscosity behavior of chitosan and *N,N,N*-Trimethylchitosan chloride salts in acid-free aqueous solution. J Macromol Sci Part A Pure Appl Chem 2006; 43: 555-72. [http://dx.doi.org/10.1080/10601320600575298]

[58] Britto D, Forato LA, Assis OB. Determination of the average degree of quaternization of N,N,N-trimethylchitosan by solid state 13C NMR. Carbohydr Polym 2008; 74: 86-91. [http://dx.doi.org/10.1016/j.carbpol.2008.01.021]

[59] Jintapattanakit A, Mao S, Kissel T, Junyaprasert VB. Physicochemical properties and biocompatibility of N-trimethyl chitosan: effect of quaternization and dimethylation. Eur J Pharm Biopharm 2008;

70(2): 563-71.
[http://dx.doi.org/10.1016/j.ejpb.2008.06.002] [PMID: 18602467]

[60] Britto D, Campana-Filho SP. A kinetic study on the thermal degradation of N,N,N-trimethylchitosan. Polym Degrad Stabil 2004; 84: 353-61.
[http://dx.doi.org/10.1016/j.polymdegradstab.2004.02.005]

[61] Snyman D, Hamman JH, Kotze AF. Evaluation of the mucoadhesive properties of N-trimethyl chitosan chloride. Drug Dev Ind Pharm 2003; 29(1): 61-9.
[http://dx.doi.org/10.1081/DDC-120016684] [PMID: 12602493]

[62] Cardile V, Frasca G, Rizza L, *et al.* Improved adhesion to mucosal cells of water-soluble chitosan tetraalkylammonium salts. Int J Pharm 2008; 362(1-2): 88-92.
[http://dx.doi.org/10.1016/j.ijpharm.2008.06.017] [PMID: 18634863]

[63] van der Merwe SM, Verhoef JC, Verheijden JH, Kotzé AF, Junginger HE. Trimethylated chitosan as polymeric absorption enhancer for improved peroral delivery of peptide drugs. Eur J Pharm Biopharm 2004; 58(2): 225-35.
[http://dx.doi.org/10.1016/j.ejpb.2004.03.023] [PMID: 15296951]

[64] Hagenaars N, Mania M, de Jong P, *et al.* Role of trimethylated chitosan (TMC) in nasal residence time, local distribution and toxicity of an intranasal influenza vaccine. J Control Release 2010; 144(1): 17-24.
[http://dx.doi.org/10.1016/j.jconrel.2010.01.027] [PMID: 20100528]

[65] Kotzé A, de Leeuw B, Lueßen H, de Boer A, Verhoef J, Junginger H. Chitosans for enhanced delivery of therapeutic peptides across intestinal epithelia: *in vitro* evaluation in Caco-2 cell monolayers. Int J Pharm 1997; 159: 243-53.
[http://dx.doi.org/10.1016/S0378-5173(97)00287-1]

[66] Borchard G, Lueßen HL, de Boer AG, Verhoef JC, Lehr C-M, Junginger HE. The potential of mucoadhesive polymers in enhancing intestinal peptide drug absorption. III: Effects of chitosan-glutamate and carbomer on epithelial tight junctions *in vitro*. J Control Release 1996; 39: 131-8.
[http://dx.doi.org/10.1016/0168-3659(95)00146-8]

[67] Atyabi F, Majzoob S, Dorkoosh F, Sayyah M, Ponchel G. The impact of trimethyl chitosan on *in vitro* mucoadhesive properties of pectinate beads along different sections of gastrointestinal tract. Drug Dev Ind Pharm 2007; 33(3): 291-300.
[http://dx.doi.org/10.1080/03639040601085391] [PMID: 17454062]

[68] Mi F-L, Wu Y-Y, Lin Y-H, *et al.* Oral delivery of peptide drugs using nanoparticles self-assembled by poly(γ-glutamic acid) and a chitosan derivative functionalized by trimethylation. Bioconjug Chem 2008; 19(6): 1248-55.
[http://dx.doi.org/10.1021/bc800076n] [PMID: 18517235]

[69] Amidi M, Pellikaan HC, de Boer AH, Crommelin DJ, Hennink WE, Jiskoot W. Preparation and physicochemical characterization of supercritically dried insulin-loaded microparticles for pulmonary delivery. Eur J Pharm Biopharm 2008; 68(2): 191-200.
[http://dx.doi.org/10.1016/j.ejpb.2007.05.007] [PMID: 17576056]

[70] Amidi M, Krudys KM, Snel CJ, *et al.* Efficacy of pulmonary insulin delivery in diabetic rats: use of a model-based approach in the evaluation of insulin powder formulations. J Control Release 2008; 127(3): 257-66.
[http://dx.doi.org/10.1016/j.jconrel.2008.01.019] [PMID: 18353483]

[71] Haas J, Ravi Kumar MN, Borchard G, Bakowsky U, Lehr C-M. Preparation and characterization of chitosan and trimethyl-chitosan-modified poly-(ε-caprolactone) nanoparticles as DNA carriers. AAPS PharmSciTech 2005; 6(1): E22-30.
[http://dx.doi.org/10.1208/pt060106] [PMID: 16353959]

[72] Sadeghi AM, Dorkoosh FA, Avadi MR, Saadat P, Rafiee-Tehrani M, Junginger HE. Preparation, characterization and antibacterial activities of chitosan, N-trimethyl chitosan (TMC) and N-

diethylmethyl chitosan (DEMC) nanoparticles loaded with insulin using both the ionotropic gelation and polyelectrolyte complexation methods. Int J Pharm 2008; 355(1-2): 299-306. [http://dx.doi.org/10.1016/j.ijpharm.2007.11.052] [PMID: 18206322]

[73] Belalia R, Grelier S, Benaissa M, Coma V. New bioactive biomaterials based on quaternized chitosan. J Agric Food Chem 2008; 56(5): 1582-8. [http://dx.doi.org/10.1021/jf071717+] [PMID: 18271546]

[74] Ruihua H, Bingchao Y, Zheng D, Wang B. Preparation and characterization of a quaternized chitosan. J Mater Sci 2012; 47: 845-51. [http://dx.doi.org/10.1007/s10853-011-5862-4]

[75] Wu J, Su Z-G, Ma G-H. A thermo- and pH-sensitive hydrogel composed of quaternized chitosan/ glycerophosphate. Int J Pharm 2006; 315: 1-11.

[76] Xiao B, Wan Y, Wang X, Zha Q, Liu H, Qiu Z, *et al.* Synthesis and characterization of N-(2-hydroxy)propyl-3-trimethyl ammonium chitosan chloride for potential application in gene delivery. Colloids Surf B Biointerfaces 2012; 91: 168-74.

[77] Cho J, Grant J, Piquette-Miller M, Allen C. Synthesis and physicochemical and dynamic mechanical properties of a water-soluble chitosan derivative as a biomaterial. Biomacromolecules 2006; 7(10): 2845-55. [http://dx.doi.org/10.1021/bm060436s] [PMID: 17025361]

[78] dos Santos DM, Bukzem AdeL, Campana-Filho SP. Response surface methodology applied to the study of the microwave-assisted synthesis of quaternized chitosan. Carbohydr Polym 2016; 138: 317-26. [http://dx.doi.org/10.1016/j.carbpol.2015.11.056] [PMID: 26794768]

[79] Desbrières J, Martinez C, Rinaudo M. Hydrophobic derivatives of chitosan: characterization and rheological behaviour. Int J Biol Macromol 1996; 19(1): 21-8. [http://dx.doi.org/10.1016/0141-8130(96)01095-1] [PMID: 8782715]

[80] Wu J, Wei W, Wang L-Y, Su Z-G, Ma G-H. A thermosensitive hydrogel based on quaternized chitosan and poly(ethylene glycol) for nasal drug delivery system. Biomaterials 2007; 28(13): 2220-32. [http://dx.doi.org/10.1016/j.biomaterials.2006.12.024] [PMID: 17291582]

[81] Alipour SM, Nouri M, Mokhtari J, Bahrami SH. Electrospinning of poly(vinyl alcohol)-water-soluble quaternized chitosan derivative blend. Carbohydr Res 2009; 344(18): 2496-501. [http://dx.doi.org/10.1016/j.carres.2009.10.004] [PMID: 19883903]

[82] Liang H, Zhou B, He L, *et al.* Fabrication of zein/quaternized chitosan nanoparticles for the encapsulation and protection of curcumin. RSC Advances 2015; 5: 13891-900. [http://dx.doi.org/10.1039/C4RA14270E]

[83] Xu Y, Du Y, Huang R, Gao L. Preparation and modification of N-(2-hydroxyl) propyl-3-trimethyl ammonium chitosan chloride nanoparticle as a protein carrier. Biomaterials 2003; 24(27): 5015-22. [http://dx.doi.org/10.1016/S0142-9612(03)00408-3] [PMID: 14559015]

[84] Huang R, Chen G, Sun M, Hu Y, Gao C. Preparation and characteristics of quaternized chitosan/poly(acrylonitrile) composite nanofiltration membrane from epichlorohydrin cross-linking. Carbohydr Polym 2007; 70: 318-23. [http://dx.doi.org/10.1016/j.carbpol.2007.04.017]

[85] Seong H-S, Whang HS, Ko S-W. Synthesis of a quaternary ammonium derivative of chito-oligosaccharide as antimicrobial agent for cellulosic fibers. J Appl Polym Sci 2000; 76: 2009-15. [http://dx.doi.org/10.1002/(SICI)1097-4628(20000628)76:14<2009::AID-APP3>3.0.CO;2-W]

[86] Rosa S, Laranjeira MC, Riela HG, Fávere VT. Cross-linked quaternary chitosan as an adsorbent for the removal of the reactive dye from aqueous solutions. J Hazard Mater 2008; 155(1-2): 253-60. [http://dx.doi.org/10.1016/j.jhazmat.2007.11.059] [PMID: 18180101]

[87] Zhao Y, Tian J-S, Qi X-H, Han Z-N, Zhuang Y-Y, He L-N. Quaternary ammonium salt-functionalized chitosan: An easily recyclable catalyst for efficient synthesis of cyclic carbonates from epoxides and carbon dioxide. J Mol Catal Chem 2007; 271: 284-9. [http://dx.doi.org/10.1016/j.molcata.2007.03.047]

[88] Badawy ME, Rabea EI, Rogge TM, *et al.* Synthesis and fungicidal activity of new N,O-acyl chitosan derivatives. Biomacromolecules 2004; 5(2): 589-95. [http://dx.doi.org/10.1021/bm0344295] [PMID: 15003025]

[89] Badawy ME, Rabea EI, Rogge TM, *et al.* Fungicidal and insecticidal activity of O-acyl chitosan derivatives. Polym Bull 2005; 54(4-5): 279-89. [http://dx.doi.org/10.1007/s00289-005-0396-z]

[90] Pavinatto A, Souza AL. Delezuk J a M, Pavinatto FJ, Campana-Filho SP, Oliveira ON. Interaction of O-acylated chitosans with biomembrane models: Probing the effects from hydrophobic interactions and hydrogen bonding. Colloids Surfaces B Biointerfaces. Elsevier BV 2014; 114: 53-9.

[91] Le Tien C, Lacroix M, Ispas-Szabo P, Mateescu M-A. N-acylated chitosan: hydrophobic matrices for controlled drug release. J Control Release 2003; 93(1): 1-13. [http://dx.doi.org/10.1016/S0168-3659(03)00327-4] [PMID: 14602417]

[92] Hirano S, Ohe Y, Ono H. Selective N-acylation of chitosan. Carbohydr Res 1976; 47(2): 315-20. [http://dx.doi.org/10.1016/S0008-6215(00)84198-1] [PMID: 1268884]

[93] Hirano S, Yamaguchi Y, Kamiya M. Novel N-saturated-fatty-acyl derivatives of chitosan soluble in water and in aqueous acid and alkaline solutions. Carbohydr Polym 2002; 48: 203-7. [http://dx.doi.org/10.1016/S0144-8617(01)00243-0]

[94] Kurita K. Controlled functionalization of the polysaccharide chitin. Prog Polym Sci 2001; 26(9): 1921-71. [http://dx.doi.org/10.1016/S0079-6700(01)00007-7]

[95] Tong Y, Wang S, Xu J, Chua B, He C. Synthesis of O,O′-dipalmitoyl chitosan and its amphiphilic properties and capability of cholesterol absorption. Carbohydr Polym 2005; 60(2): 229-33. [http://dx.doi.org/10.1016/j.carbpol.2005.01.005]

[96] Sui W, Song G, Chen G, Xu G. Aggregate formation and surface activity property of an amphiphilic derivative of chitosan. Colloids Surfaces A Physicochem Eng Asp 2005; 256: 29-33. [http://dx.doi.org/10.1016/j.colsurfa.2004.09.019]

[97] Peesan M, Sirivat A, Supaphol P, Rujiravanit R. Dilute solution properties of hexanoyl chitosan in chloroform, dichloromethane, and tetrahydrofuran. Carbohydr Polym 2006; 64: 175-83. [http://dx.doi.org/10.1016/j.carbpol.2005.11.010]

[98] Wu Y, Hisada K, Maeda S, Sasaki T, Sakurai K. Fabrication and structural characterization of the Langmuir-Blodgett films from a new chitosan derivative containing cinnamate chromophores. Carbohydr Polym 2007; 68(4): 766-72. [http://dx.doi.org/10.1016/j.carbpol.2006.08.016]

[99] Woraphatphadung T, Sajomsang W, Gonil P, *et al.* pH-Responsive polymeric micelles based on amphiphilic chitosan derivatives: Effect of hydrophobic cores on oral meloxicam delivery. Int J Pharm 2016; 497(1-2): 150-60. [Internet]. [http://dx.doi.org/10.1016/j.ijpharm.2015.12.009] [PMID: 26657271]

[100] Fan Y, Liu Y, Xi J, Guo R. Vesicle formation with amphiphilic chitosan derivatives and a conventional cationic surfactant in mixed systems. J Colloid Interface Sci 2011; 360(1): 148-53. [http://dx.doi.org/10.1016/j.jcis.2011.04.009] [PMID: 21546032]

[101] Peng H, Li W, Ning F, *et al.* Amphiphilic chitosan derivatives-based liposomes: synthesis, development, and properties as a carrier for sustained release of salidroside. J Agric Food Chem 2014; 62(3): 626-33. [http://dx.doi.org/10.1021/jf4039925] [PMID: 24372377]

[102] Sashiwa H, Shigemasa Y, Roy R, Homogeneous N. O-Acylation of Chitosan in Dimethyl Sulfoxide with Cyclic Acid Anhydrides. Chem Lett 2000; (10): 1186-7. [http://dx.doi.org/10.1246/cl.2000.1186]

[103] Van De Velde K, Kiekens P. Structure analysis and degree of substitution of chitin, chitosan and dibutyrylchitin by FT-IR spectroscopy and solid state13C NMR. Carbohydr Polym 2004; 58(4): 409-16. [http://dx.doi.org/10.1016/j.carbpol.2004.08.004]

[104] Viegas de Souza RH, Takaki M, de Oliveira Pedro R, dos Santos Gabriel J, Tiera MJ, de Oliveira Tiera VA. Hydrophobic effect of amphiphilic derivatives of chitosan on the antifungal activity against Aspergillus flavus and Aspergillus parasiticus. Molecules 2013; 18(4): 4437-50. [http://dx.doi.org/10.3390/molecules18044437] [PMID: 23591924]

[105] Lao SB, Zhang ZX, Xu HH, Jiang GB. Novel amphiphilic chitosan derivatives: Synthesis, characterization and micellar solubilization of rotenone. Carbohydr Polym 2010; 82(4): 1136-42. [http://dx.doi.org/10.1016/j.carbpol.2010.06.044]

[106] Huo M, Zhang Y, Zhou J, *et al.* Synthesis and characterization of low-toxic amphiphilic chitosan derivatives and their application as micelle carrier for antitumor drug. Int J Pharm 2010; 394(1-2): 162-73. [http://dx.doi.org/10.1016/j.ijpharm.2010.05.001] [PMID: 20457237]

[107] Harris JM, Chess RB. Effect of pegylation on pharmaceuticals. Nat Rev Drug Discov 2003; 2(3): 214-21. [http://dx.doi.org/10.1038/nrd1033] [PMID: 12612647]

[108] Knop K, Hoogenboom R, Fischer D, Schubert US. Poly(ethylene glycol) in drug delivery: pros and cons as well as potential alternatives. Angew Chem Int Ed Engl 2010; 49(36): 6288-308. [http://dx.doi.org/10.1002/anie.200902672] [PMID: 20648499]

[109] Casettari L, Vllasaliu D, Castagnino E, Stolnik S, Howdle S, Illum L. PEGylated chitosan derivatives: Synthesis, characterizations and pharmaceutical applications. Prog Polym Sci 2012; 37(5): 659-85. [http://dx.doi.org/10.1016/j.progpolymsci.2011.10.001]

[110] Goto T, Morishita M, Kavimandan NJ, Takayama K, Peppas NA. Gastrointestinal transit and mucoadhesive characteristics of complexation hydrogels in rats. J Pharm Sci 2006; 95(2): 462-9. [http://dx.doi.org/10.1002/jps.20566] [PMID: 16381013]

[111] Hu Y, Jiang H, Xu C, Wang Y, Zhu K. Preparation and characterization of poly(ethylene glycol)---chitosan with water- and organosolubility. Carbohydr Polym 2005; 61(4): 472-9. [http://dx.doi.org/10.1016/j.carbpol.2005.06.022]

[112] Mao S, Shuai X, Unger F, Wittmar M, Xie X, Kissel T. Synthesis, characterization and cytotoxicity of poly(ethylene glycol)-graft-trimethyl chitosan block copolymers. Biomaterials 2005; 26(32): 6343-56. [http://dx.doi.org/10.1016/j.biomaterials.2005.03.036] [PMID: 15913769]

[113] Casettari L, Vllasaliu D, Mantovani G, Howdle SM, Stolnik S, Illum L. Effect of PEGylation on the toxicity and permeability enhancement of chitosan. Biomacromolecules 2010; 11(11): 2854-65. [http://dx.doi.org/10.1021/bm100522c] [PMID: 20873757]

[114] Lebouc F, Dez I, Desbrières J, Picton L, Madec P-J. Different ways for grafting ester derivatives of poly(ethylene glycol) onto chitosan: related characteristics and potential properties. Polymer (Guildf) 2005; 46(3): 639-51. [http://dx.doi.org/10.1016/j.polymer.2004.11.104]

[115] Harris JM, Struck EC, Case MG, Al HE. Synthesis and Characterization of Poly(Ethylene Gycol) Derivates. J Polym Sci, Polym Phys Ed 1984; 22(2): 341-52. [http://dx.doi.org/10.1002/pol.1984.170220207]

[116] Sugimoto M, Morimotob M, Sashiwab H, Saimotob H, Shigemasabl Y. Preparation and characterization of water-soluble chitin and chitosan derivatives. Carbohydr Polym 1998; 36(1): 49-

59.
[http://dx.doi.org/10.1016/S0144-8617(97)00235-X]

[117] Deng L, Qi H, Yao C, Feng M, Dong A. Investigation on the properties of methoxy poly(ethylene glycol)/chitosan graft co-polymers. J Biomater Sci Polym Ed 2007; 18(12): 1575-89.
[PMID: 17988521]

[118] Yoksan R, Akashi M, Hiwatari K, Chirachanchai S. Controlled hydrophobic/hydrophilicity of chitosan for spheres without specific processing technique. Biopolymers 2003; 69(3): 386-90.
[http://dx.doi.org/10.1002/bip.10400] [PMID: 12833264]

[119] Ouchi T, Nishizawa H, Ohya Y. Aggregation phenomenon of PEG-grafted chitosan in aqueous solution. Polymer (Guildf) 1998; 39(21): 5171-5.
[http://dx.doi.org/10.1016/S0032-3861(97)10020-9]

[120] Fangkangwanwong J, Akashi M, Kida T, Chirachanchai S. One-pot synthesis in aqueous system for water-soluble chitosan-graft-poly(ethylene glycol) methyl ether. Biopolymers 2006; 82(6): 580-6.
[http://dx.doi.org/10.1002/bip.20511] [PMID: 16552764]

[121] Jeong Y-I, Kim D-G, Jang M-K, Nah J-W. Preparation and spectroscopic characterization of methoxy poly(ethylene glycol)-grafted water-soluble chitosan. Carbohydr Res 2008; 343(2): 282-9.
[http://dx.doi.org/10.1016/j.carres.2007.10.025] [PMID: 18035341]

[122] Gorochovceva N, Makuška R. Synthesis and study of water-soluble chitosan-O-poly(ethylene glycol)-graft-copolymers. Eur Polym J 2004; 40(4): 685-91.
[http://dx.doi.org/10.1016/j.eurpolymj.2003.12.005]

[123] Saito H, Wu X, Harris JM, Hoffman AS. Graft copolymers of poly (ethylene glycol) (PEG) and chitosan. Macromol Rapid Commun 1997; 18(7): 547-50.
[http://dx.doi.org/10.1002/marc.1997.030180703]

[124] Cho HJ, Kim G, Kwon HS, Oh YK. Functional PEGylated chitosan systems for biopharmaceuticals. In: Sarmento B, Neves J das, Eds. Chitosan-based systems for biopharmaceuticals: delivery, targeting and polymer therapeutics. Chichester: John Wiley 2012; pp. 301-17.
[http://dx.doi.org/10.1002/9781119962977.ch17]

[125] Makuška R, Gorochovceva N. Regioselective grafting of poly(ethylene glycol) onto chitosan through C-6 position of glucosamine units. Carbohydr Polym 2006; 64(2): 319-27.
[http://dx.doi.org/10.1016/j.carbpol.2005.12.006]

[126] Nishimura S-I, Kohgo O, Kurita K, Kuzuhara H. Chemospecific manipulations of a rigid polysaccharide: syntheses of novel chitosan derivatives with excellent solubility in common organic solvents by regioselective chemical modifications. Macromolecules 1991; 24(17): 4745-8.
[http://dx.doi.org/10.1021/ma00017a003]

[127] El-Sherbiny IM, Smyth HD. Controlled release pulmonary administration of curcumin using swellable biocompatible microparticles. Mol Pharm 2012; 9(2): 269-80.
[http://dx.doi.org/10.1021/mp200351y] [PMID: 22136259]

[128] Yoksan R, Matsusaki M, Akashi M, Chirachanchai S. Controlled hydrophobic/hydrophilic chitosan: colloidal phenomena and nanosphere formation. Colloid Polym Sci 2004; 282(4): 337-42.
[http://dx.doi.org/10.1007/s00396-003-0933-y]

[129] Facchinatto WM. Preparação e Caracterização de Nanosuspensões e Hidrogéis de N,O-metoxipoli (etilenoglicol)-g-quitosana para aplicação em Sistemas de Liberação de Fármacos Antitumorais. Instituto de Química de São Carlos, Universidade de São Paulo, São Carlos 2016. f. Dissertação (Mestrado em Físico-Química); 2016.

[130] Berger J, Reist M, Mayer JM, Felt O, Peppas NA, Gurny R. Structure and interactions in covalently and ionically crosslinked chitosan hydrogels for biomedical applications. Eur J Pharm Biopharm 2004; 57(1): 19-34.
[http://dx.doi.org/10.1016/S0939-6411(03)00161-9] [PMID: 14729078]

[131] Gonsalves Arlan de Assis, Melo Araujo Cleônia Roberta, Soares Nivaldo Alves. Different strategies for crosslinking of chitosan. Química Nova 2011; 34(7): 1215-23.

[132] Singh A, Narvi SS, Dutta PK, Pandey N. External stimuli response on a novel chitosan hydrogel crosslinked with formaldehyde. Bull Mater Sci 2006; 29(3): 233-8. [http://dx.doi.org/10.1007/BF02706490]

[133] Pratt DY, Wilson LD, Kozinski JA. Preparation and sorption studies of glutaraldehyde cross-linked chitosan copolymers. J Colloid Interface Sci 2013; 395: 205-11. [http://dx.doi.org/10.1016/j.jcis.2012.12.044] [PMID: 23374440]

[134] Tahtat D, Mahlous M, Benamer S, Khodja AN, Oussedik-Oumehdi H, Laraba-Djebari F. Oral delivery of insulin from alginate/chitosan crosslinked by glutaraldehyde. Int J Biol Macromol 2013; 58: 160-8. [http://dx.doi.org/10.1016/j.ijbiomac.2013.03.064] [PMID: 23567292]

[135] Gupta KC, Jabrail FH. Glutaraldehyde cross-linked chitosan microspheres for controlled release of centchroman. Carbohydr Res 2007; 342(15): 2244-52. [http://dx.doi.org/10.1016/j.carres.2007.06.009] [PMID: 17610856]

[136] Huang R-h, Chen G-h, Sun M-k, Hu Y-m, Gao C-j. Preparation and characteristics of quaternized chitosan/poly(acrylonitrile) composite nanofiltration membrane from epichlorohydrin cross-linking. Carbohydr Polym 2007; 70(3): 318-23. [http://dx.doi.org/10.1016/j.carbpol.2007.04.017]

[137] Demarchi CA, Debrassi A. A magnetic nanogel based on O-carboxymethylchitosan for antitumor drug delivery: synthesis, characterization and *in vitro* drug release. Soft Matter 2014; 10(19): 3441-50.

[138] Lu G, Sheng B, Wang G, *et al.* Controlling the degradation of covalently cross-linked carboxymethyl chitosan utilizing bimodal molecular weight distribution. J Biomater Appl 2009; 23(5): 435-51. [PMID: 18697877]

[139] Lu G, Kong L, Sheng B, Wang G, Gong Y, Zhang X. Degradation of covalently cross-linked carboxymethyl chitosan and its potential application for peripheral nerve regeneration. Eur Polym J 2007; 43(9): 3807-18. [http://dx.doi.org/10.1016/j.eurpolymj.2007.06.016]

[140] Reves BT, Bumgardner JD, Haggard WO. Fabrication of crosslinked carboxymethylchitosan microspheres and their incorporation into composite scaffolds for enhanced bone regeneration. J Biomed Mater Res B Appl Biomater 2013; 101(4): 630-9. [http://dx.doi.org/10.1002/jbm.b.32865] [PMID: 23359414]

[141] Agustina Aldana A, Gonzalez A, Strumia MC, Martinelli M. Preparation and characterization of chitosan/genipin/poly(N-vinyl-2-pyrrolidone) films for controlled release drugs. Mater Chem Phys 2012; 134(1): 317-24. [http://dx.doi.org/10.1016/j.matchemphys.2012.02.071]

[142] Baran ET, Mano JF, Reis RL. Starch-chitosan hydrogels prepared by reductive alkylation cross-linking. J Mater Sci Mater Med 2004; 15(7): 759-65. [http://dx.doi.org/10.1023/B:JMSM.0000032815.86972.5e] [PMID: 15387411]

[143] Hoffmann B, Seitz D, Mencke A, Kokott A, Ziegler G. Glutaraldehyde and oxidised dextran as crosslinker reagents for chitosan-based scaffolds for cartilage tissue engineering. J Mater Sci Mater Med 2009; 20(7): 1495-503. [http://dx.doi.org/10.1007/s10856-009-3707-3] [PMID: 19259790]

[144] Li F, Liu WG, Yao KD. Preparation of oxidized glucose-crosslinked N-alkylated chitosan membrane and *in vitro* studies of pH-sensitive drug delivery behaviour. Biomaterials 2002; 23(2): 343-7. [http://dx.doi.org/10.1016/S0142-9612(01)00111-9] [PMID: 11761153]

[145] Bao D, Chen M, Wang H, Wang J, Liu C, Sun R. Preparation and characterization of double crosslinked hydrogel films from carboxymethylchitosan and carboxymethylcellulose. Carbohydr

Polym 2014; 110: 113-20.
[http://dx.doi.org/10.1016/j.carbpol.2014.03.095] [PMID: 24906736]

[146] Muzzarelli RA. Genipin-crosslinked chitosan hydrogels as biomedical and pharmaceutical aids. Carbohydr Polym 2009; 77(1): 1-9.
[http://dx.doi.org/10.1016/j.carbpol.2009.01.016]

[147] Wang L, Stegemann JP. Glyoxal crosslinking of cell-seeded chitosan/collagen hydrogels for bone regeneration. Acta Biomater 2011; 7(6): 2410-7.
[http://dx.doi.org/10.1016/j.actbio.2011.02.029] [PMID: 21345389]

[148] Mi FL, Tan YC, Liang HC, Huang RN, Sung HW. *In vitro* evaluation of a chitosan membrane cross-linked with genipin. Journal of Biomaterials Science-Polymer Edition 2001; 12(8): 835-50.

[149] Clasen C, Wilhelms T, Kulicke WM. Formation and Characterization of Chitosan Membranes. Biomacromolecules 2006 Nov 01; 7(11): 3210-22.
[http://dx.doi.org/10.1021/bm060486x]

[150] Liu Y-F, Huang K-L, Peng D-M, Ding P, Li G-Y. Preparation and characterization of glutaraldehyde cross-linked O-carboxymethylchitosan microspheres for controlled delivery of pazufloxacin mesilate. Int J Biol Macromol 2007; 41(1): 87-93.
[http://dx.doi.org/10.1016/j.ijbiomac.2007.01.003] [PMID: 17292465]

[151] Beppu MM, Vieira RS, Aimoli CG, Santana CC. Crosslinking of chitosan membranes using glutaraldehyde: Effect on ion permeability and water absorption. J Membr Sci 2007; 301(1–2): 126-30.
[http://dx.doi.org/10.1016/j.memsci.2007.06.015]

[152] Chen A-H, Liu S-C, Chen C-Y, Chen C-Y. Comparative adsorption of Cu(II), Zn(II), and Pb(II) ions in aqueous solution on the crosslinked chitosan with epichlorohydrin. J Hazard Mater 2008; 154(1-3): 184-91.
[http://dx.doi.org/10.1016/j.jhazmat.2007.10.009] [PMID: 18031930]

[153] Fujikawa S, Yokota T, Koga K, Kumada J. The continuous hydrolysis of geniposide to genipin using immobilized beta-glucosidase on calcium alginate gel. Biotechnol Lett 1987; 9(10): 697-702.
[http://dx.doi.org/10.1007/BF01024600]

[154] Sung HW, Chen CN, Huang RN, Hsu JC, Chang WH. *In vitro* surface characterization of a biological patch fixed with a naturally occurring crosslinking agent. Biomaterials 2000; 21(13): 1353-62.
[http://dx.doi.org/10.1016/S0142-9612(00)00017-X] [PMID: 10850929]

[155] Mi F-L, Tan Y-C, Liang H-F, Sung H-W. *In vivo* biocompatibility and degradability of a novel injectable-chitosan-based implant. Biomaterials 2002; 23(1): 181-91.
[http://dx.doi.org/10.1016/S0142-9612(01)00094-1] [PMID: 11762837]

[156] Li X, Nan K, Shi S, Chen H. Preparation and characterization of nano-hydroxyapatite/chitosan cross-linking composite membrane intended for tissue engineering. Int J Biol Macromol 2012; 50(1): 43-9.
[http://dx.doi.org/10.1016/j.ijbiomac.2011.09.021] [PMID: 21983025]

[157] Li Q, Wang X, Lou X, *et al.* Genipin-crosslinked electrospun chitosan nanofibers: Determination of crosslinking conditions and evaluation of cytocompatibility. Carbohydr Polym 2015; 130: 166-74.
[http://dx.doi.org/10.1016/j.carbpol.2015.05.039] [PMID: 26076613]

[158] Sung HW, Huang DM, Chang WH, Huang RN, Hsu JC. Evaluation of gelatin hydrogel crosslinked with various crosslinking agents as bioadhesives: *in vitro* study. J Biomed Mater Res 1999; 46(4): 520-30.
[http://dx.doi.org/10.1002/(SICI)1097-4636(19990915)46:4<520::AID-JBM10>3.0.CO;2-9] [PMID: 10398013]

[159] Fiamingo A, Campana-Filho SP. Structure, morphology and properties of genipin-crosslinked carboxymethylchitosan porous membranes. Carbohydr Polym 2016; 143: 155-63.
[http://dx.doi.org/10.1016/j.carbpol.2016.02.016] [PMID: 27083355]

[160] Mi FL, Shyu SS, Peng CK. Characterization of ring-opening polymerization of genipin and pH-dependent cross-linking reactions between chitosan and genipin. Journal of Polymer Science Part a-. Polym Chem 2005; 43(10): 1985-2000.
[http://dx.doi.org/10.1002/pola.20669]

CHAPTER 6

Characterization Techniques for Chitosan and its Based Materials

Nauro da Silveira[1], **Denys A. S. Rodrigues**[1], **Tito Roberto S. Cadaval**[1], **Guilherme L. Dotto**[2] and **Luiz Antonio A. Pinto**[1,*]

[1] *Industrial Technology Laboratory, School of Chemistry and Food, Federal University of Rio Grande – FURG, Rio Grande – RS, Brazil*

[2] *Chemical Engineering Department, Federal University of Santa Maria-UFSM, Santa Maria-RS, Brazil*

Abstract: All synthesized material needs to be characterized before application. For each type of application there are a number of possible techniques that may be used in order to identify specific material properties and their behaviors during a particular test. Currently a series of chitosan based materials have been synthesized and applied in various fields of science due the versatilities of the materials and its applications. Drug delivery systems, cellular interactions, *in vivo* studies, nanoparticles encapsulated vesicular systems, wastewater treatment and recovery of precious metals, are some of the applications of chitosan and its based materials found in the scientific literature. Thus, this chapter presents a series of techniques used in the characterization of materials, searching to relate the techniques used with the scientific applications. Analyses that aim at to determine the mechanical and chemical properties of the new material are presented, as well as the ways of monitoring and evaluation of interactions between cells and materials synthesized in cell interactions and *in vivo* systems.

Keywords: Cellular interactions, *In vivo* characterization, Material characterization.

INTRODUCTION

The chitosan based materials possess applications in different technological area. In order to characterize and evaluate the performance of the new material, a series of analyses is necessary. The chitosan composites have influence in various biochemical and physiological processes involved in wound healing, because the chitosan provides antimicrobial activity, cell stimulation, and filmogenicity. In the

* **Corresponding author Luiz Antonio A. Pinto:** Industrial Technology Laboratory, School of Chemistry and Food, Federal University of Rio Grande – FURG, Rio Grande – RS, Brazil; Tel/Fax: (55) 53 32336969; E-mail: dqmpinto@furg.br

Guilherme Luiz Dotto, Sergio Paulo Campana-Filho and Luiz Antonio de Almeida Pinto (Eds)

treatment of healing lesions or ulcers, in vascular problems and systemic diseases, the association of the chitin and chitosan with a foam-like dressing can effectively induce nearly physiological repair. Many forms of chitosan nanoparticles have been produced and are suitable as drug vehicles. Results confirm the potential of this drug delivery platform and the subsequent need of a full preclinical evaluation in order to satisfy the specific regulatory demands of this mode of administration [1]. Chitosan is also a protagonist in the delivery of DNA for gene therapy, due to its unique characteristics that include polyelectrolyte complex formation with favorable complex stability values during delivery [2].

Chitosan composites result in a new class of biomaterials that possess mechanical, physicochemical, and functional properties, which have potential for use in different applications [3]. However, before the application of the chitosan material, it is necessary to identify the chemical and physical characteristics of this material using many different analytical techniques.

In some cases, it is necessary to identify the thermal characteristics of the new material, and the DSC and TGA analyses need to be evaluated. The porosity and the superficial area are fundamental information when a mass transference study is performed and, in this case, the BET analysis is imperative. When chemical modification is carried out in the chitosan polymer chains, like a crosslinked or a grafting, the modification in the functional groups can be identified using FTIR and RMN spectroscopic analyses. The removal of ions from the solvent using chitosan material as adsorbent to wastewater treatment and precious metals recovery are extensively studied in the literature.

Thus, many other characterization techniques can be utilized in the elaboration and in the application of these materials in drug delivery systems, cellular interactions, *in vivo* studies, nanoparticles encapsulated vesicular systems, wastewater treatment and precious metals recovery. In this chapter, some of the most important chitosan-based structures are concisely described with their characteristics and applications.

MATERIAL CHARACTERIZATION

Differential Scanning Calorimetry (DSC)

Differential Scanning Calorimetry (DSC) is a technique that enables accurate determination of the temperature at which structural changes occur in the sample. This can lead to develop new products with enhanced textural attributes. Thermal changes can occur as either the first or second order transitions. First order transitions can be associated with endothermic or exothermic changes, such as crystallization and melting. On the other hand, a glass transition is a second order transition, and it involves a stepped change in the heat capacity. DSC analysis has been used especially for the development of new polymers, with the idea of identifying their technological applicability, thus, overcoming their limitations [4].

Martínez-Camacho, [5] studied the thermal characteristics of the pure chitosan films and the chitosan composite in order to evaluate the antifungal activity of the same. Chitosan composite was obtained by the addition of 20% sorbitol (w/w with respect to the amount of chitosan) as a plasticizing agent, and was compared with pure chitosan film. In these assays, it was observed a single value of Tg, so it can be assumed that sorbitol and chitosan were miscible.

Another applicability of chitosan base materials is their use in the treatment of liquid effluents containing recalcitrant molecules, as colorants. With the aim to evaluate possible dye-films interactions, Dotto *et al.* [6] studied the characteristics of the chitosan film after and before the adsorptions of two food dyes from aqueous solutions. In DSC curves, it was observed an endothermic peak at 75.3 °C with enthalpy of 89.1 J g^{-1}. This peak was attributed to evaporation of residual water. For the chitosan films adsorbed with acid red 18 and adsorbed with FD&C blue no. 2, the peak was shifted to 70.3 °C (enthalpy of 62.6 J g^{-1}) and 73.7 °C (enthalpy of 64.7 J g^{-1}), respectively. These alterations suggested that the interactions between films and water were weakened after adsorption. This can be occurred due to the migration of large dye molecules into the biopolymer chains.

Thermal Gravimetric Analysis TGA

Thermal gravimetric analysis (TGA) is a thermal analysis in which changes in physical and chemical properties of the materials are measured in function of the temperature, like mass loss. It is an especially useful technique for the characterization of polymeric materials, including thermoplastics, elastomers, composites, plastic films, and coatings [7, 8].

In the biosorption of azo dyes from aqueous solution using glutaraldehyde-crosslinked chitosan, the thermal decomposition of the microparticles was analyzed by Chen *et al.* [9]. The decomposition temperatures of the microparticles prepared from three different methods were of 273.5, 224.3 and 260.9 °C. These results indicated that the microparticles formed from sodium tripolyphosphate solution were more easily decomposed than those formed from the sodium hydroxide solution. These characteristics were attributed to the decomposition of tripolyphosphate into phosphoric acid.

One of the main characteristics of an adsorbent is its surface area. In order to increase the surface area of chitosan, Li *et al.* [10] studied the preparation and characterization of new foam adsorbents of poly(vinyl alcohol)/chitosan (PVA/CS) composites, and their applied to removal of dye and heavy metal from aqueous solution. The PVA/CS composite foam exhibited well improved thermal stability in comparison to PVA foam. The TG curve of PVA foam showed three stages consecutive: the first was between 196.0 and 299.1 °C, followed by the second stage from 299.1 to 578.0 °C and the last stage from 578.0 and 798.8 °C. There was a total lost of 76.66% at last stage, and this was attributed to the thermal decomposition of PVA. The thermal gravimetric curve of PVA/CS composite foam also showed three different stages: the first stage started from 220.0 to 407.1 °C, followed by the second from 407.1 to 652.9 °C and the last stage from 652.9 to 807.5 °C. The total lost was 78.02%, due to the formation of large amounts of volatiles from the PVA and CS.

Fourier Transformer Infrared (FT-IR)

Infrared radiation refers to the part of electromagnetic spectrum between the visible and microwaves region. This spectroscopic technique allows to obtain information about the structure of the organic molecules, since each functional group have particular characteristics related to the shape, frequency and intensity in the infrared spectrum. Fourier Transformer Infrared (FT-IR) is the mostly utilized spectrometry. It converts an interferogram from the time domain in the frequency domain and produces a complete infrared spectrum [11].

One of the most important characteristic of the chitosan is the deacetylation degree, and it interferes directly in the applicability of chitosan based materials. In order to compare chitosan obtained from chitin of shrimp waste silage and commercial chitosan in flakes and films, Khan *et al.* [12] measured the deacetylation degree of the different chitosans using FT-IR analysis. The deacetylation degrees were determined by eq. (**1**) and eq. (**2**):

$$DD(\%) = 100 - \left[\left(\frac{A_{1655}}{A_{3450}} \right) \times \frac{100}{1.33} \right] \quad \textbf{(1)}$$

$$DD(\%) = 100 - \left[\left(\frac{A_{1655}}{A_{3450}} \right) \times 115 \right] \quad \textbf{(2)}$$

being A_{1655} the absorbance at 1655cm^{-1} of the amide I band, A_{3450} the absorbance at 3450cm^{-1} of the hydroxyl band and 1.33 the value of the A_{1655}/A_{3450} proportion for a completely acetylated chitosan.

Chitosan is a potentially useful material to produce films containing chitosan as a drug carrier for controlled release. In this context, Rodrigues *et al.* [13] studied *in vitro* release and characterization of chitosan films with dexamethasone. Attenuated total reflectance Fourier transformed infrared spectroscopy (ATR-FTIR) spectrums of blanks matrix chitosan films, dexamethasone, monolayer drug-loaded film and bilayer drug-loaded film were performed, in order to identify chemical reactions between the drug and the matrix. The analysis in ATR-FTIR not showed new absorption bands in both drug loaded films, leading to the conclusion that there was not chemical reaction between the drug and the matrix and dexamethasone, probably it did not lose its activity in the film loaded with drug.

Nuclear Magnetic Resonance (NMR)

NMR is another form of absorption spectrophotometry, which under specific conditions can determine a variety of information regarding the molecular structure of the polymer. The sample can absorb electromagnetic radiation in the radio frequency region governed by the characteristics of the sample. A plot of the frequency of the absorption peaks *versus* peaks intensity constitutes an NMR spectrum. The most common NMR analysis is utilizing in proton carbon spectroscopy, and this technique can be utilized to the identification of organic substances [11].

The application of chitosan in several science fields is limited due to its solubility only in dilute organic acid solution. In this way, many studies have been performed to synthesis of chitosan derivatives by chemical modification in order to increase the solubility in aqueous medium. Some examples of the applications of water-soluble chitosan derivatives are: antibacterial activity, anticoagulant activity, antioxidant activity absorption enhancer across mucosal surfaces, and gene delivery vehicles [14].

Sajomsang *et al.* [14] studied the synthesis of methylated chitosan containing aromatic moieties to use in chemoselectivity and to evaluate the effect on molecular weight. Chitosan containing 4-methylbenzyl, 4-N,N-dimethyl-aminobenzyl and 4-pyridylmethyl substituents with various extents of substitution, were synthesized. The extents of N-methylation and O-methylation were determined by proton NMR, and the influence of the sodium hydroxide concentration during methylation of N-(4-N,N-dimethylaminobenzyl) chitosan was investigated. After, the antibacterial activity of these chitosan derivatives was evaluated against *Staphylococcus aureus* (Gram-positive) and *Escherichia coli* (Gram-negative) bacteria.

In other work, Carneiro *et al.* [15] studied the use of chitosan as a coating to protect aluminum alloy 2024. The characterization of the material was performed using RMN. Different inhibitors were added in the matrix, and the interaction between inhibitor and functional groups from chitosan were investigated to evaluated the coating efficiency. The obtained systems were characterized by proton-nuclear magnetic resonance (1H NMR). The spectrum allowed the determination of the degree of substitution in the polymer chain.

X-rays Diffraction (XRD)

X-rays are electromagnetic waves and can be described as an electric field oscillating in a single plane, where the plane is perpendicular to the direction of propagation (i.e., a polarized transverse wave). The X-ray wavelengths are generally considered to lie between 0.1 A and 100 A [16].

In studies of grafting of acrylic monomers into chitosan, the XRD analysis showed that, because of the grafting onto chitosan backbone, the original crystallinity of chitosan was destroyed. The incorporation of grafting agent to the chitosan chains increased its solubility in water. The reaction conditions, such as initiator concentration, monomer concentration, reaction temperature and time, had great influence on grafting copolymerization [17].

Chitosan can be used in nerve repair, and studies showed that the fetal mouse cerebral cortex cells can grow on chitosan film. However, the improper mechanical properties of chitosan restrict its application in nerve regeneration, because chitosan is more rigid and brittle than nerve tissues. Therefore, chitosan film must be modified to improve its mechanical properties before it can be used in nerve repair [18].

Cheng *et al.* [18] developed a series of chitosan–gelatin composite films by varying the components ratio to prepare soft and elastic biomaterial, which could be used for nerve regeneration. A series of chitosan–gelatin films was prepared in different ratios. X-ray analysis showed good compatibility between these two biopolymers, and the gelatin spectrum, presented an amorphous characteristic. The crystallinity of the films decreased with increasing gelatin content. The decreased in the crystallinity was attributed to the broken hydrogen bonding in the chitosan molecules. The amino groups in chitosan forms hydrogen bonding with carboxyl groups in gelatin. This breaks the hydrogen bonding between amino groups and hydroxyl groups in chitosan polymer and, thus, results in an amorphous structure [19, 20].

Scanning Electron Microscopy (SEM) Studies

The scanning electron microscope (SEM) is utilized in diverse fields, such as materials development, metallic materials, ceramics, and polymers. A scanning electron microscope is a type of electron microscope that produces images of a sample by scanning it with a focused beam of electrons. The electrons interact with atoms in the sample, producing various signals that contain information about the samples surface topography and composition. SEM can achieve resolution better above than 1 nm, and the samples must be solid and must fit into the microscope chamber [21].

In order to obtain chitosan nanoparticles to enhance the cellular uptake of curcumin, Xiao *et al.* [21] produced nanoparticles of kafirin/carboxymethyl chitosan. SEM images of the lyophilized powders of both formulations showed that the spherical structures of the individual particles were maintained intact, but stacked in large aggregation were formed.

In the characterization of dielectrophoresis-aligned nanofibrous silk fibroin–chitosan scaffold to use in interactions with endothelial cells for tissue engineering applications, SEM were used to assess the surface properties like texture and fibril size of the scaffolds [22]. Representative SEM images of silk fibroin-chitosan scaffolds with fibrous structures are shown in Fig. (**1a-c**). Fibril sizes obtained from the SEM images are summarized in Fig. (**1d**).

Chitosan scaffold adsorbent for the removal of hazardous food dyes from aqueous solutions was studied by Esquerdo *et al* [23]. In SEM analysis, it was observed that the chitosan scaffold presented a megaporous structure with pores size from 50 to 200 μm. The chitosan scaffold showed suitable structural characteristics for adsorption purposes, with high values of pore size, porosity and specific surface area. The total adsorption capacities of the scaffold were from 788 to 3316 mg g^{-1}, at 298 K.

Transmission Electron Microscopy (TEM)

Transmission electron microscopy (TEM) is a technique in which a beam of electrons is transmitted through a thin sample. An image is formed from the interaction of the electrons transmitted through the specimen. The image is magnified and focused onto an imaging device, on a layer of photographic film, or a camera [24].

TEM is capable to provide images at a significantly higher resolution than light microscopes and this enables to examine fine detail like a single column of atoms. TEM represents a major analysis method in a range of scientific fields, as in physical, chemical and biological resources. TEM find application in cancer research, virology, materials science as well as pollution, nanotechnology, and semiconductor research. Alternate modes of use allow to observe modulations in chemical identity, crystal orientation, electronic structure and sample induced electron phase shift, as well as, the regular absorption based imaging [25].

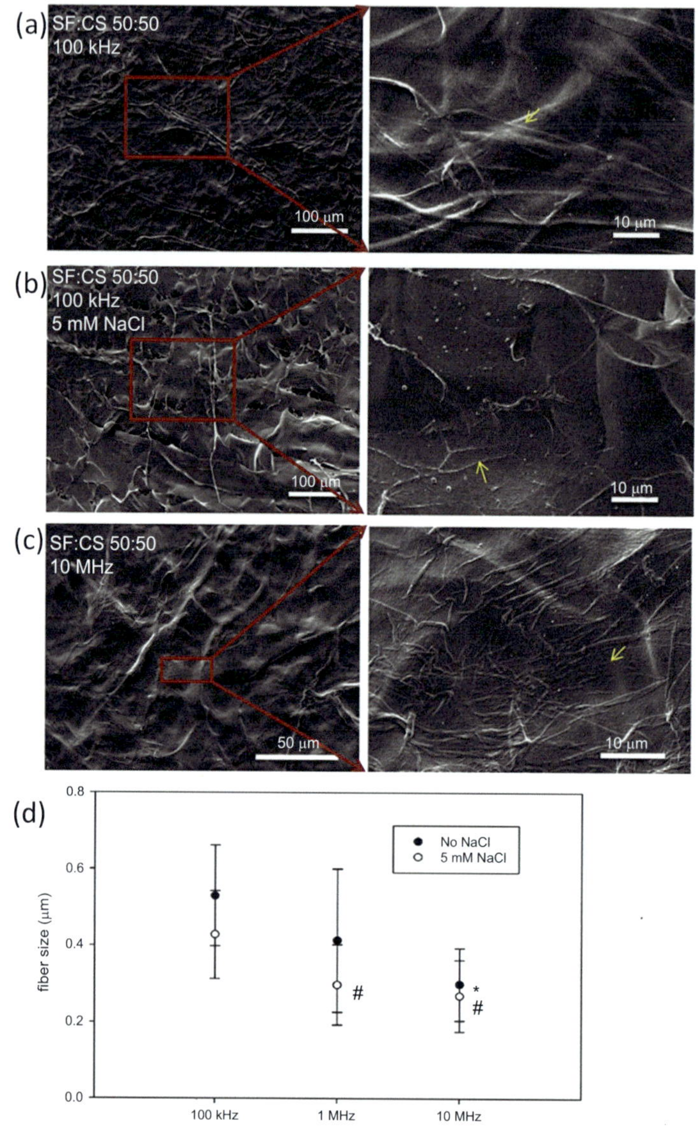

Fig. (1). (**a–c**) SEM images of scaffolds with fibrous structures (fiber structures are indicated by yellow, and samples were frozen at -80 °C after processing). (**d**): Fiber sizes measured from SEM images, comparing the scaffolds (50:50) without and with 5 mM NaCl solution [22].

The preparation and characterization (*in vitro* and *in vivo*) of chitosan nanoparticles encapsulated vesicular systems for oral immunization were evaluated by Vyas *et al.* [26]. The shape and surface morphology of the nanoparticles were observed by TEM, and the nanoparticles made exclusively of chitosan showed a solid and consistent structure, whereas those loaded with antigen exhibited a compact core, presumably, of the antigen surrounded by a coating. The vesicular formulations were of shape spherical, and the encapsulated nanoparticles can be clearly seen as black dense zones inside the vesicles in TEM images of the formulations in Fig. (**2a**).

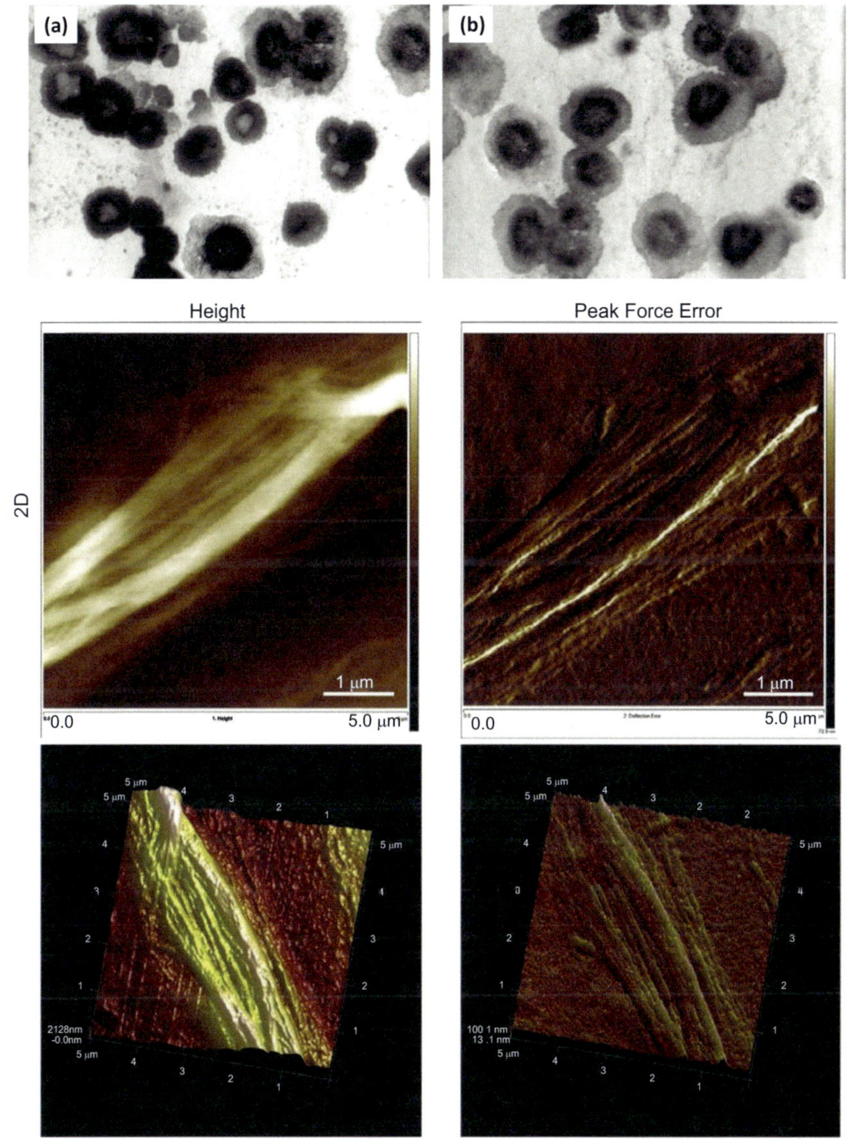

Fig. (2). (**a**) TEM image of chitosan nanoparticles encapsulated in liposomes (×5500), and (**b**) niosomes (×6000) [26]. (**c**) AFM images of scaffolds with nanofibrous structures. A 5 µm × 5 µm area was scanned (50:50) scaffolds were fabricated at a DEP frequency of 10 MHz in the presence of 5 mM NaCl [22].

The use of Methotrexate-loaded chitosan nanogels for drug delivery to brain was studied by Azadi *et al.* [27]. The nanoparticles were prepared and after *in vitro* characterized, and their transport across blood–brain barrier was investigated in animals. The ionic gelation method was used to preparation of drug-loaded nanogels. The size evaluation of samples by TEM showed spherical shapes. In addition, the apparent vacuole assembly of the nanogels as a typical behavior of the samples was noteworthy. It was observed that, the appropriate size and the spherical shape were the favorable factors for the way of the particles input to the brain parenchyma *via* endocytosis.

BET

The BET method or Multimolecular Adsorption Theory is a mathematical theory with purpose of describing the physical adsorption of gas molecules on a solid surface, and it is used as basis for an important analytical technique for measuring specific surface area of a material. The method of BET is considered an extension of the Langmuir theory, developed by Irving Langmuir in 1916 [28, 29].

This analysis enables the determination of surface area, pore volume and distribution of meso and micropores, and it can be performed employing nitrogen. In this method, the gas passes over a cooled sample at the temperature of liquid nitrogen under pressure. The N_2 physically adsorbed produces a change in output composition, measured by a detector. On heating the N_2 loss of the sample cell is desorbed and the peak area is proportional to the mass of N_2 desorbed. By the gas volume desorbed and using the BET equation determines the volume of nitrogen required to cover the adsorbent with a monolayer surface. This equipment is essential for the textural characterization of materials, such as, silica, zeolites, polymers, scaffolds and catalysts [30].

In order to remove the anionic dye reactive blue 19 from aqueous medium, Mirmohseni [31] produced chitosan hollow fibers using the dry-wet spinning technique. The BET surface area of chitosan hollow fibers was found of 48 m^2 g^{-1} based on N_2 adsorption isotherms. The specific surface areas of chitosan beads and flakes were of 14.1 and 11.8 m^2 g^{-1}, respectively. These results confirm that chitosan hollow fibers as biosorbent have highest specific surface area compared with the chitosan beads and flakes [32, 33].

A new type of grafting chitosan was synthesized by Dai *et al.* [34] using 2-hydroxyethyltrimethyl ammonium chloride. The adsorption of Cr (VI) on new chitosan was studied, and the effect of different factors on the adsorption and the adsorption mechanism were considered. The surface areas of pure and grafted chitosan were measured with the BET method. The pure chitosan showed 1.45 m^2 g^{-1} of surface area and 5.10 nm of average pore diameter. The grafted chitosan surface area was of 1.75 m^2 g^{-1} and the average pore diameter of 6.82 nm.

Atomic-force Microscopy (AFM)

AFM is a scanning probe microscopy, with resolution on the order of fractions of a nanometer. The information are recorded by a probe which touch the surface of the sample with piezoelectric elements, which facilitating precise movements of the electronic command from scanning.

This technique provides information about force measurement and imaging. In force measurement can determine the forces between the probe and the sample as a function of their distance, and can be applied to obtain a force spectroscopy. In acquisition of images, the force generated between the sample and the probe enable to form a three-dimensional image of the sample surface at a high resolution. An AFM image is a simulated image based on the height of each point of the surface and, where each coordinate in the plane of the surface has a height h [35].

In the characterization of nanofibrous–chitosan scaffold to apply in interactions with endothelial cells for tissue engineering applications studied by Dunne *et al.* [22], the new material was characterized with atomic force microscopy (AFM) in order to analyze the scaffold topography. The interaction of human umbilical vein endothelial cells with scaffolds was evaluated using AFM to measure the apparent elastic modulus of the cells. AFM images of chitosan based material were performed at 10 MHz in the presence of 5 mM NaCl revealed fibril bundles consisting of fine nanofibers (77.96 ± 4.69 nm) (Fig. **2c**). In the characterization of scaffold apparent elastic modulus, when the loading force was of 0.1 nN, the apparent elastic modulus was of 39.9 ± 22.4 kPa. As the force increased to 1.0 nN and above, the apparent modulus also increased.

The assembly of kafirin/carboxymethyl chitosan nanoparticles to enhance the cellular uptake of curcumin was studied by Huang *et al.* [21]. The surface morphology and three-dimensional AFM images were evaluated for kafirin based colloidal particles containing curcumin and kafirin/carboxymethyl-chitosan nanoparticles. Spherical particles were observed for both formulations. Compared with kafirin based colloidal particles containing curcumin, kafirin/carboxymethyl-chitosan nanoparticles showed larger individual particles with smoother surface. The result suggested that kafirin based colloidal particles containing curcumin showed particles size in the range from 125 to 246 nm with mean a particles size of 193 nm. The values for kafirin/carboxymethyl-chitosan nanoparticles were from 165 to 285 nm and the mean particles size of 225 nm.

Confocal Laser Scanning Microscope

Confocal laser scanning microscopy (CLSM) is an optical technique for increasing optical resolution by means of adding a spatial hole placed at the confocal plane of the lens, in order to remove the out-of-focus light [36]. It enables the reconstruction of three-dimensional structures from the obtained images by a set of images at different depths. This technique can be utilized *in vivo* experiments and in materials science.

CLSM is a scanning imaging technique in which the resolution obtained is best explained by comparing with another scanning technique, like the scanning electron microscope (SEM), and has the advantage of not requiring a probe as in an AFM. The distance from the objective lens to the surface is comparable to the conventional optical microscope.

Electrostatic interaction between chitosan-modified nanoparticle and tumor were evaluated by Yang *et al.* [37]. They studied the direction paclitaxel (drug that prevents uncontrolled cell division), in the lung tumor treatment in mice by intravenous administration of chitosan-modified paclitaxel-loaded nanoparticles. To investigate the distribution profile of coumarin 6 *in vivo*, the nanoparticles were administered to mice *via* the tail vein at a dosage of 10 mg kg^{-1}. Thereafter, the lung tissues slices were directly placed in a sample holder, covered with a coverslip glass, and observed under confocal laser scanning microscopy, and the adsorption amount of coumarin 6 was further quantified. According to the confocal microscopy results in Fig. (**3a**), cellular uptake of C-NPs was much greater compared with U-NPs at all pHs examined, and increased significantly as the pH in the medium decreased for both cell lines.

In an attempt to increase the cellular uptake of genistein, Yang *et al* [38] studied the use of lipid carriers into human lens epithelial cells by chitosan hydrochlorides coatings. This study attempted to increase the cellular uptake of developed genistein loaded nanostructured lipid carriers into human lens epithelial cells by chitosan hydrochlorides coatings, when applied in post lens capsule treatment, and to provide further understanding of the uptake and anti-proliferation mechanisms. The uptake of various particle sizes was evaluated and visualized by confocal laser scanning microscopy (CLSM), showing a size-dependent manner. The uptake was visualized for particles of different sizes using CLSM to support the quantitative uptake measurements.

Fluorescence Microscopy

A fluorescence microscope is equipment that uses fluorescence and phosphorescence to study properties of organic or inorganic substances, and it refers to any microscope which uses fluorescence to generate an image. Many fluorescent stains have been developed for a range of biological molecules. Some of them own intrinsic fluorescence and bind to the molecule of interest in order to identify it [39].

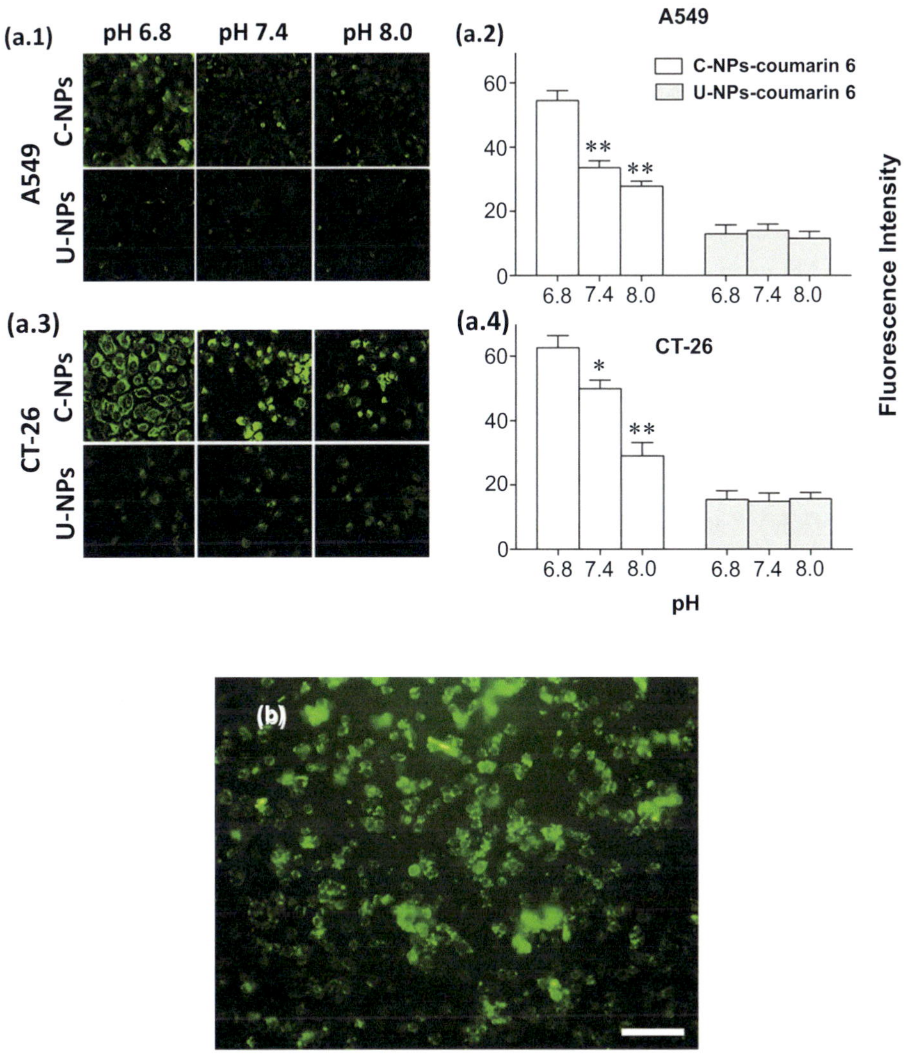

Fig. (3). (**a**) AFM image Confocal Representative confocal microscopic images of A549 (a.1) and CT-26 cells (a.3), following incubation for 6 h with chitosan-modified coumarin 6-loaded PLGA nanoparticles (C-NPs-coumarin 6) and unmodified coumarin 6-loaded PLGA nanoparticles (U-NPs-coumarin 6) at pH 6.8, 7.4 or 8.0. The fluorescence intensities of A549 (a.2) and CT 26 cells (a.4) are shown as graphs (n = 3). $^{*}P < 0.05$, $^{**}P < 0.001$ [37]. (**b**) Fluorescence images of HepG2 cells aggregated with FITC-labeled chitosan-AA. Scale bar: 100 mm [40].

Chitosan was evaluated by Liu *et al.* [40] as an inter-cellular linker to accelerate multicellular spheroid generation in hydrogel scaffold. It was developed a method to chemically aggregate cells, which involved a first treatment of cells with $NaIO_4$ to produce the surface aldehyde functionalities, and then, incubation with acrylic acid-modified chitosan. Biocompatibility test indicated that the acrylic acid-modified chitosan was non-toxic to cells, even at a high concentration. The appearance of the cells cultured in the scaffold was observed *in situ* by an inverted fluorescence microscope. The aggregates induced by acrylic acid-modified chitosan formed by the fluorescent linker were imaged using fluorescence microscope. As shown in Fig. (**3b**), most of the linkers were located at the surface of the cells. The result indicated that acrylic acid-modified chitosan has a high affinity with the cell surface.

Acharya *et al.* [41] produced a bimodal molecular imaging probe based on magneto-fluorescent nanocomposite encapsulated by chitosan. The material showed biocompatibility, with visualization of *in vitro* specific cancer cells and the *in vivo* lung tissues. The multifunctional hybrid nanocomposite material, consisted of chitosan encapsulated iron oxide as contrasting agent, a fluorescent probe and podophyllotoxin as anticancer drug. These nanoparticles were synthesized and characterized. The *in vitro* studies were performed with human oral cancer (KB), rat glioma (C6) and human lung adenocarcinoma epithelial (A549) cells. The intracellular presences of these nanocomposites were confirmed by both, fluorescence imaging and Perl's Prussian blue staining. The studies *in vivo* fluorescence imaging and Prussian blue staining confirmed the utility of these nanocomposites as dual mode imaging probe.

Dynamic Light Scattering (DLS)

The dynamic light scattering (DLS) is a noninvasive technique that allows measurement of size and size distribution of molecules and particles, in the region of from micrometers up until smaller than 1 nm [42]. In general, the dynamic light scattering is used in the characterization of particles, emulsions and molecules dissolved or dispersed in a liquid medium. The Brownian motion of particles or molecules in suspension or solution leads to the laser light be scattered to different extents. The analysis of these intensity fluctuations is proportional to the speed of Brownian motion, and thus, the particle size can be determined using the Stokes-Einstein relationship. DLS can also be used to analyze the behavior of complex fluids, such as concentrated polymer solutions [43].

Zeta Potential

Almost all the materials, macroscopic or particulates, in contact with a liquid can acquire an electric charge on its surface. This charge can appear due to the adsorption of ions on the particle surface or by dissociation of ionogenic groups on its surface. The net charge on the particle surface causes influence in the form that the ions are distributed in vicinity of the particle, leading to increase of the concentration of the contour ions near to the surface. Thus, an electrical double layer is formed at the particle interface with the liquid [44].

The zeta potential is a useful indicator to predict and control the stability of nanostructures. The value of the zeta potential is directly proportional to the nanoparticle stability, since the charged particles repel each other, inhibiting the natural tendency of aggregation.

The surface charge too affects the nanoparticles performance on the bio-distribution, which has a significant role in the targeting to the specific tissue in drug delivery systems. Thus, the zeta potential of nanoparticles not only determines its stability, but also influences the transfection efficiency by the effectiveness of its interaction with cell membranes [45].

Tijana Ristić *et al.* [46] researched the effect of chitosan nanoparticles onto Lactobacillus cells. In this study, chitosan and trimethyl chitosan solutions were utilized to produce nanoparticles using the ionic gelation method. Their characterization was focused on determining of the antimicrobial properties when utilized in the combat against the common pathogenic microorganisms and *Lactobacillus spp*. These microorganisms are usually found in resident microbiota in the vaginal and gastrointestinal tract. The average sizes of synthesized nanoparticles determined with DLS measurements was of 232.2 ± 2.5 nm and 239.9 ± 2.6 nm for chitosan and trimethyl chitosan nanoparticles, respectively, and the polydispersity index were of 0.168 and 0.175, respectively. Both types of nanoparticles exhibited positive zeta potential values at pH 3, which was attributed to the presence of protonated amino groups on the particles surface. Zeta potential values of 40.0 ± 4.90 mV for chitosan nanoparticles and 33.0 ± 4.48 mV for trimethyl chitosan nanoparticles, also indicate that the nanoparticle dispersions are stable.

In other work, Shagholani *et al.* [47] studied the synthesis and application of magnetite nanoparticles to drug delivery systems. The nanoparticles were obtained by coprecipitation under ultrasonication followed by coating with chitosan. Polyvinyl alcohol (PVA) was combined with chitosan to coating the magnetite nanoparticles. The combination occurred by hydrogen binding and ionic cross-linking of the amino and hydroxyl groups of chitosan and PVA respectively.

The hydrodynamic sizes of the Fe_3O_4 and chitosan-Fe_3O_4 nanoparticles were determined using dynamic light scattering analysis. The size distribution for the different nanoparticles is presented in Fig. (**4a**). In both samples, the particle size distribution showed only one peak with a low polydispersity index, suggesting that the nanoparticles not had much aggregation in the solutions. Mean hydrodynamic sizes for Fe_3O_4 and chitosan-Fe_3O_4 were of 38.2 and 26.7 nm, respectively. The decreasing of diameter from Fe_3O_4 to chitosan-Fe_3O_4 was attributed to surface charge of the chitosan molecules placed onto the surface of Fe_3O_4 nanoparticles. This surface charge decreased the agglomeration of nanoparticles in aqueous medium. The zeta potential of the nanoparticles was measured to investigate the surface charge of the nanoparticles. Fe_3O_4 nanoparticles had a zeta potential of +36.4 mV that was positive in deionized water. This was attributed to ammonia molecules which remained from the synthesis process adsorbed onto Fe_3O_4 nanoparticles. The increasing in positive values of the zeta potential for chitosan-Fe_3O_4 value (+48.1 mV) confirmed the functionalization of amino groups on the Fe_3O_4 surface by chitosan molecules. These positive values facilitates binding of nanoparticles with RNA and DNA plasmids in order to form complex magnetic nanoparticles for targeting genes into cells [47]

Solid lipid nanoparticles for oral drug delivery produced using chitosan coating were studied by Wang *et al.* [48]. In this work, stability, controlled delivery, mucoadhesion and cellular uptake of the nanoparticles were evaluated. The low stability of solid lipid nanoparticles under acidic condition resulted in large aggregation in gastric environment, limiting their application as oral delivery systems. A series of solid lipid nanoparticles were prepared to investigate the effects of surfactant/cosurfactant and chitosan coating on their physicochemical properties and cellular uptake. Four different formulations of solid lipid nanoparticles were performed. The particles size was expressed as the hydrodynamic diameter measured by a dynamic light scattering instrument. The particles sizes were between 427.7 ± 52.3 and 252.5 ± 24.8 nm, and the polydispersity index of solid lipid nanoparticles were between 0.24 ± 0.02 and 0.13 ± 0.03.

X-ray Photoelectron Spectroscopy (XPS)

X-ray photoelectron spectroscopy (XPS) is a quantitative spectroscopic technique which determines the elementary composition, empirical formula, chemical state and electronic state of the elements which constitutes the sample. XPS spectrums are obtained by irradiating with a beam of X-rays in the sample, and the simultaneously measuring of the kinetic energy and the number of electrons that escape from the material surface being analyzed. XPS can be used to measure elemental composition of the surface, obtain empirical formula of pure materials, identify contamination in the surface of the sample, the chemical or electronic state of each element in the surface and uniformity of elemental composition across the surface.

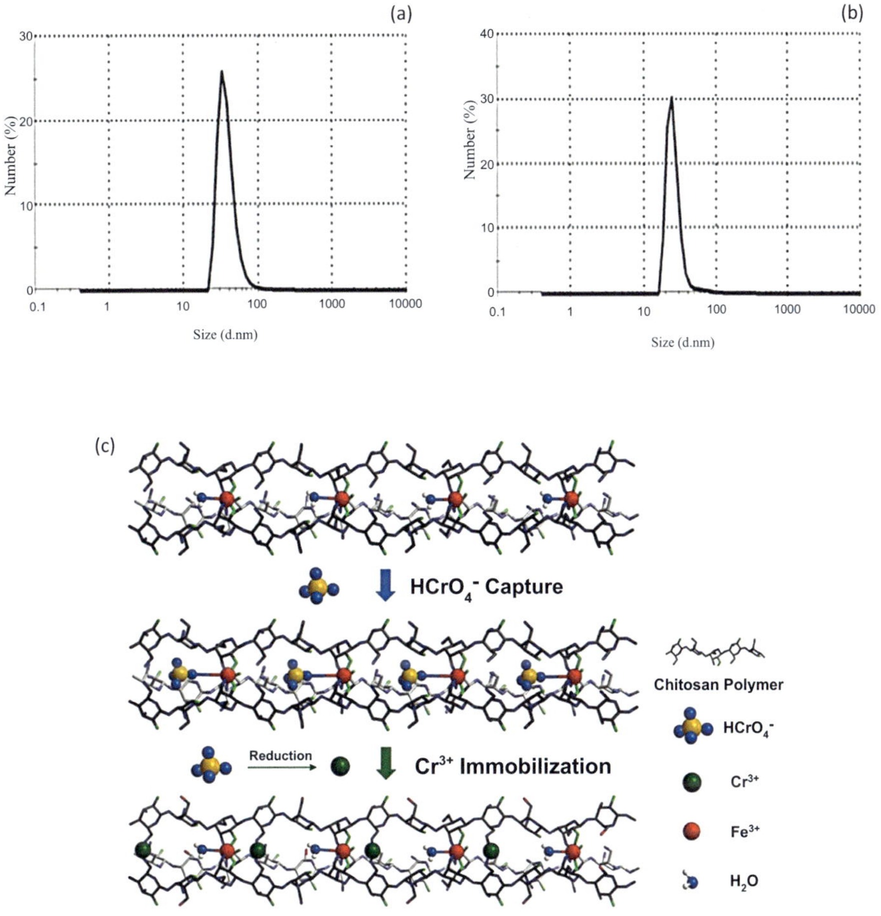

Fig. (4). Size distribution by number obtained by DLS; (**a**) Fe_3O_4, (**b**) Ch- Fe_3O_4 nanoparticles [47], and (**c**) the proposed adsorption-coupled reduction mechanism [51].

In order to achieve low concentrations of heavy metals in wastewater, many researches regarding the chitosan as adsorbent were performed. However, few information is given about the properties of the resulting complexes. This information can bring a better understanding of the mechanisms involved in metal binding to the polymer. XPS is a powerful technique to investigate how metal ions bind in the adsorbent surface [49].

Chitosan membrane containing copper nanoparticles is a potential adsorbent for chromate and vanadate removal from aquatic environment. X-ray photoelectron spectroscopy (XPS) was performed by Godoi *et al.* [49] to investigate the influence of copper nanoparticles on Cr (VI) and V (V) sorption processes. XPS spectra obtained after the ions adsorption showed that, Cr (VI) and V (V) interacted of different form with chitosan membrane containing copper nanoparticles. Both species had similar reduction potentials: Cr(VI)/Cr(III), E^0 = +1.33 V and V(V)/V(IV), E^0 = +1.00 V against Cu(II)/Cu(I), E^0 = +0.16 V. Cr(VI) was reduced to Cr(III), while Cu(I) was oxidized to Cu(II). On the other side, in the case of vanadate adsorption, the interactions between vanadate species and chitosan amine groups were more significant than the redox reactions involving copper.

In other work, Vieira *et al.* [50] studied the adsorption of copper, chromium and mercury ions onto natural and crosslinked (glutaraldehyde and epichlorohydrin) chitosan matrices, once the diverse functional groups present in this biopolymer can induce different adsorption mechanisms. The X-ray photoelectron spectroscopy (XPS) showed that these metals linked to the glutaraldehyde-crosslinked chitosan, differently from those other two types of matrices. Hence, amino group availability and the formation of new structures, such as imino bonds are the key factors. The Copper (II) stabilization was found to be poor in glutaraldehyde-crosslinked chitosan. However, Hg (II) ion presents higher adsorption capacity in this kind of matrix. Chromium (VI) was reduced in all three matrices. In this case, chromium (VI) is probably not well stabilized by the functional groups of these polymers and can undergo the action of their reducing groups.

Metal–chitosan complexes stood out as an effective adsorbent for the removal of Cr (VI) from water solution. However, whether this kind of complex can reduce Cr (VI) to less-toxic Cr (III) and what are the mechanisms of detoxification processes are still unknown. In order to explain these questions, Shen *et al.* [51] used the X-ray photoelectron spectroscopy (XPS), due to high sensitivity for probing the structure of matter at the local atomic scale around almost any specific element in the periodic table. In this study, despite the high adsorption efficiency (maximum adsorption capacity of 173.1 mg g^{-1} in 10 min), the significant improvement in the Cr (VI) reduction by chitosan–Fe (III) complex compared with the normal crosslinked chitosan was demonstrated. In addition, the structure of chitosan–Fe (III) complex and its functional groups in relation to Cr (VI) were detoxified. The XPS spectra indicated that the primary alcoholic function on C-6 served as an electron donor during Cr (VI) reduction and was oxidized to a carbonyl group. Overall, the possible mechanism proposed for the adsorption, coupled to reduction for the removal of Cr (VI) by the chitosan–Fe (III) complex is presented in Fig. (**4b**).

CONFLICT OF INTEREST

The authors confirm that they have no conflict of interest to declare for this publication.

ACKNOWLEDGEMENTS

The authors are grateful to the Brazilian agencies CAPES and CNPq for the financial support.

REFERENCES

[1] Jayakumar R, Prabaharan M, Nair SV, Tamura H. Novel chitin and chitosan nanofibers in biomedical applications. Biotechnol Adv 2010; 28(1): 142-50.

[2] Muzzarelli C, Stanic V, Gobbi L, Tosi G, Muzzarelli RA. Spray-drying of solutions containing chitosan together with polyuronans and characterisation of the microspheres. Carbohydr Polym 2004; 57(1): 73-82.

[3] Agrawal P, Strijkers GJ, Nicolay K. Chitosan-based systems for molecular imaging. Adv Drug Deliv Rev 2010; 62(1): 42-58.

[4] Wunderlich B. Thermal Analysis. New York: Academic Press 1990; pp. 137-40.

[5] Martinez-Camacho AP, Cortez-Rocha MO, Ezquerra-Brauer JM, Graciano-Verdugo AZ. Rodriguez-Félix F, Castillo-Ortega MM. Chitosan composite films: Thermal, structural, mechanical and antifungal properties. Carbohydr Polym 2010; 82(2): 305-15.

[6] Dotto GL, Moura JM, Cadaval TR, Pinto LA. Application of chitosan films for the removal of food dyes from aqueous solutions by adsorption. Chem Eng J 2013; 214: 8-16.

[7] Coats AW, Redfern JP. Thermogravimetric Analysis: A Review. Analyst (Lond) 1963; 906-24.

[8] Tikhonov NA, Arkhangelsky IV, Belyaev SS, Matveev AT. Carbonization of polymeric nonwoven materials. Thermochim Acta 2009; 486(1–2): 66-70.

[9] Chen AH, Chen SM. Biosorption of azo dyes from aqueous solution by glutaraldehyde-crosslinked chitosans. J Hazard Mater 2009; 172(2–3): 1111-21.

[10] Li X, Li Y, Zhang S, Ye Z. Preparation and characterization of new foam adsorbents of poly(vinyl alcohol)/chitosan composites and their removal for dye and heavy metal from aqueous solution. Chem Eng J 2012; 183: 88-97.

[11] Silverstein M. Robert, Webster X Francis KJD. Spectrometric Identification of Organic Compounds. Organic Chemistry 2005; pp. 1-550.

[12] Khan TA, Peh KK, Ch'ng HS. Reporting degree of deacetylation values of chitosan: The influence of analytical methods. J Pharm Pharm Sci 2002; 5(3): 205-12.

[13] Rodrigues LB, Leite HF, Yoshida MI, Saliba JB, Junior AS, Faraco AA. *In vitro* release and characterization of chitosan films as dexamethasone carrier. Int J Pharm 2009; 368(1–2): 1-6.

[14] Sajomsang W, Tantayanon S, Tangpasuthadol V, Daly WH. Synthesis of methylated chitosan containing aromatic moieties: Chemoselectivity and effect on molecular weight. Carbohydr Polym 2008; 72(4): 740-50.

[15] Carneiro J, Tedim J, Ferreira MG. Chitosan as a smart coating for corrosion protection of aluminum alloy 2024: A review. Prog Org Coat 2015; 89: 348-56.

[16] Cullity BD. Elements of X-ray diffraction. Addison-Wesley Publishing Company Inc. 1978.

[17] Rinaudo M. Ã MR. Chitin and chitosan: Properties and applications. Prog Polym Sci 2006; 31(7): 603-32.

[18] Cheng M, Deng J, Yang F, Gong Y, Zhao N, Zhang X. Study on physical properties and nerve cell affinity of composite films from chitosan and gelatin solutions. Biomaterials 2003; 24(17): 2871-80.

[19] Lee YM, Nam SY, Kim JH. Pervaporation of water–ethanol through poly(vinyl alcohol)/chitosan blend membrane. Polym Bull 1992; 29: 423-8.

[20] Kim JH, Lee YM. Synthesis and properties of diethylaminoethyl chitosan. Polymer (Guildf) 1993; 34(9): 1952-7.

[21] Xiao J, Nian S, Huang Q. Assembly of kafirin/carboxymethyl chitosan nanoparticles to enhance the cellular uptake of curcumin. Food Hydrocoll 2015; 51: 166-75.

[22] Dunne LW, Iyyanki T, Hubenak J, Mathur AB. Characterization of dielectrophoresis-aligned nanofibrous silk fibroin-chitosan scaffold and its interactions with endothelial cells for tissue engineering applications. Acta Biomater Acta Materialia Inc 2014; 10(8): 3630-40.

[23] Esquerdo VM, Cadaval TR, Dotto GL, Pinto LA. Chitosan scaffold as an alternative adsorbent for the removal of hazardous food dyes from aqueous solutions. J Colloid Interface Sci 2014; 424: 7-15.

[24] Ernst R. The early development of electron lenses and electron microscopy. Hirzel Press 1980; pp. 1-140.

[25] Hawkes P. The beginnings of electron microscopy. Academic Press 1985; pp. 1-633.

[26] Jain S, Kumar Sharma R, Vyas SP. Chitosan nanoparticles encapsulated vesicular systems for oral immunization: preparation, in-vitro and in-vivo characterization. Jpp 2006; 58: 303-10.

[27] Azadi A, Hamidi M, Rouini MR. Methotrexate-loaded chitosan nanogels as "Trojan Horses" for drug delivery to brain: Preparation and *in vitro/in vivo* characterization. Int J Biol Macromol 2013; 62: 523-30.

[28] Langmuir I. The adsorption of gases on plane surfaces of glass, mica and platinum. J Am Chem Soc 1918; 40(9): 1361-403.

[29] Brunauer S, Emmett PH, Teller E. Adsorption of gases in multimolecular layers. J Am Chem Soc 1938; 60(1): 309-19.

[30] Allen T. Particle size measurement. Springer 1996; Vol. 2: p. 39.

[31] Mirmohseni A. Chitosan hollow fibers as effective biosorbent toward dye: preparation and modeling. Bioresour Technol 2012; 121: 212-20.

[32] Zhang Q, Deng S, Yu G, Huang J. Removal of perfluorooctane sulfonate from aqueous solution by crosslinked chitosan beads: Sorption kinetics and uptake mechanism. Bioresour Technol 2011; 102(3): 2265-71.

[33] Wu FC, Tseng RL, Juang RS. Enhanced abilities of highly swollen chitosan beads for color removal and tyrosinase immobilization. J Hazard Mater 2001; 81(1–2): 167-77.

[34] Dai J, Ren FL, Tao C. Adsorption of Cr(VI) and speciation of Cr(VI) and Cr(III) in aqueous solutions using chemically modified chitosan. Int J Environ Res Public Health 2012; 9(5): 1757-70.

[35] Lang KM, Hite DA, Simmonds RW, McDermott R, Pappas DP, Martinis JM. Conducting atomic-force microscopy for nanoscale tunnel barrier characterization. Rev Sci Instrum 2004; 75(8): 2726-31.

[36] Pawley JB. Handbook of biological confocal microscopy. Springer US Press 2006; pp. 1-985.

[37] Yang R, Shim WS, De Cui F, *et al.* Enhanced electrostatic interaction between chitosan-modified PLGA nanoparticle and tumor. Int J Pharm 2009; 371(1–2): 142-7.

[38] Zhang W, Liu J, Zhang Q, *et al.* Enhanced cellular uptake and anti-proliferating effect of chitosan hydrochlorides modified genistein loaded NLC on human lens epithelial cells. Int J Pharm 2014; 471: 118-26.

[39] Cremer C, Cremer T. Considerations on a laser-scanning-microscope with high resolution and depth of field. Microsc Acta 1978; 81(1): 31-44.

[40] Liu Y, Guan Y, Zhang Y. Chitosan as inter-cellular linker to accelerate multicellular spheroid generation in hydrogel scaffold. Polym 2015; 77: 366-76.

[41] Walia S, Sharma S, Markand Kulurkar P, Patial V, Acharya A. A bimodal molecular imaging probe based on chitosan encapsulated magneto-fluorescent nanocomposite offers biocompatibility, visualization of specific cancer cells *in vitro* and lung tissues *in vivo.* Int J Pharm 2016; 498(1–2): 110-8.

[42] Berne BJ, Pecora R. Dynamic light scattering. Courier Dover Publications 2000. ISBN 0-486-41155-9

[43] Pecora R. Doppler shifts in sight scattering from pure liquids and polymer solutions. J Chem Phys 1964; 40: 1604-14.

[44] Greenwood R, Kendall K. Electroacoustic studies of moderately concentrated colloidal suspensions. J Eur Ceram Soc 1999; 19(4): 479-88.

[45] Kievit FM, Veiseh O, Bhattarai N, *et al.* PEI–PEG–chitosan-copolymer-coated iron oxide nanoparticles for safe gene delivery: synthesis, complexation, and transfection Adv. Funct Mater 2009; 19: 2244-51.

[46] Ristić T, Lasič S, Kosalec I, Bračič M, Fras-Zemljič L. The effect of chitosan nanoparticles onto Lactobacillus cells. React Funct Polym 2015; 97: 56-62.

[47] Shagholani H, Ghoreishi SM, Mousazadeh M. Improvement of interaction between PVA and chitosan *via* magnetite nanoparticles for drug delivery application. Int J Biol Macromol 2015; 78: 130-6.

[48] Luo Y, Teng Z, Li Y, Wang Q. Solid lipid nanoparticles for oral drug delivery: Chitosan coating improves stability, controlled delivery, mucoadhesion and cellular uptake. Carbohydr Polym 2015; 122: 221-9.

[49] de Godoi FC, Rodriguez-Castellon E, Guibal E, Beppu MM. An XPS study of chromate and vanadate sorption mechanism by chitosan membrane containing copper nanoparticles. Chem Eng J 2013; 234: 423-9.

[50] Vieira RS, Oliveira MLM, Guibal E, Rodriguez-Castellon E, Beppu MM. Copper, mercury and chromium adsorption on natural and crosslinked chitosan films: An XPS investigation of mechanism. Colloids Surfaces A Physicochem Eng Asp 2011; 374(1–3): 108-4.

[51] Shen C, Chen H, Wu S, *et al.* Highly efficient detoxification of Cr (VI) by chitosan–Fe(III) complex: Process and mechanism studies. Journal of Hazardous Materials 2013; 244– 245: 689-97.

CHAPTER 7

Application of Chitosan Based Materials for Dyes/Metals Removal

Tito Roberto S. Cadaval[1], **Mery Luiza G. Vieira**[1], **Alisson S. Camara**[1], **Guilherme L. Dotto**[2] and **Luiz A. A. Pinto**[1,*]

[1] *Industrial Technology Laboratory, School of Chemistry and Food, Federal University of Rio Grande - FURG, Rio Grande-RS, Brazil*

[2] *Environmental Processes Laboratory, Chemical Engineering Department, Federal University of Santa Maria-UFSM, Santa Maria-RS, Brazil*

Abstract: Application of chitosan based materials in wastewater treatment has received considerable attention in last years. The most frequently applied technique for this purpose is the adsorption operation, which typically use activated carbon as adsorbent in the industrial wastewater treatment systems, because it has a large specific surface area. However, there has been an increasing interest in utilizing adsorbents from renewable sources for dyes/metals, removal from wastewater. Chitosan based materials has been widely used to remove heavy metals and dyes due to its high content of amino ($-NH_2$) and hydroxyl (-OH) functional groups, which have high activity as adsorption sites. This chapter presents some applications of chitosan based materials for dyes/metals, removal from aqueous medium, providing information obtained using batch and continuous adsorption systems. The effects of parameters in the adsorption capacity, such as chitosan characteristics, process variables, chemistry interaction adsorbent/adsorbate, equilibrium, kinetics and thermodynamics are presented and discussed. The chapter also presents the equilibrium and kinetic models most commonly used and the thermodynamic studies reported for adsorption onto different chitosan based materials. Several examples are discussed to demonstrate the versatility of the material and its application as an adsorbent. Desorption performance and the structural parameters of the polymer in relation to regeneration as well as the development of new adsorbents are presented.

Keywords: Adsorption, Chitosan, Dyes, Materials, Metals.

INTRODUCTION

The increasing demand for new materials for the removal of substances, like synthetic dyes and metal ions, from industrial liquid waste has led many

[*] **Corresponding author Luiz A. A. Pinto:** Industrial Technology Laboratory, School of Chemistry and Food, Federal University of Rio Grande – FURG, Rio Grande – RS, Brazil; Tel/Fax: (55) 53 32336969; E-mail: dqmpinto@furg.br

Guilherme Luiz Dotto, Sergio Paulo Campana-Filho and Luiz Antonio de Almeida Pinto (Eds)

researchers to evaluate the possibility to use renewable based materials to uptake these elements [1, 2]. Currently, several adsorbents have been used to remove different types of dyes and metal ions from wastewater especially activated carbon, clays and biopolymers [3].

Polysaccharides, as a class of natural macromolecules, have the tendency to be extremely bioactive and are generally derived from agricultural feedstock, as cellulose, starch, pectin, or crustacean shell wastes, as chitin and chitosan [4]. In terms of availability, chitin is the second most important natural polymer in the world [5], it is also obtained from fungi and bacteria [6]. Chitosan is obtained on an industrial scale by the alkaline deacetylation of chitin [7]. Both chitin and chitosan are linear copolymers of D-GlcN and D-GlcNAc distributed randomly and which are linked fully in the β-1,4-configuration, generating in an organizing rigid and unbranched structure. The abundance of hydroxyl groups (1 primary hydroxyl at C-6 and 1 secondary hydroxyl at C-3) and highly reactive amino group (at C-2) or its N-acetyl counterpart (wholly in chitin) with concomitant tendency for intra and intermolecular hydrogen bonds results in the formation of linear aggregates with extensive crystallinity. The main sources exploited are marine crustaceans, mainly shrimp and crabs. In particular, the development of chitosan-based materials as useful adsorbent polymeric matrices is expanding in area of adsorption science [8]. Chitin and chitosan are biopolymers with immense structural possibilities for chemical and mechanical modifications to generate novel properties, functions and applications. Chitosan is a good adsorbent to remove various kinds of anionic and cationic dyes as well as heavy metal ions [9, 10].

Chemical modifications that lead to the formation of chitosan derivatives, grafting chitosan and chitosan composites have gained much attention, and have been extensively studied and widely reported in the literatures. Application of chitinous products in wastewater treatment has received considerable attention in recent years [11 - 14]. This chapter provides information relevant on the application of chitosan materials for removing dyes and heavy metal ions from aqueous medium. Adsorption capacity and the experimental conditions have been presented, including the parameters that might be involved during adsorption operation. The uptake performance, the process variables, the chitosan characteristics, the chemistry of the dye and the solution conditions used in batch and continuous systems are evaluated. Studies on the biosorption capacity, equilibrium and kinetic models, and the thermodynamic studies reported for biosorption onto chitosan and chitosan based materials systems are presented.

INFLUENCE OF PROCESS VARIABLES

Adsorption is the phenomenon in which the spontaneous adhesion of molecules present in a fluid (adsorbate) to a solid surface (adsorbent) is observed. The degree of adsorption depends on the thermodynamic conditions, adsorbent surface and interaction between adsorbent and adsorbate. This occur due to the existence of unbalanced forces on the solid surface, leading to the attraction of the molecules present in the fluid in contact with the solid adsorbent for a certain time [15].

The adsorption phenomena are classified as the responsible forces in two types: chemical adsorption and physical adsorption. The chemical adsorption or chemisorption, is so called because in this process occurs an effective exchange of electrons between the solid and the adsorbed molecule, resulting in the following characteristics: forming a monolayer on the solid surface, irreversibility and release of a considerable amount of energy (in the order of a chemical reaction). Physical adsorption is a reversible phenomenon, generally observed when occurs the deposition of more than one layer of adsorbate onto adsorbent surface. The forces acting on physical adsorption are identical to the cohesive forces that operate in liquid state, solid and gas. The released energies are relatively low and rapidly reach the equilibrium [15, 16].

Regarding the removal of metallic ions and synthetic dyes from aqueous medium, the adsorption appears as an alternative method, due to its ease of operation and efficiency over conventional methods, which are expensive or technically complicated. Moreover, it is a versatile method that has wide application range and an interesting relation cost / benefit, especially when using adsorbents from industrial waste. Usually, the adsorption processes in batch or continuous systems are studied in relation to their equilibrium conditions, kinetic, thermodynamic and mechanisms involved [1 - 3, 17, 18].

For a better understanding of a process that uses adsorption operation must be designed studies in order to elucidate the relevant factors in the system, such as adsorption capacity parameters and physicochemical parameters. In this context, authors have developed works for the adsorption of metal ions and synthetic dyes using models isotherms and kinetics and thermodynamic models [19 - 21], seeking more detailed explanations of the adsorptive processes and how they can be applied. The distribution of adsorbate between solid phase and liquid is reached by varying the initial concentration of the adsorbate, C_o, keeping the solution volume, V, and adsorbent mass, m constant. The residual concentration of the adsorbate solution is used to calculate the adsorption capacity of the mass balance using eq. (**1**).

$$q = \frac{C_0 - C_f}{m} V \quad (1)$$

where C_0 and C_f are initial and final concentrations of the adsorbate, respectively, V is the volume of solution and m is the mass of adsorbent. The removal percentage (R) can be determined by eq. (**2**)

$$R\% = \frac{C_0 - C_f}{C_0} 100 \quad (2)$$

Adsorption Isotherms

The equilibrium study in the adsorption provides important information to evaluate the affinity and adsorption capacity of an adsorbent in relation to the adsorbate. This is one of the most important criteria in selecting a suitable adsorbent. In adsorption processes, equilibrium is established when the adsorbate concentrations in the liquid phase and the solid phase do not change over time. This behavior can be studied through mathematical models called adsorption isotherms [15, 22].

The adsorption isotherms are curves representing the adsorption capacity, q (adsorbent mass / adsorbent mass) *versus* the residual concentration of the adsorbate in solution at equilibrium C_e, under specific experimental conditions. The adsorption capacity in equilibrium (q_e) can be determined by eq. (**3**):

$$q_e = \frac{V(C_0 - C_e)}{m} \quad (3)$$

where C_0 is the initial solute concentration in liquid phase (mg L^{-1}), C the concentration of solute at equilibrium (mg L^{-1}), V is the solution volume (L) and m is the mass of adsorbent (g). Ruthven (1984) presents five types of characteristic adsorption isotherms as shown in Fig. (**1**) [15].

The isotherms of the type I exhibit a convex plateau, which is associated with the formation of a monomolecular layer in microporous or nonporous adsorbents, in which the pore sizes are not much larger than the molecular diameter of the adsorbate. Types II and III describe multilayer adsorption usually associated to adsorbents with heterogeneous pore sizes. Isotherms of type IV are for the adsorption of two layers or which occurs in much larger pores than the molecular diameter of the adsorbate. The curves of the type V describe behavior similar to type IV, considering strong and / or weak interactions between the adsorbent and the adsorbate.

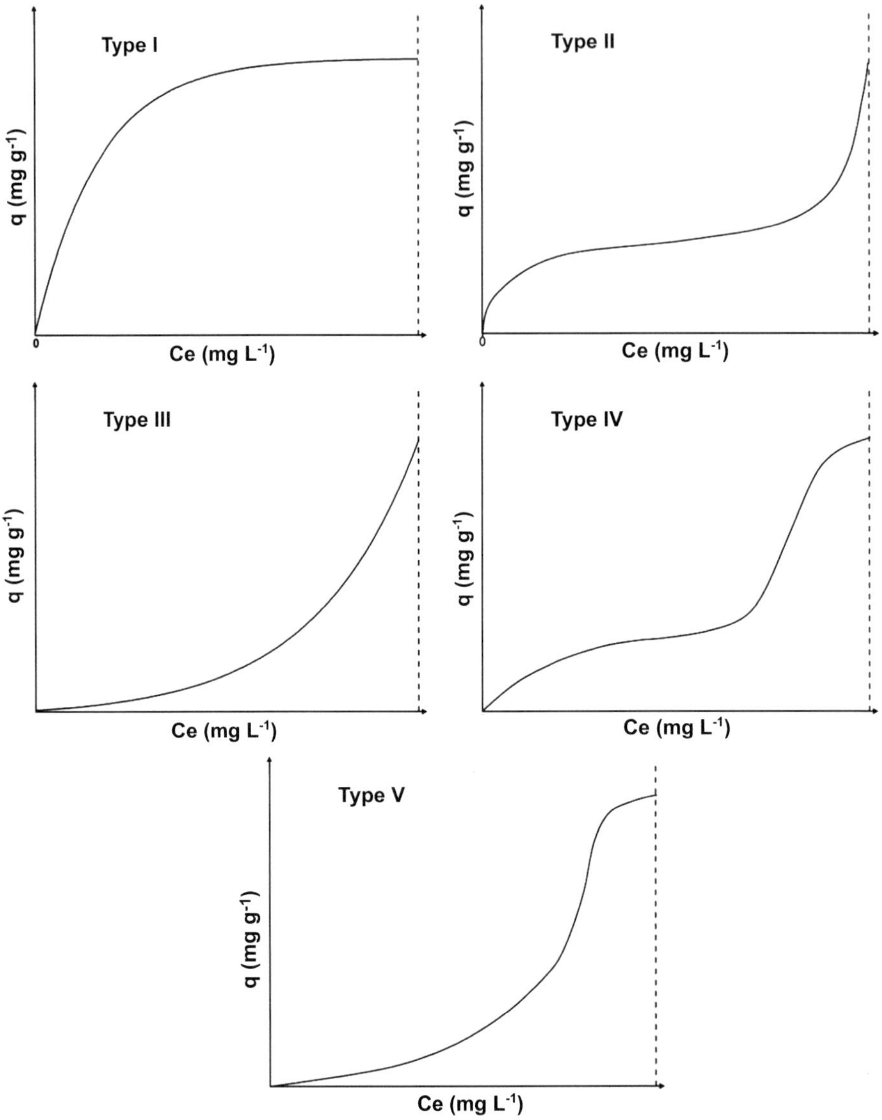

Fig. (1). Adsorption isotherms characteristics [15].

There are several isotherms models available to analyze experimental data and to describe the adsorption equilibrium, where the most used are the Henry, Langmuir, Freundlich, BET, Toth, Temkin, Redlich-Peterson, Sips, Frumkin, Harkins-Jura, Halsey, Henderson e Dubinin-Radushkevich [1, 9, 19, 23]. The following are presented the most common isotherms in the removal of metal ions and synthetic dyes by chitosan in aqueous medium.

Henry Isotherm

The Henry isotherm is used when the surface of the adsorbent can be considered uniform and the diluted solution, so that the molecules show almost no effect of interaction with neighboring molecules. The relationship between the amount of adsorbate in the fluid phase and the amount of adsorbate on the solid phase is linear, with a proportionality constant called Henry constant and can be represented by eq. (**4**) [15]:

$$q_e = K_H C_e \tag{4}$$

where q_e is the adsorption capacity in equilibrium (mg g^{-1}) and C_e the concentration of the liquid phase at equilibrium (mg L^{-1}).

Langmuir Isotherm

The Langmuir isotherm model was proposed in 1918 for the adsorption of gases at metal surfaces. This model assumes that an adsorbent has specific sites, homogeneous and identical adsorption energy, and provides a coating of monolayer on the external surface of the adsorbent. Thus, when a molecule reaches a certain place no further adsorption can occur there. Since the adsorbent has a finite capacity to adsorb a substance, the saturation of the monolayer (with $Ce \rightarrow \infty$) can be represented by eq. (**5**) [15, 24]:

$$q_e = \frac{q_m k_L C_e}{1 + k_L C_e} \tag{5}$$

where, q_m is the maximum adsorption capacity of the monolayer (mg g^{-1}), k_L is the Langmuir constant (L mg^{-1}), q_e and C_e are the adsorption capacity (mg g^{-1}) and the equilibrium concentration (mg L^{-1}), respectively.

Freundlich Isotherm

The Freundlich isotherm is an empirical equation used for heterogeneous systems where heterogeneity is characterized by the factor 1/*n*, eq. (**6**) [25].

$$q_e = k_F C_e^{1/n} \tag{5}$$

where k_F is the constant Freundlich ((mg g^{-1}) (L $mg^{-1})^{1/n}$) and $1/n$ is the heterogeneity factor.

BET Isotherm

The BET (Brunauer, Emmett e Teller) isotherm is an extension of the Langmuir theory for monolayer adsorption to the multilayer adsorption, when the increase of adsorbate-adsorbate interaction and multilayer formation occurs due to secondary adsorption at a given site, and can be represented by eq. (**7**) [15, 26]:

$$q_e = \frac{q_{BET} K_1 C_e}{(1-K_2 C_e)(1-K_2 C_e + K_1 C_e)} \quad \textbf{(6)}$$

where, q_{BET} is the monolayer adsorption capacity (mg g^{-1}), K_1 and K_2 are the BET constants (L mg^{-1}).

Thermodynamics of Adsorption

The adsorption thermodynamics is determined using the thermodynamic equilibrium coefficients, obtained at different temperatures and concentrations, in order to verify the energy intensity of adsorption. The adsorption characteristics of a system can be expressed in thermodynamic parameters, such as the variation in Gibbs free energy ($\Delta G°$), variation of enthalpy ($\Delta H°$) and variation of entropy ($\Delta S°$). These values indicate if the process is spontaneous, endothermic or exothermic, and provide information about the system randomness. According to thermodynamics, $\Delta G°$ can be calculated by eq. (**8**) [27 - 29].

$$\Delta G^0 = -RT \ln(\rho_w K_D) \quad \textbf{(7)}$$

where, K_D is the thermodynamic equilibrium constant (L g^{-1}), ρ_w is the water density (g L^{-1}), T is the temperature (K) and R is the universal gas constant (kJ mol^{-1} K^{-1}). Usually its value is obtained from the best fit isotherm parameters [27 - 29].

According to thermodynamics, the variation in Gibbs free energy corresponds to the difference between the variation in enthalpy of adsorption ($\Delta H°$) and the product of entropy variation ($\Delta S°$) in the operation temperature. Thus, applying this concept to eq. (**7**), the enthalpy change (ΔH^0) (kJ mol^{-1}) and entropy change (ΔS^0) (kJ mol^{-1} K^{-1}) were determined by Van't Hoff's plot.

$$\ln(\rho_w K_D) = \frac{\Delta S^0}{R} - \frac{\Delta H^0}{RT} \quad \textbf{(8)}$$

By fit the experimental data with eq. (**8**) is possible to determine the slope $\Delta H/RT$ and an interception $\Delta S/R$ [28, 29].

Kinetics of Adsorption

The kinetic study is fundamental in an adsorption operation, as it controls the process efficiency, provides information about the rate which the reactions occur, and the factors that influence the adsorption rate [30]. This group of models include of the pseudo-first order, pseudo-second order and the model Elovich [30]. The adsorption capacity *versus* time (q_t) can be determined by eq. (**9**)

$$q_t = \frac{V(C_0 - C_t)}{m} \tag{9}$$

where C_t is the adsorbate concentration in liquid phase at time t (mg L^{-1}).

Pseudo-First Order Model

It was proposed by Lagergren in 1898, based on the Newton's law cooling. This model assumes that the adsorption occurs as a result of a concentration gradient between the adsorbent surface and the solution, and can be expressed according to eq. (**10**):

$$\frac{dq_t}{dt} = k_1(q_e - q_t) \tag{10}$$

where, k_1 is the rate constants of pseudo-first order model, in (min^{-1}), q_t e q_e are the adsorption capacity at time t and in the equilibrium, respectively (mg g^{-1}), and t is the time (min).

Solving eq. (**10**) by separable variables, considering q_t=0 at t=0 and q_t=q_t for t=t, and rearranging, the pseudo-first order kinetic can be represented by eq. (**11**) [31]:

$$q_t = q_1(1 - \exp(-k_1 t)) \tag{11}$$

where, q_1 is the theoretical values for the adsorption capacity (mg g^{-1}).

Pseudo-Second Order Model

The kinetic model of pseudo-second order equation presents both effects, internal as well as external, of mass transfer, and is generally suitable for chemisorption process. This model can be expressed according to eq. (**12**):

$$\frac{dq_t}{dt} = k_2(q_e - q_t)^2 \quad \textbf{(12)}$$

where, k_2 is the rate constants of pseudo-second order models, in (g mg^{-1}min^{-1}), and t is the time (min).

Solving eq. (**12**) by separable variables, considering q_t=0 in t=0 and q_t= q_t em t=t, and rearranging, the pseudo-second order kinetic can be represented by eq. (**13**) [32]:

$$q_t = \frac{t}{(1/k_2 q_2^2) + (t/q_2)} \quad \textbf{(13)}$$

and q_2 is the theoretical value for the adsorption capacity (mg g^{-1}).

Elovich Model

When the adsorption process involves chemisorption on solid surface, and the adsorption rate decreases with time due to coverage of the surface layer, The Elovich model is one of the most used and is shown in eq. (**13**) [33]:

$$q_t = \frac{1}{a}\ln(1 + abt) \quad \textbf{(14)}$$

where, a is the initial velocity due to dq/dt with q_t=0 (mg g^{-1} min^{-1}) and b is the desorption constant of the Elovich model (g mg^{-1}).

Weber e Morris Model

The elucidation of the mechanisms that occur in the adsorption phenomenon is of fundamental importance, to check how the interactions occur between the adsorbent and the adsorbate. In general the mechanisms involved in the adsorption of the adsorbate by an adsorbent are: external mass transfer, intraparticle diffusion, and chemical reaction [15].

To identify these mass transfer mechanisms, is generally used theory of Weber and Morris (1963). According to these authors, plotting a graph of adsorption capacity as a function of square root of time, it is possible to identify different linear portions wherein each representing a distinct mechanism of mass transfer [26, 34]. The first linear portion can be attributed to external mass transfer mechanism, and the second is relative to the intraparticle diffusion. The Weber and Morris model is shown in eq. (**14**):

$$q_t = k_{di} t^{1/2} + C_i \tag{14}$$

Batch and Continuous Systems

Batch adsorption assays are used easily in the laboratory for the treatment of small volume of effluents, but less convenient to use on industrial scale, where large volumes of wastewater are continuously generated. Batch adsorption provides certain preliminary information such as pH for maximum adsorption, maximum initial adsorbate concentration, and particle size for optimum adsorption, and approximate time for adsorption, as well as, the adsorption capacity of the adsorbent [35]. However, in fixed bed, the adsorbate is continuously in contact with a given quantity of fresh adsorbent, thus providing the required concentration gradients between adsorbent and adsorbate for adsorption. Fixed bed operations are widely used in pollution control processes, such as for the removal of ions or organic compounds, by adsorption [36]. This system involves percolation of wastewater through percolating material. During the flow of the wastewater through the percolator, the wastewater is purified by physicochemical processes. The design and theory of fixed-bed adsorption systems focuses on establishing the shape of the breakthrough curve and its velocity through the bed. Breakthrough curves and bed volumes are usually used in the evaluation of the performance of a fixed-bed column, being the total removal percentage, R, given by eq. (**15**) [37].

$$R = \frac{\int_0^{t_{total}} \left(1 - \frac{C_t}{C_0}\right) dt}{t_{total}} 100 \tag{15}$$

ADSORPTION USING CHITOSAN BASED MATERIALS

The chitosan corresponds to polymers with more than 75% deacetylated content, and the deacetylation of all the chain is difficult to be obtained due to the depolymerization of the structure. Copolymers with various deacetylation degree and molecular weight are commercially available. Chitosan is very interest because their percentage of amino groups when compared to the synthetically substituted in other synthetics biopolymers. Due to these amino groups, chitosan is soluble in acid solutions and chemically versatile. Most of the properties of chitosan can be related to its cationic nature, which to possibility an elaboration of many different chitosan based materials from raw structure in order to apply in adsorption operation [1 - 3, 31, 38].

Currently, the preparation and uses of chitosan and its derivatives (chitosan-based materials) are an important research area. Most researches demonstrate that decreasing the average particles size increases the adsorption capacity. However, the flake and powder forms show low BET surface area and low porosity, resulting in high pressure drops in fixed beds. Furthermore, when chitosan is used in powder form, centrifugation or filtration is necessary for the phase separation after adsorption operation, and the regeneration and reuse of the adsorbent becomes difficult. Consequently, many works have been performed to develop chitosan based materials with characteristics like large surface area, high porosity, regeneration capacity, easy phase separation, high adsorption capacity, insolubility in acid media and better mechanical properties [11, 23, 31, 39].

Some of the new materials that utilize chitosan as a starting material are: films, membranes, nanofibers, sponges, gels beads and nanoparticles. In some cases chitosan is supported on inert materials [23, 37, 38, 40 - 42]. The utilization of these new materials presents many advantages in terms of applicability to a wide variety of operation configurations. However, there is a need of better standardization of the production process to be able to prepare reproducible polymers having the same characteristics. Changes in the specifications of the polymer may significantly change its adsorption performance. These modifications in chitosan polymer can be performed by physical or chemical methods. Different reviews of chitosan-based materials have been reported in the adsorption of metal and dyes from aqueous medium. Following are the most commonly performed modifications in the chitosan polymers, in order to improve its use as adsorbent in the removal of dyes and metal ions.

Physical Modifications

The control of the polymer form can be helpful for designing the adsorption operation. The material can be modified physically, in order to obtain differently polymer forms, such as, films, nanoparticles, porous structures and nanofibers for water treatment.

Chitosan Film

Chitosan has been successfully used by several researchers as an adsorbent for the capture of metals and dyes dissolved in the aqueous solutions. However, when the polymer is utilized in the powder form, it is very difficult to remove the adsorbent from aqueous phase after the adsorption, as well as, to regeneration and reuse it. The separation of the powder chitosan from the aqueous medium can be performed using filtration or centrifugation. Among the chitosan based materials, the chitosan films are an attractive way to remove pollutants from aqueous solutions, mainly due to their good mechanical properties. Moreover, the films can be separated from the liquid medium by a simple grating, and can be reused in a more appropriate [9, 13, 17, 31, 43].

Recently, chitosan films were applied for the removal of lead, chromium, cooper and mercury, vanadium and dyes from aqueous solutions. Desorption and the reuse of dyes and heavy metals by chitosan film were evaluated in literature [10, 23]. Cadaval *et al.* [23] studied the adsorption and desorption of vanadium ions onto chitosan films. Authors performed five cycles of adsorption/desorption, keeping similar values for adsorption capacity and mechanical proprieties of the films. The Fig. (**2c**) presents the chitosan film during the adsorption of a food dye.

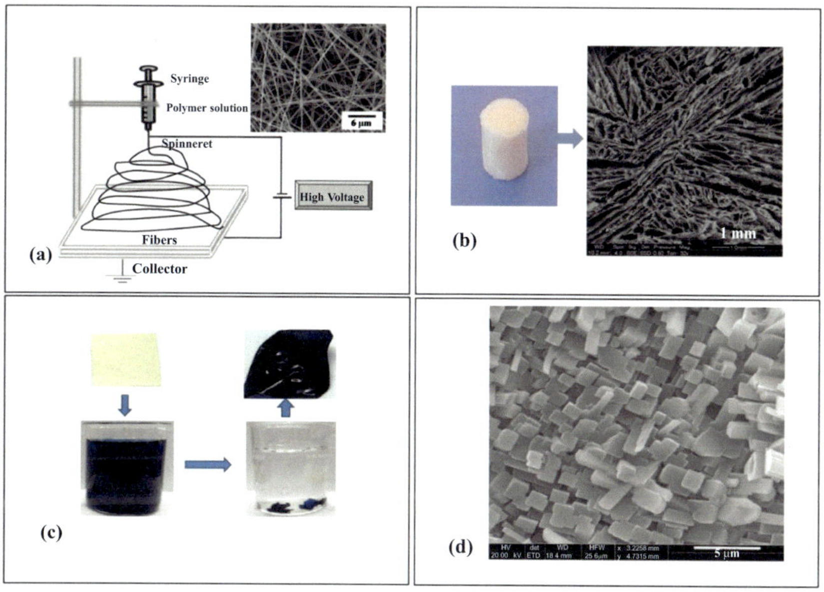

Fig. (2). Different physical modifications of the chitosan-based materials: (**a**) nanofibers; (**b**) scaffold; (**c**) films; (**d**) nanochitosan.

Porous Structures

The adsorption capacity of chitosan depends on its physical and structural parameters, such as, crystallinity, surface area, porosity and particles size. Thus, the development of chitosan scaffold Fig. (**2b**) with high values of specific surface area and porosity is an alternative to improve its characteristics as adsorbent.

Allouche *et al.* [38] produced a porous structure of chitosan in order to remove mercury from water solution. Granulated chitosan was hydrated and the solution was prepared by acetic acid (1%, w/w). The solutions were poured in molds that were maintained −80° C in freezer. The mercury adsorption capacity on the chitosan structure reached 350 mg g^{-1}.

Esquerdo *et al.* [11] produced a chitosan scaffold, of megaporous structure, as an alternative structure of the polymer to remove six different food dyes from aqueous solutions. The chitosan scaffold was prepared using chitosan sample (in the powder form) dissolved in 1.0% acetic acid solution at room temperature. The chitosan solution was homogenized at 10,000 rpm, and it was maintained at 193 K during 48 h in an ultra–freezer. After, it was carried out in freeze drying at 219 K for 48 h under vacuum of 44 mmHg. The chitosan scaffold presented pores sizes from 50 to 200 μm, porosity of 92.2 ± 1.2% and specific surface area of 1135 ± 2 m^2 g^{-1}. The chitosan megaporous scaffold showed good structural characteristics and high adsorption capacities for the six food dyes (788-3316 mg g^{-1}).

Chitosan Nanofibers

Nanofibers are defined as solid fibers with diameters in the range from 1 to 1000 nm. Methods such as the sol–gel, chemical vapor deposition and thermal oxidation, have been used to prepare nanofibers. However, the electrospinning is the most adopted method to produce nanofibers from chitosan (Fig. **2a**). The viscosity is a key parameter in electrospinning since it is related to the degree of entangled of the chitosan polymer chain in the solution, and limits its flow during the production of nanofibers [8, 44]. Electrospinning can be influenced by several other variables, such as the operating voltage, the distance between the collection plate and the needle tip, the temperature, the pressure, and the flow rate. The applied voltage determines the degree of electrostatic interaction forces that induce the expulsion of a polymer jet. In some cases, an increase in fiber diameter can be observed with increased voltage [44, 45]. The electrospun nanofibers due to its high specific surface area and high porosity with fine pores present excellent ability for removing heavy metal ions. Chitosan nanofibers prepared by the electrospinning operation, have been used like adsorbent for the removal heavy metals and dyes from aqueous systems [40, 46].

Chitosan in Fixed Bed

The application of chitosan as adsorbent in dynamic systems using fixed bed columns have some peculiarities, among these, one should observe the characteristics of the particles. These must have density, shape and size in order to give mechanical strength to the material. The main objective is to have a derivative of chitosan that enables adsorption in fixed bed column without drawbacks to the system, such as, high pressure drop or column obstruction. Thus, it has been reported in the literature the chitosan materials obtained through coating techniques [47, 48].

To deposit coatings appropriately on each substrate there is availability of numerous techniques. When emphasis is placed on the coating structure and properties of the deposition, the coating techniques can be classified into atomistic deposition, deposition of particulates, mass coating and surface modification [37].

In general, the atomistic deposition techniques consist in the formation of a film formed by atoms. The atoms are deposited by condensing on the substrate, followed by nucleation and growth. In the particle deposition techniques occur the melting of the particles, after the solidification or the sintering, resulting in a microstructure. The mass coating consists only of the application of large quantities of coating on the material surface; while, the surface modification techniques changes the composition or properties of material, by thermal, mechanical or chemical treatments. The coating technique most commonly used for immobilization of chitosan material is mass coating. This technique stands out among others because it is a technique of simple operation, in which the geometry of the substrate to be coated can be variable [37, 49 - 51].

Many adsorbents were developed by coating chitosan, such as: ceramic alumina, polyvinyl chloride beads, sand, fly ash, glass beads and bentonite beads. These adsorbents showed a high potential to remove metals and dyes from wastewater. The preference for the application of this technique is in the higher performance, small variation in the weight, size and shape of the particles. In addition, the possibility of change in the formulation of the coating, such as the incorporation of plasticizer and solvent. The plasticizer has the purpose of giving flexibility to the final film, while the solvent is responsible for ensuring controlled deposition of film forming materials on the substrate surface [37, 50, 52 - 54].

Despite significant advances in coating techniques, some aspects are still a technological challenge. There is need for more efforts in understanding the coating foundations, such as the control of the chemical structure of the layer, the adhesion of the layer on a substrate, and deposition rate. Thus, the study of coating substrates with chitosan by the coating techniques is characterized as an important advance for the adsorption of ions and dyes.

Nanomaterials

The development of nanomaterials has advanced in recent years and the use of nanochitosan has been receiving attention from the scientific community for application in wastewater treatment [55, 56].

Chitosan in the powder form shows low surface area ($< 20\ m^2\ g^{-1}$), and the use of nanochitosan is an interesting alternative because there is a considerable increase in the surface area, which in theory improves adsorbent capacity. Many studies have compared the adsorption performance of chitosan and nanochitosan, where the results showed that there is an increase in the adsorption capacity when nanochitosan was used [56, 57]. The Fig. (**2d**) shows the scanning electron microscopy of the nanochitosan structure for wastewaters treatment [56].

Another advantage in using nanochitosan is related to phenomena that depend directly on the particle size, such as, chemical, electronic, magnetic and mechanical properties, and also size effect (quantum effect) [57]. Therefore the use of nanochitosan show advantages when utilized for modification (grafting and cross-linked), because use in nano form reduces the effect caused by these modifications, which decrease the amount of groups (-NH_2) and (-OH) available, leading to a decrease in the adsorption capacity. The utilization of nanochitosan in adsorption processes has shown promising results, but there is need for more studies before this technology becomes available for commercial application [55].

Chemical Modifications

The chemical modification of chitosan can be justified by two reasons: (a) avoid the solubilization of the chitosan polymer when the adsorption or desorption is performed in acidic solutions; (b) improve the adsorption characteristics (increase of adsorption capacities or selectivity) [7]. For example, the Vakili group [58, 59], performed efforts to remove the reactive blue 4 dye from aqueous media, using modified chitosan beads. In one work, the adsorption behavior of chitosan (CS) beads modified with 3-aminopropyl triethoxysilane (APTES) was investigated. The results revealed that the adsorption capacity of modified CS beads (433.77 mg/g) was 1.37 times higher than that of unmodified CS beads (317.23 mg/g) [58]. In another work, the adsorption performance of chitosan (CS) hydrogel beads was investigated after impregnation of CS with hexadecylamine (HDA) as a cationic surfactant [59]. It was found that the adsorption capacity was increased by 1.43 times from 317 mg/g (CS) to 454 mg/g (CS/HDA). Examples of grafting (a) and crosslinking (b) reactions with chitosan are illustrated in Fig. (**3**).

Cross-linking

The crosslinked are an interesting way of chitosan derivatives materials that can be utilized in the material elaboration such as beads and gels. They are produced by cross-covalent bonds that connect the polymeric chains of chitosan. In this chemical modification, the crosslinking agents are molecules with at least two reactive functional groups that allow the formation of bridges between polymer chains [1, 2, 60].

The reticulation degree depends on many variables, however, the amount of reticulate agent is the crucial factor, and it formation is favored when the chitosan molecular weight and the temperature are increased. Considering that the covalent bonds occur in the polymer units deacetylated chitosan, the deacetylation degree also directly influences the number of possible interconnections. However, for this same reason, crosslinking causes a significant decrease in the adsorption capacity of the polymer, since the amine groups are unavailable, especially in the case of adsorption through chelation mechanisms. This chemical modification also decreases the accessibility to internal sites of the material and leads to a loss in the flexibility of the polymer chains [7, 60, 61].

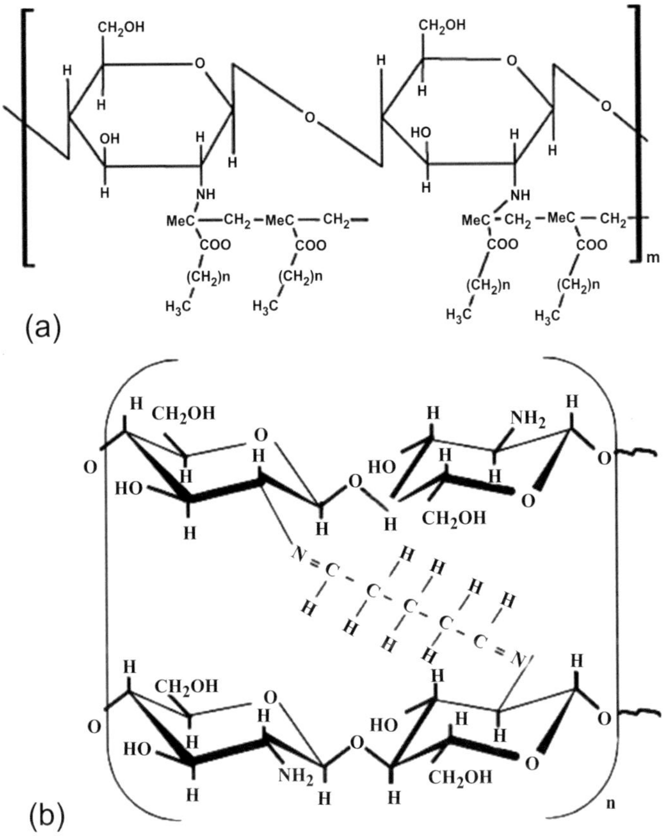

Fig. (3). Example of grafting (**a**) and crosslinking (**b**) of chitosan [60, 66].

The cross-linking may be performed using different cross-linking agents such as glutaraldehyde, 1,1,3,3-tetramethoxypropane, oxidized β-cyclodextrin polyaldehyde, ethyleneglycol diglycidyl ether and tri-polyphosphate [7, 62].

Grafting Reactions

The grafting consists in the insertion of new functional groups on the chitosan backbone. In the water treatment application, the main objectives of the grafting in chitosan polymers in are [7]: (a) increase the density of adsorption sites; (b) change the pH range for metal/dye adsorption; (c) increase the adsorption sites and the adsorption selectivity.

If the chitosan is in gel form, the water content leads to the reduction of the adsorption sites density. Thus, it's necessary to use grafting of new amino groups in order to compensate these restrictions and to keep the adsorption potential. For this propose, many substances have been successfully grafted on chitosan by different processes. For example, in order to change the more adequate pH to remove ions from aqueous medium, grafted of carboxylic functions has been regarded as an interesting change for increasing the adsorption capacity. Carboxylic acids have also been grafted on chitosan, in order to promote chelating derivatives for the uptake of cations [7, 63 - 66].

INTERACTION MECHANISMS

There are many researches associated to remove contaminants from liquid effluents using chitosan based material like adsorbent. However, only a few of them presents the chemical interactions in the adsorption mechanisms.

In the adsorption of metal and dyes ions, the amino and hydroxyl groups are the responsible by interactions however the main reactive sites are the amino groups. These reactive groups may interact with the adsorbate through different ways, which depends on the characteristics of the ions, pH of the solution and the matrix effect. The free electron on nitrogen amino group may interact with cations at pH close to neutrality or weak acidity. On the other hand, when the chitosan is in acid solution, the protonation of amine groups occurs and, consequently, there is a potential for the anions capture. The ions adsorption onto chitosan polymer occurs predominantly by two mechanisms, the chelation and the electrostatic attraction, depending on the composition of the solution and the pH of the medium, since these parameters can affect the protonation of the polymer and the speciation of the ions (when referring to metal) [1, 2, 7, 67, 68].

Chelation

The chelation mechanism is usually recurrent in the adsorption of metal ions. The adsorption of metal ions by chitosan polymer occurs predominantly due the formation of a chitosan-metal ion complex. The nitrogen and oxygen electrons present in the amino group and the hydroxyl groups can involve the transition metal ions, in some coordination form through the covalent bonds. Two hypotheses have been proposed in order to interpret the chelation form involved in the ion-polymer interaction: (a) the bridge model, and (b) the pendant model. In the bridge model, metal ions are bound with several amine groups from the same chain or from different chains, *via* inter or intra molecular bond, while the opposite occurs in the pendant model, in which a metal ion is bound to one amine group. The Fig. (**4**) presents some formation of chitosan chelates with copper ions (a, b) and the possible structure of chitosan-Zn complexes (c, d,e) [7, 67].

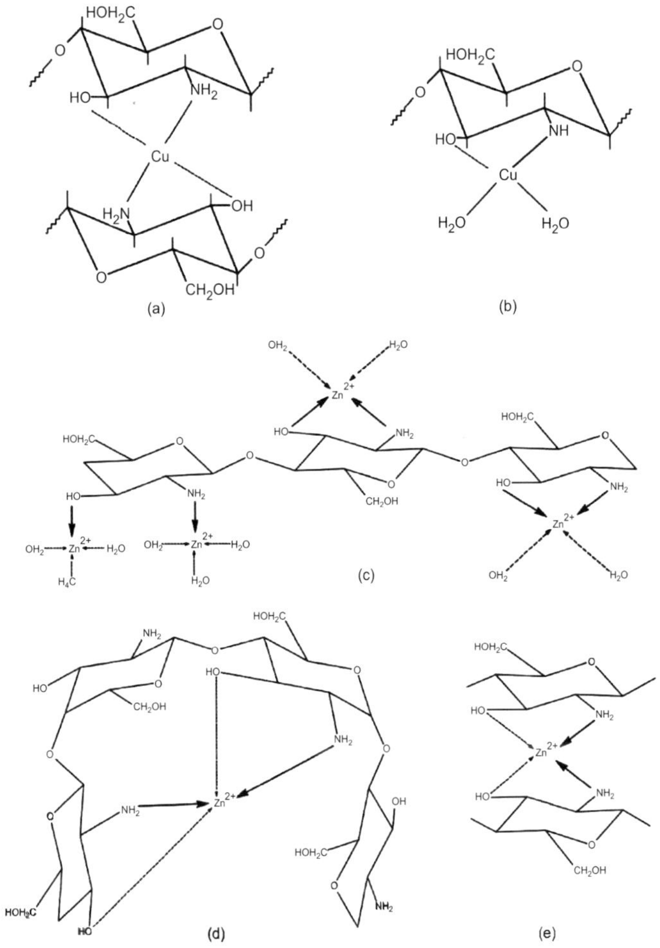

Fig. (4). Formation of chitosan chelates with copper ions (**a, b**), and the possible structure of chitosan-Zn complexes (**c, d, e**) [67].

Many parameters can affect the interaction between the chitosan-metal ion complex formation, however, probably more than one form of complexation can to occur and coexist in the same environment adsorption. The range of parameters affecting chelate formation includes: pH effect, ion selectivity, polymer chain and deacetylation degree.

pH Effect

Many studies have demonstrated that pH change leads to a change in the metal speciation, the number of protonated amino groups and, consequently, in metal ion adsorption capacity [69, 70]. Previous work indicated that the Cr(VI) adsorption capacity was sensitive to pH, however, the Cu(II) adsorption capacity was not influenced in the range pH 3 to 5 [71]. Another study showed that a low pH favored the protonation of the amino sites decreasing the copper-chelating, and this suggests that at a neutral pH more copper ions will be adsorbed [72].

Ion Selectivity

The selectivity of the chitosan polymer in a heterogeneous solution has been a limited focus on adsorption studies. The phenomena that can occurs due the pH influence, concentration range and different speciation of metals, leads to the difficult interpretation of the adsorption operation. At low pH, the chitosan protonation in the adsorption of metal anions is predominant in relation to the uptake of metal cations. Usually, chitosan has no affinity for alkaline and alkaline-earth metals and is possible to separate it from transition metals.

Rhazi *et al.* [73], in study of the adsorption of ions mixtures, showed the selectivity rule as follow: Cu(II) ≥ Hg(II) > Zn(II) > Cd(II) > Ni(II) ≥ Co(II); and studies performed by Vold *et al.* [71] confirmed the selectivity preferential of chitosan toward copper ions. They found also that the affinity of chitosan for cations not depends on the physical form of the polymer, and that the selectivity not depends on the size and hardness of considered ions. The preference of interaction can be the change through the chemical modification by grafting, to change the sites and the adsorption mechanism.

Polymer Chain and Deacetylation Degree

The adsorptive characteristics of chitosan depend on its degree of deacetylation, the distribution of acetyl group, chain length, and molecular mass distribution. The polymer chain flexibility can be changed using different molecular weight, degree of deacetylation and change of pH. The fraction of deacetylation is the major parameter to influence complexation [74], since it directly affects the density of amino groups on the surface of chitosan and, therefore, the capacity and mechanism of complex formation [67]. Previous work demonstrated the importance of chain length for the efficient complexation of copper ions [73]. Cadaval *et al.* [9] studied chitosan with different deacetylation degrees to remove Cr (VI) from aqueous solution by adsorption. The results showed that the deacetylation degrees increase the increase the adsorption capacity and the more appropriate conditions for Cr (VI) adsorption by chitosan were 95% deacetylation degree and pH 3, since $HCrO_4^-$ adsorption occurs by electrostatic interaction with protonated amino groups.

Ion Exchange/Electrostatic Attraction

The cationic character of amine groups in acid solutions is controlled by the pKa of the polymer, which is directly proportional to the degree of deacetylation of the chitosan chain. These properties are very important for the understanding of the electrostatic interactions. Many studies have focused on the adsorption of metal anions and anionic dyes on chitosan and its derivative materials. In neutral pH, about 50% of amine groups are protonated and available for the adsorption of anions [2]. However, the existence of free amine groups can lead to the chelation of cations. As the pH decreases, the protonation of amine groups increases. The ions adsorbed per mole of protonated amine groups can be determined plotting the logarithm of the distribution coefficient *versus* the pH, and the curve slope can be used for its determination [75]. The distribution coefficient is defined as the ratio of adsorption capacity and the equilibrium concentration in the solution. The distribution coefficient reaches a maximum before it to decrease in low pH. This behavior can be explained by a competitor effect of the anions brought about by the dissociation of the acid used for pH control. The more appropriated pH is frequently found around pH 2-4 [75]. Below this limit value, usually a large excess of competitor anions limits the adsorption efficiency. This electrostatic attraction may occur by direct interaction with free anions [76, 77].

Dotto *et al.* [10] used chitosan films to remove anionic food dyes from aqueous solutions by adsorption. They found that electrostatic interactions occurred between the negatively charged groups of the dyes (sulfonate or carboxylate) and protonated amino groups of the chitosan matrix. Cadaval *et al.* [78] evaluated the adsorption of food azo dyes onto chitosan films and showed, in a thermodynamic evaluation, that the magnitude of enthalpy change was consistent with electrostatic interactions due the interactions that occurred between the negatively charged groups of the dyes (sulfonate groups) and protonated amino groups of the chitosan film. Esquerdo *et al.* [11] studied chitosan scaffold like an alternative adsorbent for the removal of five hazardous food dyes from aqueous solutions. They inferred that the electrostatic interactions could have occurred between the food dyes and the chitosan scaffold.

DESORPTION AND RECYCLING

Chitosan based materials present many characteristics, such as their adsorption capacity, selectivity and the reuse possibility. The regeneration is very important since the reuse of the adsorbent is related to the economic viability of the operation, reducing the process costs and opening the possibility of recovering the contaminant. For this purpose, it is desirable to remove the contaminants and apply the chitosan derivative adsorbent for another adsorption cycle. Sometimes the recovery of adsorbent that is connected to the adsorbate by physisorption (like anionic dyes in electrostatic interaction) can be achieved by using an alkaline solution. In alkaline medium, the amino groups are deprotonated and the electrostatic interaction is broken. Desorption capacity is increased with a pH increase. Rêgo *et al.* [13] in desorption assays of food dyes showed that the chitosan film maintained its adsorption capacity by two cycles and the NaOH solution 0.50 mol L^{-1} was the eluent and the desorption were carried out in 5 min. In the first two cycles, the films maintained their physical characteristics and the same adsorption capacity. Chatterjee *et al.* [79] performed desorption of textile dye from the chitosan beads by changing the pH of the solution and showed that the beads could be reused five times keeping the mechanical properties and adsorption capacity.

In the case of the metal cations adsorption, the chelation mechanism is very sensitive to pH and usually adsorption does not occur at low pH. Therefore, a simple change in the pH of the solution may reverse the reaction. It would also be possible to use a strong chelating agent, such as EDTA, but this method is more complex because the eluents contain the chelated metal, which is difficult to separate. The method of pH change has also been successfully used for desorbing metal anions from loaded sorbent; however, in this case, it was necessary to increase the pH using sodium hydroxide (or ammonium hydroxide) [80]. Molybdate and vanadate were completely recovered from loaded phases using sodium hydroxide solutions. The adsorbent can be successfully reused for at least 10 cycles with no significant decrease in the adsorption capacities [81]. Cadaval *et al.* [23] studied the reuse of chitosan films for vanadium ion adsorption, and desorption cycles were carried out using NH_4Cl solution (0.01 mol L^{-1}). Six adsorption-desorption cycles were possible, maintaining similar adsorption capacity and mechanical properties of chitosan film. Hu *et al.* [82] reported that the dyes adsorbed on chitosan nanoparticles could be desorbed in an alkaline medium but not in a neutral medium. Lima *et al.* [83] reported that grafted chitosan can be regenerated with aqueous solutions containing sodium chloride without the use of organic solvents or pH modification.

CONFLICT OF INTEREST

The authors confirm that they have no conflict of interest to declare for this publication.

ACKNOWLEDGEMENTS

The authors are grateful to the Brazilian agencies CAPES and CNPq for the financial support.

REFERENCES

[1] Crini G, Badot PM. Application of chitosan, a natural aminopolysaccharide, for dye removal from aqueous solutions by adsorption processes using batch studies: a review of recent literature. Prog Polym Sci 2008; 33: 399-447.
[http://dx.doi.org/10.1016/j.progpolymsci.2007.11.001]

[2] Guibal E. Interactions of metal ions with chitosan-based sorbents: a review. Separ Purif Tech 2004; 38: 43-74.
[http://dx.doi.org/10.1016/j.seppur.2003.10.004]

[3] Wan-Ngah WS, Teong LC, Hanafiah MA. Adsorption of dyes and heavy metal ions by chitosan composites: a review. Carbohydr Polym 2011; 83: 1446-56.
[http://dx.doi.org/10.1016/j.carbpol.2010.11.004]

[4] Ramesh HP, Tharanathan RN. Carbohydrates the renewable raw materials of high biotechnological value. Crit Rev Biotechnol 2003; 23(2): 149-73.
[http://dx.doi.org/10.1080/713609312] [PMID: 12889744]

[5] Rinaudo M. Chitin and chitosan: properties and applications. Prog Polym Sci 2006; 31: 603-32.
[http://dx.doi.org/10.1016/j.progpolymsci.2006.06.001]

[6] Tharanathan RN, Kittur FS. Chitin the undisputed biomolecule of great potential. Crit Rev Food Sci Nutr 2003; 43(1): 61-87.
[http://dx.doi.org/10.1080/10408690390826455] [PMID: 12587986]

[7] Guibal E. Interactions of metal ions with chitosan-based sorbents: a review. Sep Sci Technol 2004; 38: 43-74.

[8] Muzzarelli RA. Nanochitins and nanochitosans, paving the way to eco-friendly and energy-saving exploitation of marine resources. In: Matyjaszewski K, Möller M, Eds. Polymer Science: A Comprehensive Reference. 10 ed.. Amsterdam: Elsevier BV 2012; pp. 153-64.

[9] Cadaval-Jr TR, Camara AS, Dotto GL, Pinto LA. Adsorption of Cr (VI) by chitosan with different deacetylation degrees. Desalination Water Treat 2013; 51: 7690-9.
[http://dx.doi.org/10.1080/19443994.2013.778797]

[10] Dotto GL, Moura JM, Cadaval TR, Pinto LA. Application of chitosan films for the removal of food dyes from aqueous solutions by adsorption. Chem Eng J 2013; 214: 8-16.
[http://dx.doi.org/10.1016/j.cej.2012.10.027]

[11] Esquerdo VM, Cadaval TR Jr, Dotto GL, Pinto LA. Chitosan scaffold as an alternative adsorbent for the removal of hazardous food dyes from aqueous solutions. J Colloid Interface Sci 2014; 424: 7-15.
[http://dx.doi.org/10.1016/j.jcis.2014.02.028] [PMID: 24767491]

[12] Dotto GL, Cunha JM, Calgaro CO, Tanabe EH, Bertuol DA. Surface modification of chitin using ultrasound-assisted and supercritical CO_2 technologies for cobalt adsorption. J Hazard Mater 2015; 295: 29-36.
[http://dx.doi.org/10.1016/j.jhazmat.2015.04.009] [PMID: 25880046]

[13] Rêgo TV, Cadaval TR Jr, Dotto GL, Pinto LA. Statistical optimization, interaction analysis and desorption studies for the azo dyes adsorption onto chitosan films. J Colloid Interface Sci 2013; 411: 27-33.
[http://dx.doi.org/10.1016/j.jcis.2013.08.051] [PMID: 24112836]

[14] Allouche FN, Guibal E, Mameri N. Preparation of a new chitosan-based material and its application for mercury sorption. Colloids Surf A Physicochem Eng Asp 2014; 446: 224-32.
[http://dx.doi.org/10.1016/j.colsurfa.2014.01.025]

[15] Ruthven DM. Principles of adsorption and adsorption processes. New York: John Wiley & Sons 1984.

[16] Suzuky M. Adsorption Engineering. Tokyo: Kodansha 1990.

[17] Vieira RS, Oliveira ML, Guibal E, Rodríguez-Castellón E, Beppu MM. Copper, mercury and chromium adsorption on natural and crosslinked chitosan films: An XPS investigation of mechanism. Colloids Surf A Physicochem Eng Asp 2011; 374: 108-14.
[http://dx.doi.org/10.1016/j.colsurfa.2010.11.022]

[18] Reddy MC, Sivaramakrishna L, Reddy AV. The use of an agricultural waste material, Jujuba seeds for the removal of anionic dye (Congo red) from aqueous medium. J Hazard Mater 2012; 203-204: 118-27.
[http://dx.doi.org/10.1016/j.jhazmat.2011.11.083] [PMID: 22209325]

[19] Aydin YA, Aksoy ND. Adsorption of chromium on chitosan: Optimization, kinetics and thermodynamics. Chem Eng J 2009; 151: 188-94.
[http://dx.doi.org/10.1016/j.cej.2009.02.010]

[20] Dotto GL, Pinto LA. Adsorption of food dyes onto chitosan: optimization process and kinetic. Carbohydr Polym 2011; 84: 231-8.
[http://dx.doi.org/10.1016/j.carbpol.2010.11.028]

[21] Kannamba B, Reddy KL, AppaRao BV. Removal of Cu(II) from aqueous solutions using chemically modified chitosan. J Hazard Mater 2010; 175(1-3): 939-48.
[http://dx.doi.org/10.1016/j.jhazmat.2009.10.098] [PMID: 19942344]

[22] Blázquez G, Calero M, Hernáinz F, Tenorio G, Martín-Lara MA. Equilibrium biosorption of lead (II) from aqueous solutions by solid waste from olive-oil production. Chem Eng J 2010; 160: 615-22.
[http://dx.doi.org/10.1016/j.cej.2010.03.085]

[23] Cadaval-Jr TR, Dotto GL, Seus ER, Mirlean N, Pinto LA. Vanadium removal from aqueous solutions by adsorption onto chitosan films. Desalination Water Treat 2016; 57: 16583-91.
[http://dx.doi.org/10.1080/19443994.2015.1079741]

[24] Langmuir I. The constitution and fundamental properties of solids and liquids. J Am Chem Soc 1916; 38: 2221-95.
[http://dx.doi.org/10.1021/ja02268a002]

[25] Freundlich HM. Über die adsorption in lösungen. Z Phys Chem 1906; 57: 385-471.

[26] Piccin JS, Gomes CS, Feris LA, Gutterres M. Kinetics and isotherms of leather dyes adsorption by tannery solid waste. Chem Eng J 2012; 183: 30-8.
[http://dx.doi.org/10.1016/j.cej.2011.12.013]

[27] Milonjic SK. A consideration of the correct calculation of thermodynamic parameters of adsorption. J Serb Chem Soc 2007; 72: 1363-7.
[http://dx.doi.org/10.2298/JSC0712363M]

[28] Liu Y. Is the free energy change of adsorption correctly calculated? J Chem Eng Data 2009; 54: 1981-5.
[http://dx.doi.org/10.1021/je800661q]

[29] Zhou X, Liu H, Hao J. How to calculate the thermodynamic equilibrium constant using the langmuir equation. Adsorpt Sci Technol 2012; 30: 647-9.
[http://dx.doi.org/10.1260/0263-6174.30.7.647]

[30] Qiu H, Pan LL, Zhang QJ, Zhang W, Zhang Q. Critical review in adsorption kinetic models. J Zhejiang Univ Sci A 2009; 10: 716-24.
[http://dx.doi.org/10.1631/jzus.A0820524]

[31] Moura JM, Gründmann DD, Cadaval-Jr TR, Dotto GL, Pinto LA. Comparison of chitosan with different physical forms to remove Reactive Black 5 from aqueous solutions. J Environ Chem Eng 2016; 4: 2259-67.
[http://dx.doi.org/10.1016/j.jece.2016.04.003]

[32] Skodras G, Diamantopoulou I, Pantoleontos G, Sakellaropoulos GP. Kinetic studies of elemental

mercury adsorption in activated carbon fixed bed reactor. J Hazard Mater 2008; 158(1): 1-13. [http://dx.doi.org/10.1016/j.jhazmat.2008.01.073] [PMID: 18321645]

[33] Wu FC, Tseng RL, Juang RS. Characteristics of elovich equation used for the analysis of adsorption kinetics in dye chitosan systems. Chem Eng J 2009; 150: 366-73. [http://dx.doi.org/10.1016/j.cej.2009.01.014]

[34] Weber WJ, Morris JC. Kinetics of adsorption of carbon from solutions. J Sanitary Eng Div Proceed Am Soc Civil Eng 1963; 89: 31-59.

[35] Auta M, Hameed BH. Chitosan-clay composite as highly effective and low-cost adsorbent for batch and fixed-bed adsorption of methylene blue. Chem Eng J 2014; 237: 352-61. [http://dx.doi.org/10.1016/j.cej.2013.09.066]

[36] Futalan CM, Kan CC, Dalida ML, Pascua C, Wan MW. Fixed-bed column studies on the removal of copper using chitosan immobilized on bentonite. Carbohydr Polym 2011; 83: 697-704. [http://dx.doi.org/10.1016/j.carbpol.2010.08.043]

[37] Vieira ML, Esquerdo VM, Nobre LR, Dotto GL, Pinto LA. Glass beads coated with chitosan for the food azo dyes adsorption in a fixed bed column. J Ind Eng Chem 2014; 20: 3387-93. [http://dx.doi.org/10.1016/j.jiec.2013.12.024]

[38] Allouche FN, Guibal E, Mameri N. Preparation of a new chitosan-based material and its application for mercury sorption. Colloids Surf A Physicochem Eng Asp 2014; 446: 224-32. [http://dx.doi.org/10.1016/j.colsurfa.2014.01.025]

[39] Gupta VK, Suhas. Application of low-cost adsorbents for dye removal a review. J Environ Manage 2009; 90(8): 2313-42. [http://dx.doi.org/10.1016/j.jenvman.2008.11.017] [PMID: 19264388]

[40] Beheshti H, Irani M, Hosseini L, Rahimi A, Aliabadi M. Removal of Cr (VI) from aqueous solutions using chitosan/MWCNT/Fe_3O_4 composite nanofibers-batch and column studies. Chem Eng J 2016; 284: 557-64. [http://dx.doi.org/10.1016/j.cej.2015.08.158]

[41] Lu Y, Chang PR, Zheng P, Ma X. Porous 3D network rectorite/chitosan gels: Preparation and adsorption properties. Appl Clay Sci 2015; 107: 21-7. [http://dx.doi.org/10.1016/j.clay.2015.01.020]

[42] Vakili M, Rafatullah M, Salamatinia B, *et al.* Application of chitosan and its derivatives as adsorbents for dye removal from water and wastewater: a review. Carbohydr Polym 2014; 113: 115-30. [http://dx.doi.org/10.1016/j.carbpol.2014.07.007] [PMID: 25256466]

[43] Tao Y, Ye L, Pan J, Wang Y, Tang B. Removal of Pb(II) from aqueous solution on chitosan/TiO(2) hybrid film. J Hazard Mater 2009; 161(2-3): 718-22. [http://dx.doi.org/10.1016/j.jhazmat.2008.04.012] [PMID: 18495341]

[44] Zhao LM, Shi LE, Zhang ZL, *et al.* Preparation and application of chitosan nanoparticles and nanofibers. Braz J Chem Eng 2011; 28: 353-62. [http://dx.doi.org/10.1590/S0104-66322011000300001]

[45] Sencadas V, Correia DM, Areias A, *et al.* Determination of the parameters affecting electrospun chitosan fiber size distribution and morphology. Carbohydr Polym 2012; 87: 1295-301. [http://dx.doi.org/10.1016/j.carbpol.2011.09.017]

[46] Mirmohseni A, Seyed Dorraji MS, Figoli A, Tasselli F. Chitosan hollow fibers as effective biosorbent toward dye: preparation and modeling. Bioresour Technol 2012; 121: 212-20. [http://dx.doi.org/10.1016/j.biortech.2012.06.067] [PMID: 22858488]

[47] Barron-Zambrano J, Szygula A, Ruiz M, Sastre AM, Guibal E. Biosorption of Reactive Black 5 from aqueous solutions by chitosan: column studies. J Environ Manage 2010; 91(12): 2669-75. [http://dx.doi.org/10.1016/j.jenvman.2010.07.033] [PMID: 20724062]

[48] Wan-Ngah WS, Teong LC, Toh RH, Hanafiah MA. Utilization of chitosan-zeolite composite in the removal of Cu(II) from aqueous solution: Adsorption, desorption and fixed bed column studies. Chem Eng J 2012; 209: 46-53. [http://dx.doi.org/10.1016/j.cej.2012.07.116]

[49] Wan MW, Kan CC, Rogel BD, Dalida ML. Adsorption of copper (II) and lead (II) ions from aqueous solution on chitosan-coated sand. Carbohydr Polym 2010; 80: 891-9. [http://dx.doi.org/10.1016/j.carbpol.2009.12.048]

[50] Popuri SR, Vijaya Y, Boddu VM, Abburi K. Adsorptive removal of copper and nickel ions from water using chitosan coated PVC beads. Bioresour Technol 2009; 100(1): 194-9. [http://dx.doi.org/10.1016/j.biortech.2008.05.041] [PMID: 18614363]

[51] Torres MA, Beppu MM, Santana CC. Characterization of chemically modified chitosan microspheres as adsorbents using standard proteins (bovine serum albumin and lysozyme). Braz J Chem Eng 2007; 24: 325-36. [http://dx.doi.org/10.1590/S0104-66322007000300003]

[52] Boddu VM, Abburi K, Talbott JL, Smith ED, Haasch R. Removal of arsenic (III) and arsenic (V) from aqueous medium using chitosan-coated biosorbent. Water Res 2008; 42(3): 633-42. [http://dx.doi.org/10.1016/j.watres.2007.08.014] [PMID: 17822735]

[53] Wen Y, Tang ZR, Chen Y, Gu YX. Adsorption of Cr (VI) from aqueous solutions using chitosan-coated fly ash composite as biosorbent. Chem Eng J 2011; 175: 110-6. [http://dx.doi.org/10.1016/j.cej.2011.09.066]

[54] Calagui MJ, Senoro DB, Kan CC, Salvacion JW, Futalan CM, Wan MW. Adsorption of indium(III) ions from aqueous solution using chitosan-coated bentonite beads. J Hazard Mater 2014; 277: 120-6. [http://dx.doi.org/10.1016/j.jhazmat.2014.04.043] [PMID: 24802798]

[55] Tan KB, Vakili M, Horri BA, *et al.* Adsorption of dyes by nanomaterials: Recent developments and adsorption mechanisms. Separ Purif Tech 2015; 150: 229-42. [http://dx.doi.org/10.1016/j.seppur.2015.07.009]

[56] Sivakami MS, Gomathi T, Venkatesan J, Jeong HS, Kim SK, Sudha PN. Preparation and characterization of nano chitosan for treatment wastewaters. Int J Biol Macromol 2013; 57: 204-12. [http://dx.doi.org/10.1016/j.ijbiomac.2013.03.005] [PMID: 23500442]

[57] Akbari H, Mehrabadi A, Torabian A. Determination of nanofiltration efficiency in arsenic removal from drinking water. Iran J Environ Health Sci Eng 2010; 7: 273-8.

[58] Vakili M, Rafatullah M, Ibrahim MH, Abdullah AZ, Salamatinia B, Gholami Z. Chitosan hydrogel beads impregnated with hexadecylamine for improved reactive blue 4 adsorption. Carbohydr Polym 2016; 137: 139-46. [http://dx.doi.org/10.1016/j.carbpol.2015.09.017] [PMID: 26686114]

[59] Vakili M, Rafatullah M, Salamatinia B, Ibrahim MH, Abdullah AZ. Elimination of reactive blue 4 from aqueous solutions using 3-aminopropyl triethoxysilane modified chitosan beads. Carbohydr Polym 2015; 132: 89-96. [http://dx.doi.org/10.1016/j.carbpol.2015.05.080] [PMID: 26256328]

[60] Harish-Prashanth KV, Tharanathan RN. Chitin/chitosan: modifications and their unlimited Application potential an overview. Trends Food Sci Technol 2007; 18: 117-31. [http://dx.doi.org/10.1016/j.tifs.2006.10.022]

[61] Erosa MS, Medina TI, Mendoza RN, Rodriguez MA, Guibal E. Cadmium sorption on chitosan sorbents: kinetic and equilibrium studies. Hydrometallurgy 2001; 61: 157-67. [http://dx.doi.org/10.1016/S0304-386X(01)00166-9]

[62] Capitani D, De Angelis AA, Crescenzi V, Masci G, Segre AL. NMR study of a novel chitosan-based hydrogel. Carbohydr Polym 2001; 45: 245-52. [http://dx.doi.org/10.1016/S0144-8617(00)00255-1]

[63] Ruiz M, Sastre AM, Guibal E. Pd and Pt recovery using chitosan gel beads. II. Influence of chemical modifications on sorption properties. Sep Sci Technol 2002; 37: 2385-403. [http://dx.doi.org/10.1081/SS-120003519]

[64] Crini G, Martel B, Torri G. Adsorption of C.I. Basic Blue 9 on chitosan-based materials. Int J Environ Pollut 2008; 34: 451-65. [http://dx.doi.org/10.1504/IJEP.2008.020809]

[65] Gaffar MA, El-Rafie SM, El-Tahlawy KF. Preparation and utilization of ionic exchange resin *via* graft copolymerization of beta-CD itaconate with chitosan. Carbohydr Polym 2005; 56: 387-96. [http://dx.doi.org/10.1016/j.carbpol.2004.01.007]

[66] Konaganti VK, Kota R, Patil S, Madras G. Adsorption of anionic dyes on chitosan grafted poly(alkyl methacrylate)s. Chem Eng J 2010; 158: 393-401. [http://dx.doi.org/10.1016/j.cej.2010.01.003]

[67] Gerente C, Lee VK, Le Cloirec P, McKay G. Application of chitosan for the removal of metals from wastewaters by adsorption—mechanisms and models review. Crit Rev Environ Sci Technol 2007; 37: 41-127. [http://dx.doi.org/10.1080/10643380600729089]

[68] Gotoh T, Matsushima K, Kikuchi K. Adsorption of Cu and Mn on covalently cross-linked alginate gel beads. Chemosphere 2004; 55(1): 57-64. [http://dx.doi.org/10.1016/j.chemosphere.2003.10.034] [PMID: 14720547]

[69] Guibal E, Saucedo I, Roussy J, Le Cloirec P. Uptake of uranyl ions by new sorbing polymers—Discussion of adsorption isotherms and pH effect. React Polym 1994; 23: 147-56. [http://dx.doi.org/10.1016/0923-1137(94)90015-9]

[70] Wu FC, Tseng RL, Juang RS. Role of pH in metal adsorption from aqueous solutions containing chelating agents on chitosan. Ind Eng Chem Res 1999; 38: 270-5. [http://dx.doi.org/10.1021/ie980242w]

[71] Schmuhl R, Krieg HM, Keizer K. Adsorption of Cu(II) and Cr(VI) ions by chitosan: Kinetics and equilibrium studies. Water SA 2001; 27: 1-8.

[72] Inoue K, Baba Y, Yoshizuka K. Adsorption of metal-ions on chitosan and cross-linked copper(II) - complexed chitosan. Bull Chem Soc Jpn 1993; 66: 2915. [http://dx.doi.org/10.1246/bcsj.66.2915]

[73] Rhazi M, Desbrieres J, Tolaimate A, *et al.* Influence of the nature of the metal ions on the complexation with chitosan. Application to the treatment of liquid waste. Eur Polym J 2002; 38: 1523-30. [http://dx.doi.org/10.1016/S0014-3057(02)00026-5]

[74] Vold IM, Varum KM, Guibal E, Smidsrod O. Binding of ions to chitosan-selectivity studies. Carbohydr Polym 2003; 54: 471-7. [http://dx.doi.org/10.1016/j.carbpol.2003.07.001]

[75] Inoue K, Yoshizuka K, Baba Y. New developments in ion exchange. Materials, fundamentals, and applications, in: Proceedings of the International Conference on Ion Exchange. Tokyo, Japan, Kodansha: ICIE'91 1991; pp. 543-8.

[76] Lasko CL, Hurst MP. An investigation into the use of chitosan for the removal of soluble silver from industrial wastewater. Environ Sci Technol 1999; 33: 3622-6. [http://dx.doi.org/10.1021/es980443r]

[77] Kawamura Y, Yoshida H, Asai S, Tanibe H. Recovery of $HgCl_2$ using polyaminated highly porous chitosan beads: effect of salt and acid. J Chem Eng of Jpn 1998; 31: 1. [http://dx.doi.org/10.1252/jcej.31.1]

[78] Cadaval TR, Dotto GL, Pinto LA. Equilibrium isotherms, thermodynamics, and kinetic studies for the

adsorption of food azo dyes onto chitosan films. Chem Eng Commun 2015; 202: 1316-23.
[http://dx.doi.org/10.1080/00986445.2014.934449]

[79] Chatterjee S, Chatterjee S, Chatterjee BP, Das AR, Guha AK. Adsorption of a model anionic dye, eosin Y, from aqueous solution by chitosan hydrobeads. J Colloid Interface Sci 2005; 288(1): 30-5.
[http://dx.doi.org/10.1016/j.jcis.2005.02.055] [PMID: 15927558]

[80] Guzman J, Saucedo I, Revilla J, Navarro R, Guibal E. Vanadium interactions with chitosan: influence of polymer protonation and metal speciation. Langmuir 2006; 18: 1567-73.
[http://dx.doi.org/10.1021/la010802n]

[81] Milot C. Adsorption de molybdate VI sur billes de gel de chitosane-application au traitement d'effluents molydifères. PhD dissertation Université de Montpellier 1998.

[82] Hu ZG, Zhang J, Chan WL, Szeto YS. The adsorption of acid dye onto chitosan nanoparticles. Polymer (Guildf) 2006; 47: 5831-7.
[http://dx.doi.org/10.1016/j.polymer.2006.05.071]

[83] Lima IS, Ribeiro ES, Airoldi C. The use of chemically modified chitosan with succinic anhydride in the methylene blue adsorption. Quim Nova 2006; 29: 501-6.
[http://dx.doi.org/10.1590/S0100-40422006000300018]

CHAPTER 8

Application of Chitosan Based Materials for Drug Delivery

Poliana P. Lopes[1,*], Nathalie B. Barroca[2], Ana Luísa Daniel-da-Silva[3] and Bárbara J. Leite Ferreira[3]

[1] *Chemical Engineering Department, Federal University of Santa Maria, Santa Maria, Brazil*

[2] *School of Mechanical and Aerospace Engineering, Queen's University Belfast, United Kingdom*

[3] *CICECO-Aveiro Institute of Materials, Department of Chemistry, University of Aveiro, Aveiro, Portugal*

Abstract: One of the main challenges in drug delivery field is the development of formulations or devices designed to allow the transport of a suitable amount of the active pharmaceutical substance to the site of action in the body, and improve its efficacy by controlling the rate, time and place of release. The distribution of the drug to tissues different than the target place is unnecessary, wasteful, and a potential cause of toxicity. Chitosan, a versatile hydrophilic polysaccharide derived from chitin, is a non-toxic, biocompatible and biodegradable polymer which has attracted significant scientific interest, for its fundamental role in the progress of drug delivery technology. This natural biopolymer has inspired the development of safe and effective drug carrier systems due to its unique physicochemical and biological characteristics. The cationic character, exclusive of the chitosan, is attributed to the primary amino groups in the main backbone, which make its surface positively charged under acidic conditions; an important requirement that has been responsible for the enhancement of the load capability, tailoring the release profile of the drug and increasing the solubility. This chapter focuses on the use of chitosan-based therapeutic systems, including particles, hydrogel, film and scaffolds, through various anatomical routes by which the drug may be introduced in human body, such as oral, nasal, ocular, transdermal and local administration.

Keywords: Biomedical applications, Chitosan, Drug carrier, Hydrogel, Membrane, Nanoparticles, Route of administration.

INTRODUCTION

Chitin, a naturally abundant polysaccharide, is well known to be the main component of the exoskeleton of crustaceans and insects, which, by deacetylation,

* **Corresponding author Poliana P. Lopes:** Chemical Engineering Department, Federal University of Santa Maria, 9B, Santa Maria, 97.105-900, RS, Brazil; Tel/Fax: +55 (55) 32208448; E-mail: poliana.lopes@ufsm.br

Guilherme Luiz Dotto, Sergio Paulo Campana-Filho and Luiz Antonio de Almeida Pinto (Eds)

can lead to chitosan, although this reaction rarely is 100% complete, normally a partial N-deacetylation of chitin occurs [1, 2]. Therefore, chitosan is a linear copolymer consisting of glucosamine (β(1-4)-linked 2-amino-2-deoxy-D-glucose) and N-acetyl-glucosamine (2-acetamido-2-deoxy-D-glucose)), showing a rigid crystalline structure through inter- and intramolecular hydrogen bonding [3 - 5].

The history of chitosan dates back to the 19th century, when Charles Rouget boiled chitin in a concentrated alkali solution and found a new product soluble in the organic acid in 1859 [3, 6]. Later research confirmed that the product obtained was deacetylated chitin and therefore named it chitosan [7]. Chitin shows poor solubility in aqueous solution and organic solvents, limiting its use for practical applications whereas, chitosan is more suitable for bioapplications [4, 8].

Despite the fact that chitosan is not extensively present in the environment, it can be easily obtained by chemical or enzymatic hydrolysis of a natural polymer: the chitin. From a functional point of view, chitosan is quite a unique material; its intrinsic properties are so singular and valuable that it has inspired a wide variety of academic and industrial research for pharmaceutical and medical applications, especially in the field of controlled drug delivery [9 - 12]. The interest in this polymer originates in its positive facets such as biocompatibility, biodegradability, low toxicity, mucoadhesion, stimuli response and versatile biological activities as antimicrobial activity and low immunogenicity [3, 13 - 16].

The presence of reactive amino (NH_2) group in the chitosan structure makes this polymer the only polysaccharide that possesses high density of positive charges, thus the multiple free amino groups can be protonated in dilute acid solutions with pH below 6.0 creating NH_3^+ groups, which readily bind to negatively charged molecules or materials [4, 17].

Some remarkable properties of chitosan as biomaterial, previously mentioned, are based on its cationic character. The mucoadhesive properties, for example, can be explained by the electrostatic interaction developed by the presence of negatively charged sialic acid residues in the mucin [5, 18, 19]. The mechanism responsible for permeation enhancing effect is also linked to the positive charges, which seem to interact with the negative part of the cell membrane resulting in a structural reorganization and an opening of the tight junction proteins [10, 20, 21]. The analgesic effect can also be related to the polycationic nature of chitosan, once the amino groups of the D-glucosamine residues can protonate in the presence of proton ions that are released in the inflammatory reaction [22, 23]. Finally, the haemostatic activity where the positive charges of polymer have strong affinity to bond with red blood cells, activates the platelets and forms a gel-like coagulum which promotes hemostasis and seals bleeding of vessel [5, 24, 25].

Modification of chitosan can be made to improve its sensitivities to the environmental changes. Various innovative chitosan derivatives have been obtained by chemical modification, such as acylation, tosylation, alkylation or silylation, using the reactive activities of hydroxyl and amino groups [26 - 29]. In particular, the primary amino group has nucleophilic property and could generate reactive sites for many others functional compounds. The association of bioactive molecules to the chitosan provide versatile materials with specific functionality, allowing an easy modification of its biological and physical properties [2, 30 - 32].

The oral administration of chitosan has shown to result on its partial hydrolysis by lysozymes in human serum [33] and digestive enzymes in the gastrointestinal tract that generate monomers, chitooligomers and glucosamines. Among poly-saccharides, chitosan is the unique in terms of susceptibility to enzymatic depolymerization, cationicity, supply of cell-activating oligomers, and supply of *N*-acetylglucosamine for rebuilding of other biopolymers [34]. The degradation by lysozymes is very relevant as they exist in various human body fluids and tissues with concentrations from 4 to 13 mg/L in serum and from 450 to 1230 mg/L in tears [35].

Chitosan degradation is dependent on the deacetylation degree (DD), wherein a higher deacetylation leads to slower degradation [36]. Chitosan matrices with various degrees of N-acetylation were subjected to *in vitro* lysozymic degradation. Chitosan matrices with DD of 93.5%, 81.7%, and 71.7% experienced very slow degradation process, while those with DD of 62.4%, 56.1%, and 52.6% underwent significant degradation [35]. This has been used to control the degradation rate of chitosan formulations for specific uses. For instance, the biocompatibility and degradation rate of chitosan based nerve guidance channel were tuned by the acetylation degree [37]. Chitosan degradation can also be varied by carboxymethylation, bimodal molecular weight distribution of chitosan [38] or thiolation [39]. Particular attention should be paid to cross-linkers as they affect the degradation profile. Particularly, noncrosslinked chitosan films were digested by both pancreatic and colonic enzymes, while glutaraldehyde crosslinked ones were resistant to both enzyme systems and tripolyphosphate crosslinked ones were resistant only to pancreatic digestion [40].

Regarding the *in vivo* degradation, no abnormal symptom was observed with several animals when chitosan was orally administered at a dosage of 0.7-0.8 g/kg body weight/day for up to 239 days or intravenously at a dosage of 4.5 mg/kg body weight/day for 7-11 days. Also, chitosan was about 39-79% digested in rabbits and almost completely by hens and broilers [41].

With respect to biodistribution, *in vitro* and *in vivo* toxicity of chitosan, many studies have been conducted to evaluate its cytotoxicity, haemocompatibility, immunotoxicity, and biocompatibility. Comprehensive reviews on chitosan biodistribution and toxicity can be found [3, 13]. When preparing chitosan based formulations for drug delivery, modifications such as cross-linking, copolymerization, grafting and polyelectrolytes complexation have the potential to change its biodistribution and also its potential toxicity. As it will be detailed in the present review, most of the new chitosan-based formulations for drug delivery, use *in vitro* and *in vivo* models to assess their drug delivery ability but still lack an appropriate toxicity evaluation. Longer term effect on cells and tissues should be the subject of thorough research.

DRUG DELIVERY SYSTEMS

Drug delivery technology had a breakthrough in utilization over the last six decades. This progression began in 1952 with the introduction of the first sustained release formulation [42]. Until 1980s, it focused on developing macroscopic devices for oral and transdermal sustained release systems and establishing controlled drug release mechanisms [43]. The following three decades were dedicated to the investigation of self-regulated drug delivery systems, long-term depot formulations, and nanotechnology-based delivery systems, but in this period has not been as successful in generating clinical products [44 - 47]. Emerging strategies lead to the third generation drug delivery systems (from 2010) that consist in a targeted delivery, like anticancer drug delivery to tumor and siRNA delivery to target cells [48 - 52].

Drug delivery can be defined by the approach, technology, and system for transporting a pharmaceutical compound within the body in order to safely achieve the desired therapeutic effect. This process includes the administration and release of a therapeutic agent and its subsequent transport across the biological membranes to the site of action [45, 53]. Thus, it is interesting to note that the fundamental concept of delivery systems is heavily connected with the dosage form and route of administration. A suitable drug delivery system ensures that the active drug is available at the site of action for the correct time and duration.

Drugs may be administered directly to the organ affected by disease (local) [9, 54, 55] or delivered by systemic routes [56, 57], where it reaches to the circulatory system and blood, so that the entire body is affected. Although there is no formal classification of systemic routes of administration, generally the categories of enteral, parenteral and others, described by Mackichan and Doshi [58] are accepted. In enteral route the drug is placed in the gastrointestinal tract and then it absorbs to the blood. Parenteral is the introduction of medications into the body by routes other than the gastrointestinal tract but basically the term is applied to injection of substances by subcutaneous, intramuscular, intravenous, and intra-arterial routes. Systemic routes that are not classified as either enteral or parenteral typically involve absorption into the blood through a mucous membrane being also called topical route [59, 60].

Considering the dosage form, a conventional or traditional drug delivery system considers the use of multiple daily dosing of medication for treatment of disease, which quantities of drug are introduced into the system at specified intervals of time [61]. Normally this system leads to wide swings in concentration of drug in plasma (serum drug concentrations), where initially a rapid increase of drug levels in the body (peak) occurs due to the immediate release of the most of the therapeutic content after administration and then a decline sharp (valley) happens within a short period as indicated in the Fig. (**1a**). For some drugs the sharp fluctuations often cause unacceptable side-effects at the peak, followed by inadequate therapy at the valley [1, 62].

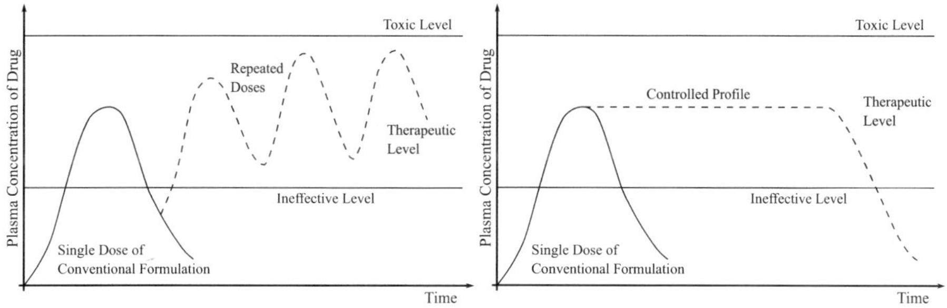

Fig. (1). Hypothetical plasma concentration profile for conventional dosage form (**a**) and for controlled release dosage form (**b**).

Controlled delivery systems are designed to promote some control of the drug in body, creating a desired environment with minimum side-effects and prolonged efficacy, enhancing the safety, efficacy and reliability of drug therapy [1, 17, 32]. This system can be graphically represented and compared to conventional formulation in Fig. (**1b**). In other words, the system attempts to control the drug concentration on target tissue or cells, maintaining a constant plasma concentration throughout the treatment period and reducing the frequency of drug administration.

There are many recent papers about chitosan-based therapeutic systems in literature focused on the administration route such as oral [27, 63], nasal [64, 65], ocular [19], gastrointestinal [66, 67] and transdermal [10, 45, 68, 69]. However, few publications relate to morphologies of the delivery vehicle, thus this chapter is based on geometries and design of chitosan drug carrier like particles, hydrogels, films and scaffolds.

PARTICULATE DRUG CARRIERS

Chitosan-based carriers in the particulate forms, including nanoparticles, microparticles and hydrogel beads, can be prepared using many different methods, which have been extensively reviewed [11, 70 - 72]. Mini-emulsion, covalent and ionic cross-linking, coacervation/precipitation and self-assembly are the most commonly used methods to prepare chitosan carriers (Table **1**). The preparative method will influence several properties of the chitosan carrier, such as particle size, surface charge, stability and loading efficiency.

Table 1. Examples of chitosan-based particulate carriers prepared using distinct methods.

Formulation	Drug	Method of Preparation	Hydrodynamic Diameter	Envisaged Administration Route	Reference
Chitosan/Alginate	Gatifloxacin	Complex coacervation	200-500 nm	Ocular	[77]
Chitosan/Alginate	Sulfasalazine	Complex coacervation	*ca.* 1 mm	Oral	[78]
Chitosan/PGA	Insulin	Complex coacervation	50 nm	Oral	[79]
N-octyl-O-glycol chitosan	Paclitaxel	Self-assembly	190-250 nm	Intravenous	[80]
Acylated carboxymethyl chitosan	Doxorubicin	Self-assembly	25 nm	Intravenous	[81]
Cholesterol-modified carboxymethyl chitosan	Paclitaxel	Self-assembly	210 nm	Intravenous	[82]
Genipin crosslinked Chitosan	Flurbiprofen	Emulsion/Cross-linking	5-10 μm	Intra-articular	[75]
Chitosan/TPP	pDNA	Ionic gelation	150 nm-2 μm	Intravenous	[83]
Chitosan/TPP	SiRNA	Ionic gelation	280-700 nm	Intravenous	[84]

Micro- and nanoparticulate chitosan based drug carriers have been prepared using water-in-oil microemulsions by emulsifying the aqueous solution of chitosan in the oil phase. A suitable surfactant is used to stabilize the aqueous droplets. Control of chitosan particle size can be achieved by controlling the size of the aqueous droplets. The addition of a cross-linker to the emulsion to promote gelation and impart desired mechanical strength to the particles is required. Chitosan crosslinking can be achieved chemically using covalent crosslinkers or physically by ionotropic gelation. Due to the presence of amine functional groups, chitosan can be easily covalently crosslinked using dialdehydes such as glutaraldehyde. Indeed, early work describing chitosan nanoparticles for drug delivery was based on glutaraldehyde crosslinked chitosan [73]. However, the later discovery of the negative effects of glutaraldehyde crosslinking on cell viability has prompt the search for less toxic alternative crosslinkers, such as genipin [74, 75], a natural occurring product extracted from Gardenia jasminoides fruits, with toxicity ca. 10,000 times lower than glutaraldehyde [76].

Taking advantage of the cationic nature of chitosan in acidic conditions, physical crosslinking can be achieved by simple electrostatic interaction with anionic species. Chitosan particles prepared by ionotropic gelation of chitosan with the polyanion tripolyphosphate (TPP) have been largely explored for pharmaceutical applications [83, 84]. The particles are prepared simply by adding dropwise a chitosan acidic solution to a solution of TPP or vice-versa. The size and surface charge of the particles can be tuned by controlling the chitosan to TPP weight ratio [85]. Alternatively, TPP can be employed to crosslink chitosan particles prepared *via* emulsion [86].

Chitosan microparticles can be also prepared using simple and complex coacervation. Simple coacervation is attained by blowing a chitosan solution into an alkaline precipitating medium. The size of the particles is affected by air pressure or spray- nozzle diameter, while the concentration of chitosan and its deacetylation degree affect the strength and porosity of the beads [87]. Glutaraldehyde is often added to improve the mechanical strength of the particles. In complex coacervation, particulate systems are produced by mixing chitosan with anionic polyelectrolytes [16]. The formation of the particles is typically driven by electrostatic interactions. Examples of polyelectrolytes used for preparing chitosan-based carriers using this technique are nature-occurring polymers such as alginate [77, 78] and κ-carrageenan [88] and synthetic polymers such as poly(γ-glutamic acid) [79]. The net charge ratio and molecular weight of the polyelectrolytes influence the physicochemical properties of the particulate systems, namely the mechanical stability, swelling behavior and dissolution profile.

Self-assembled chitosan nanoparticles have gained increased attention in recent years [12]. These particles are obtained by self-assembly of amphiphilic chitosan derivatives at concentrations larger than the critical micelle concentration. The micelle formation is typically driven by inter- and intramolecular hydrophobic interactions and hydrogen bonds, mainly in order to minimize interfacial free energy. Because the main molecular chain of chitosan is hydrophilic, the hydrophilic/hydrophobic balance of chitosan must be adjusted, usually by chemical modification with hydrophobic moieties such as alkyl [80], acyl [89] and cholesterol [82]. The molecular weight of chitosan, the nature and extent of chemical modification are the key factors affecting the self-assembly of chitosan derivatives. The micelles exhibit nanocapsule shape with hydrophobic core that can accommodate less soluble drugs and hydrophilic outer structure that provides solubility in aqueous environments. Owing to their unique core-shell architecture, nanoscale size and thermodynamic stability, chitosan micelles are very promising carriers for poorly water-soluble substances such as anticancer drugs [80 - 82].

Chitosan nanoparticles offer good biocompatibility and biodegradability. As drug delivery systems, they have attracted increasing attention for their wide applications in, for example, loading protein drugs, gene drugs, and anticancer chemical drugs, and *via* various routes of administration including oral, nasal, intravenous, and ocular [10, 90]. Owing to reduced dimensions, nanoparticles are capable of passing through biological barriers *in vivo* and delivering drugs to the lesion site to enhance efficacy. Nanoparticles can easily pass from blood vessels into tumor tissues because of leaky and irregularly dilated blood vessels at the tumor site, and can be retained at tumor sites owing to poor lymphatic drainage. This phenomenon of selective accumulation of nanoparticles near tumor tissues is termed the enhanced permeability and retention (EPR) effect. The mechanism being responsible for the permeation enhancing effect of chitosan nanoparticles is also based on the positive charges of the polymer, which seems to interact with the cell membrane resulting in a structural reorganization of tight junction-associated proteins [20].

Chitosan formulated in nanoparticles and capsules has shown great potential for encapsulating doxorubicin, which is a chemotherapeutic agent frequently used as a frontline against a variety of cancers [91]. Although effective in the regression of a number of carcinomas, the administration of free doxorubicin may induce serious side-effects such as cardiomyopathy and myelosuppression, since only a small amount of drug administrated reaches the tumor target site. In view of this, chitosan and derivatives are being used to build carriers that deliver doxorubicin more effectively to target site and reduce its systemic side-effects. Among several chitosan based doxorubicin carriers reported so far, glycol-chitosan nanoparticles have shown the best encapsulation efficiency of 97%, corresponding to a doxorubicin content up to 39% [91, 92] which might be a better option as far as chitosan DDSs are concerned. These carriers were prepared by self-assembly of doxorubicin conjugated glycol-chitosan followed by physical entrapment of doxorubicin. *In vivo* studies in mice indicate that after administration, glycol chitosan nanoparticles labeled with fluorescein isothiocyanate accumulate passively in the tumour tissue, probably *via* the EPR effect [92] and were minimally detected in the heart and lungs [93], which is promising for prevention of doxorubicin-associated cardiac toxicity.

Conventional chemotherapy often fails in cancer treatment in part due to tumor complexity and heterogeneity and development of drug resistance. In an attempt to overcome these limitations, clinicians are adopting multiple or multi-stage chemotherapeutic regimens. In this context Yoon *et al* have reported that glycol-chitosan nanoparticles are effective platforms for the encapsulation and target delivery of both doxorubicin and siRNA [94]. Regardless of the different properties of the therapeutics encapsulated, these carriers exhibited similar physicochemical properties, including size, surface properties and pH sensitive behavior. *In vivo* results in the treatment of solid tumor in mice suggest that the use of glycol-chitosan nanoparticles for the sequential treatment with doxorubicin and siRNA could successfully overcome the chemo-drug resistance problem in cancer treatment.

Glycol-chitosan nanoparticles have been also extensively explored in the development of more sophisticated platforms for cancer treatment, with responsive properties or merging therapy and diagnosis (theranostic) capabilities. Examples include echogenic glycol chitosan-based nanocapsules for cancer-targeted ultrasound-triggered delivery of doxorubicin and docetaxel drugs [95], or systems for cancer therapy and medical imaging in a single platform [96 - 98].

Chitosan based carriers offer a number of advantages for oral delivery of therapeutic macromolecules such as peptides and proteins [63, 99]. Chitosan protects loaded macromolecules against acidic denaturation and enzymatic degradation in the gastrointestinal tract and, due to its inherent mucoadhesive feature, it prolongs the intestinal residence time of therapeutics. Moreover, chitosan enhances the absorption of hydrophilic macromolecules by the intestinal epithelium. However, the effectiveness of chitosan as a mucoadhesive agent and an absorption enhancer is limited to intestinal areas where chitosan does not precipitate, *i.e.* where the pH is below the pKa of chitosan. The use of chitosan derivatives allows in principle to overcome these drawbacks since thiolated chitosan has improved mucoadhesive capability [100] and quaternized chitosan displays enhanced solubility and absorption enhancement capability in the intestinal pH environments [101].

Several studies on the use of chitosan nanoparticles for oral insulin delivery have been reported [99]. Indeed, oral delivery of insulin would be a non-invasive alternative much more comfortable for patients than repeated subcutaneous injections which is the standard treatment. Very recently Lopes and co-workers [102] have reported the fabrication of dual chitosan/albumin-coated alginate/ dextran sulfate nanoparticles for enhanced oral delivery of insulin, a pH responsive carrier with clinical potential for the therapy of type 1 diabetes mellitus. Dual chitosan/albumin coating of the nanoparticles demonstrated augmented intestinal interactions with the intestinal cells in comparison to the uncoated system, resulting in a higher permeability of insulin.

Chitosan has excellent biocompatibility and biodegradation, which has led to an increasing interest on this polymer as a non-viral carrier for gene drug delivery [90]. Naked therapeutic genes are rapidly degraded by nucleases and cannot be internalized into cells effectively, showing low transfection efficiency. Therefore, efficient gene carrier systems are needed for the successful delivery of therapeutic genes [48, 103]. These carriers must protect gene molecules from enzymatic degradation, allow internalization by the cells and release the genes in the therapeutic target, *i.e.* the cytoplasm for siRNA or the nucleus for DNA. Glucosamine unit of chitosan is protonated in slightly acidic conditions, which favors an efficient complexation of nucleic acids to chitosan nanoparticles by electrostatic interaction. Hence, two major strategies for the encapsulation of nucleic acids have been reported: the simple complexation by mixing chitosan and the nucleic acids [104] and the ionic gelation usually involving TPP as crosslinking agent [105]. The later allows physical entrapment of the nucleic acids into the chitosan nanoparticle, in addition to electrostatic interactions [53, 106]. Structural characteristics of chitosan such as molecular weight and deacetylation degree strongly influence its ability to form complexes with nucleic acids [106]. Noteworthy, chitosan still presents several limitations for the application of gene delivery. Poor water solubility of chitosan at physiological pH is critical and the pH-dependence of chitosan properties also influences nucleic acid binding capacity. Moreover, chitosan presents poor transfection efficiency, compared to other cationic polymers. Strategies to improve the efficacy of chitosan-based gene delivery systems by tailoring of chitosan's structure have emerged [48, 106]. Carriers prepared from chitosan derivatives, namely chitosan grafted to other polymers such as PEI and PEG have been developed for enhancing transfection efficiency and to increase solubility at neutral pH [106]. Recently Liao and co-workers [107] have shown that the incorporation of negatively charged poly(γ-glutamic acid) (γPGA) into chitosan complexes enhances the cellular uptake of these complexes, *via* a γPGA-specific receptor- mediated pathway, ultimately increasing their transfection efficiency markedly. The feasibility of using this ternary chitosan/γPGA-based system for DNA or siRNA transfer was demonstrated.

Stimuli Responsive Drug Delivery Systems

A challenge in developing more effective treatments is ensuring site-specific targeting of drug carriers and controlled drug release in order to prolong the permanence of therapeutics on the target, while minimizing the undesirable side effects of drugs. Furthermore, personalized medicine treatments could benefit greatly from the development of a discontinuous (triggered) drug release in response to a specific stimulus, tailored according to the specific needs of the patient. Owing to impressive progress in materials science and pharmaceutics a broad range of chitosan-based carriers with stimuli responsive properties has been designed. These carriers exhibit a sharp transition in physicochemical properties upon a small change in the surrounding environment and encapsulated pharmaceuticals may be released as result of such trigger. Stimuli may be either endogenous, such as changes in pH, ionic strength and enzyme concentration or redox potential gradients, or exogenous, such as variations in light, temperature, magnetic field gradients, ultrasounds or electric pulses [15, 108].

Chitosan has pH-sensitive properties owing to hydroxyl and amine groups present in its backbone. Aiming to obtain chitosan carriers with improved pH-tunable responsiveness, chitosan is very often chemically modified to incorporate ionizable carboxyl groups or synthetic polymers (*e.g.* poly(acrylamide) (PAAm), poly(acrylic acid) (PAA) or poly(methacrylic acid) (PMAA)) that swell or collapse upon pH variations [109]. The choice of using pH-sensitivity is mainly due to the fact that, there are pH differences between many tissues and cellular compartments of the human body. For example, pH responsiveness is a useful approach to design carriers for oral administration owing to the broad range in pH values throughout the gastrointestinal tract, ranging from pH 2 in the stomach to pH 7 in the colon [67]. Extracellular pH is lower in many tumors than in the corresponding normal tissue and the difference is attributed to the higher rate of aerobic and anaerobic glycolysis in the cancer cells. This gradient has been used to design pH-responsive tumor-targeted drug release systems based on pH-sensitive polymers [110, 111]. For instance, doxorubicin-encapsulated pH responsive Pluronic F127 nanoparticles surface-decorated with chitosan successfully reduced the size of tumors in an orthotopic human breast tumor model [112]. The system was engineered to release the doxorubicin when the nanoparticles are localized in the acidic tumor microenvironment. Chitosan provided enhanced capability of the nanocarrier in binding to CD44 receptors overexpressed on the cancer stem-like cells.

Considering the difference of redox potential existing between the intracellular and extracellular environment, which are reductive and oxidative respectively due to difference in glutathione concentration, redox-sensitive nanocarriers are a promising approach for intracellular delivery [108]. Stephen and co-workers developed a redox-responsive chitosan-based carrier for intracellular delivery of O^6-benzylguanine for the treatment of brain tumor [113]. Redox sensitiveness was owned to surface disulfide bridges formed by the oxidation of sulfhydryl groups introduced by chemical modification of chitosan. Controlled, localized release of O^6-benzylguanine was achieved under reductive intracellular conditions due to reduction of disulfide bonds by glutathione.

Thermal-sensitive hydrogels are one of the most predominantly studied classes of stimuli-responsive polymers for drug delivery applications. Thermosensitive chitosan based polymers are obtained by blending or co-polymerization of chitosan and derivatives with Thermosensitive polymers, such as N-substituted polyacryalmides, polyvinyl alcohol and methylcellulose. The resulting carriers undergo changes in the physico-chemical properties in response to changes in the surrounding temperature, usually in the form of hyperthermic stimuli at the target tissue. Poly(N-isopropylacrylamide) (PNIPAAM) is the most common monomer used to impart thermo-responsiveness to drug carriers, since it is non-toxic and its lower critical solution temperature (LCST) is close to physiological temperature and easily tunable. Recently, Wang and co-workers have reported a responsive chitosan-based nanogel built with chitosan grafted to PNIPAAM and crosslinked with N-methylenebisacrylamide with improved release of the drug 10-hydroxycamptothecine at 37°C [114]. The resulting carrier was sensitive to both temperature and pH surrounding changes. Responsive carriers can be engineered to react to temperature decrease. For example, Zhang *et al* have reported the development of thermal-sensitive Pluronic F127-chitosan core/shell nanocapsules with sharp volume changes and increased shell permeability at temperatures below physiological temperature and negligible cytotoxicity [115]. These carriers can be potentially used for drug delivery combined with cryotherapy.

Near-infrared (NIR) light is harmless and deeply penetrates in the living tissues, hence being an attractive stimulus source for triggered drug release. NIR light sensitive chitosan-based carriers were obtained owing to the introduction of carbon nano-structures with strong absorption in NIR range, such as carbon dots [116] and reduced graphene oxide [117]. These carbon materials generate heat very effectively upon NIR irradiation and acted as light activated nanoheaters to induce network changes of thermosensitive chitosan-derived nanogels, and trigger the release of doxorubicin molecules entrapped within the polymer carrier. Ultrasound waves are also appealing as a stimulus for triggered-release. Besides being non-invasive, the tissue penetration depth is easily regulated by tuning frequency, duty cycles and time of exposure [15]. Echogenic glycol chitosan-based nanocapsules have been developed for cancer-targeted ultrasound-triggered delivery of doxorubicin and docetaxel drugs [95]. Ultrasound waves could trigger the release of the drugs from chitosan capsules through thermal and mechanical effects generated by cavitation phenomena.

Magnetic-sensitive carriers are typically obtained by incorporating iron oxide nanoparticles. Magnetite (Fe_3O_4) and maghemite (γ-Fe_2O_3) are the most extensively investigated magnetic nanoparticles for drug delivery applications due to their particular magnetic properties (e.g. superparamagnetism) and no severe toxicity. Under a permanent magnetic field, magnetic carriers can be guided to the biological target to provide improved drug accumulation. Furthermore, superparamagnetic iron oxide nanoparticles are able to generate heat upon the application of an alternating magnetic field (AMF) that can be used to trigger drug release from a thermosensitive matrix. Chitosan-based ferrogels containing magnetic iron oxide nanoparticles have shown improved heating response to external time-varying magnetic fields, with potential for application as externally controlled carriers [118]. More recently, Mahdavinia and co-workers have observed that the amount of diclofenac released from carboxymethyl chitosan/carrageenan magnetic beads was increase by applying an external AMF [119].

Electrical-responsive chitosan based materials have been investigated mostly for use in microsensor and actuator applications and studies envisaging their application in drug delivery are still scarce. Zhao *et al.* have demonstrated that the controlled ibuprofen release from chitosan hydrogel films with embedded mesoporous silica nanoparticles coating titanium surfaces can be adjusted by electrical potentials [120].

Multifunctional Drug Delivery Systems

The limitations of standard diagnostic and therapeutic strategies have prompted the development of multifunctional carriers. The need for methods of early-diagnosis and more effective treatments is particularly relevant for diseases such as cancer, which is still a significant cause of mortality worldwide. In this context emerging approaches have been proposed, that take advantage of the synergistic therapeutic effect of combining distinct therapies in a single system. A different approach, often designed as theranostics, involves the integration of therapy and non-invasive diagnostic imaging into a single platform. By combining imaging with therapy, theranostics can be valuable to monitor drug delivery, drug release and drug efficacy. As described above, chitosan nanoparticles have significant advantages to carry drugs and other cargo in the body. Recent nanotechnology advances provides the opportunity to develop multifunctional chitosan-based carriers [121]. The inclusion of inorganic nanoparticles confers new functionalities to the resulting chitosan composite nanoparticles. This strategy allows to extend application of chitosan carriers to other biomedical contexts such as medical imaging and hyperthermia.

Table **2** provides selected examples of multifunctional chitosan-based nano-carriers reported in the last 5 years with indication of therapeutic and diagnostic functionalities, and corresponding nanomaterial composition. For example, platforms integrating targeted delivery of chemotherapeutic drugs and enhanced magnetic resonance imaging (MRI) of tumors have been reported [113, 122]. The carriers could provide the monitoring of biodistribution and targeting ability of the drug carriers for the tumor by non-invasive MRI owing to the presence of superparamagnetic iron oxide nanoparticles which are a T2 MRI contrast agent. Incorporation of chitosan endowed improved colloidal stability and solubility to the carriers [122] and redox-responsive drug release behavior [113]. In another study *in vivo* body distribution and tumor-targeting ability of glycol chitosan nanoparticles could be monitored by optical imaging techniques owing to particle surface labeling with Cy5.5, a near-infrared fluorescence dye [96].

Table 2. Examples of multifunctional chitosan-based nanocarriers reported in the last 5 years with indication of therapeutic and diagnostic functionalities, and corresponding nanomaterial composition.

Description	Therapeutics	Diagnostic (Imaging)	Reference
Chitosan functionalized magnetic graphene	Drug delivery Gene Therapy	Magnetic Resonance Imaging (MRI)	[122]
Magnetic iron oxide coated with chitosan	Drug Delivery	Magnetic Resonance Imaging (MRI)	[113]
Glycol chitosan NPs labeled with Cy5.5 NIR dye	Drug Delivery*	Optical Imaging	[96]
Glycol-chitosan-coated gold nanoparticles (GC-AuNPs) conjugated to fibrin-targeting peptides	Drug Delivery*	Computed tomography	[97]
Au nanoclusters/chitosan NP with suicide gene	Gene Therapy	Optical Imaging	[123]
Photosensitizer, protoporphyrin IX (PpIX), conjugated glycol chitosan (GC) nanoparticles (PpIXeGCeNPs)	Photodynamic Therapy (PDT)	Photodynamic imaging (PDI)	[98]
Echogenic glycol chitosan nanoparticles	Ultrasound Triggered Drug Delivery (Doxorubicin and Docetaxel)	Ultrasound Imaging	[95]
Fe_3O_4/chitosan core/shell nanoparticles	Drug Delivery Magnetic Hyperthermia	--	[124]
Au-nanorods/chitosan composite nanoparticles	Chemotherapy NIR Photo-thermal Therapy	--	[125]

(Table 2) contd.....

Description	Therapeutics	Diagnostic (Imaging)	Reference
Au-nanorods/chitosan composite nanoparticles loaded with the photosensitizer Indocyanine green	NIR Photo-thermal Therapy Photodynamic Therapy (PDT)	--	[126]
PEG-Chitosan@Carbon Dots Hybrid Nanogels	NIR triggered drug release pH triggered drug release	Two-photon fluorescent imaging (TPF)	[116]

*potential application (not tested)

In alternative to conventional dyes, Au nanoclusters [123] and carbon dots [116] have been also used to impart fluorescent properties to chitosan carriers. The fluorescent property of the carrier helped in the evaluation of the delivery of doxorubicin [116] and suicide gene [123] to cancer cells, by performing fluorescence microscopy. Aiming to overcame the intrinsic limitations of each imaging technique, researchers have developed chitosan carriers merging dual-modality imaging capabilities, MRI and fluorescence imaging, using glycol chitosan, Cy5.5 dye, and superparamagnetic iron oxide nanoparticles [127]. The resulting nanoparticles combine the high sensitivity of near infrared fluorescence imaging with high resolution and deep penetration capability to visualize soft tissues of MRI modality, with potential for cancer detection in earlier stages.

X-ray computerized tomography (CT) is an efficient and widespread image diagnostic technique. Owing to their remarkable properties, including high X-ray absorption coefficient, low cytotoxicity and tailored surface chemistry, gold nanoparticles have gained recent attention as an alternative to conventional iodine-based X-ray contrast agents for CT [128]. Having these properties in mind, fibrin-targeted glycol-chitosan gold nanoparticles were developed for detection and quantification of cerebral thrombi using CT imaging and monitor the thrombolytic effect of tissue plasminogen activator treatment [97]. Since fibrin is a main component of thrombus, targeting capability was achieved by conjugation of fibrin-targeting peptides to surface amine groups of chitosan.

Min and co-workers have recently reported the development of echogenic glycol chitosan-based nanoparticles that concurrently execute cancer-targeted ultrasound (US) imaging, which is a technique that offers excellent spatial and temporal resolution, and US-triggered drug delivery [95]. For the preparation of these carriers, a bioinert US gas precursor (perfluoropentane-PFP) was encapsulated into the particles using oil in water emulsions. The resulting particles comprised a PFP/hydrophobic drug (doxorubicin or docetaxel) inner core coated with a hydrophilic chitosan-based outer shell. The enhanced cellular uptake of these carriers by tumor cells was ascribed to an effect of protonation of amine groups of chitosan at the tumoral pH that is slightly acidic.

Glycol chitosan nanoparticles were also conjugated with a photosensitizer, protoporphyrin IX (PpIX), to develop a theranostic carrier for photodynamic imaging and therapy, simultaneously [98]. Photodynamic therapy is becoming more important in cancer therapy and consists in the generation of cytotoxic singlet oxygen by irradiation of a photosensitizer with light of an appropriate wavelength, to kill tumor cells through apoptosis or necrosis. Due to the fluorescence properties of the PpIX photosensitizer, the carrier provided a signal useful for diagnosis *via* photodynamic imaging. The hydrophobic PpIX photosensitizer was covalently linked to the glycol chitosan polymer. Hence, in aqueous medium, stable nanoparticles comprising a dense PpIX core coated with glycol chitosan were formed *via* self-assembly. The resulting particles displayed an on/off behavior, which is a valuable property to reduce unintended cytotoxicity, with therapeutic activity being restored after cellular uptake and irradiation.

Besides theranostic platforms, systems combining multi-therapies in one single carrier have been also proposed. For example, the incorporation of magnetic iron oxide nanoparticles (*e.g.* magnetite-Fe_3O_4) yields magnetically responsive nanocarriers that, besides appropriate drug delivery capabilities owing to chitosan shell, are able to generate heat upon an alternating magnetic gradient and can therefore further induce the death of cancer cells by magnetic hyperthermia [124]. An alternative therapy for locally producing cytotoxic heat for solid tumors treatment is photothermal therapy, that uses gold or carbon based photothermal agents to convert NIR light into heat very efficiently. In this context chitosan/Au nanorods (NRs) composite nanocarriers loaded with cisplatin were successfully prepared for combining near-IR photothermal therapy (PTT) and chemotherapy [129]. Due to a synergistic effect of chemo- and photothermal therapy, these systems provided an improved therapeutic efficacy in the treatment of H22 tumor tissue in mice, when compared with chemotherapy or photothermal treatment alone. Chitosan/Au NRs composite particles were also loaded with indocyanine green (ICG), a photosensitizer approved by FDA, to develop carriers for NIR-light triggered PTT/PDT combined therapy [126]. A sufficient ICG and AuNR accumulation at the tumor site was achieved though EPR effect of chitosan nanospheres. NIR-light triggered both the generation of local heat and reactive oxygen species by Au NRs and ICG, respectively. *In vitro* cytotoxicity and *in vivo* antitumor efficiency examinations demonstrated that these systems had significantly superior antitumor effect than PDT or PTT alone.

HYDROGELS AS DRUG CARRIERS

Hydrogels are three-dimensional and hydrophilic polymer networks that have the ability to swell in aqueous solutions and retain large amounts of fluids in the swollen state [130]. Chitosan-based hydrogels have been extensively developed in the past decades for drug delivery and tissue engineering applications and have been the subject of a few comprehensive reviews [5, 9, 67, 131 - 133].

Chitosan is easily transformed into hydrogel due to its ability to dissolve in dilute acid aqueous solutions. Chitosan hydrogels can be obtained by coagulation in alkaline media. But for most applications, chitosan must be crosslinked in order to achieve better resistance against acids, alkalis, etc. Also, physical and chemical cross-linking allow the tuning of specific features and properties that lack to pure chitosan formulations. In physical gels, polymer chains are held together by molecular entanglements, non-covalent bonds, such as Van der Waals interactions, ionic interactions, hydrogen bonding, hydrophobic interactions, traces of crystallinity and multiple helices. In chemical gels, polymer chains are connected by covalent bonds. These covalent links are developed by the addition of crosslinkers such as gluteraldehyde, formaldehyde, genipen, N,N-methylenebis(acrylamide) or by ionizing radiations. Most of the cross-linkers have an inherent toxicity so in the chemical gels, a purification step must be performed to remove the unreacted crosslinkers [9, 67, 131, 134]. Hydrogels exhibiting thermosensitivity and pH sensitivity can also be produced with or without physical cross-linkers, that either provide hydrophobic or electrostatic associations [134]. Among physical cross-linkers, sodium citrate and tripolyphosphate provide electrostatic interactions, β-glycerophosphate (β-GP) provide hydrophobic ones.

Many approaches have been developed to tune the conditions of drug release of chitosan-based hydrogels. This can be done by combining chitosan with other natural compounds or synthetic polymers. Hence, novel formulations are being developed that exhibit advantages of both components. Furthermore, the combination can occur *via* simple physical mixing, *via* the formation of polyelectrolytes, *via* graft-copolymerization with pH sensitive or thermosensitive polymers and *via* the formation of interpenetrating polymeric networks. We report in this review the most recent works on chitosan-based gels that have been specifically investigated to meet the specific requirements of ocular, nasal, oral, and transdermal drug delivery formulations.

Ocular Drug Delivery

Current options for ocular conditions often present issues related to poor drug availability due to the protective mechanisms of the eye. Common therapeutics are in the form of eye drops, but the natural tear drainage and blinking action of the eye result in a 10-fold reduction in the drug concentration within 4-20 min. However, the daily instillation required for efficacy often lead to undesirable side-effects that may cause treatment adherence failure and persistence of ocular conditions [135, 136]. The use of *in situ* forming gels is a promising strategy to surpass this problem. Ophthalmic *in situ* gels are applied to the eye as liquids, and undergo gelation upon exposure to physiological conditions, causing an increase in the ocular residence time, *via* enhanced viscosity, bioavailability and mucoadhesive properties [137, 138].

Chitosan was proven to be well tolerated by the corneal surface after topical administration of chitosan solutions regardless the deacetylation degree (DD) (between 53 and 87%), concentration and molecular weight of chitosan in the formulations [139]. Also, it was pointed to enable an increased precorneal drug residence times by increasing solution viscosity due to its pseudoplastic and viscoelastic properties that can be varied by tuning the degree of deacetylation [140] and by its mucoadhesive properties, characterized by the ability to form ionic interaction between its positively charged amino groups and the negatively charged sialic acid residues of mucin [141, 142]. Comparing to commercial collyrium (Tobrex®), 3-fold increase of the corneal residence time was achieved in chitosan based formulations [139].

Hybrid polymer networks have been the subject of some research for ophthalmic drug delivery. Acrylic acid moieties were immobilized onto chitosan chains through amidic bonds to allow the modified chitosan (AA chitosan) to participate in radical polymerization with water soluble comonomers without the need of cross-linkers. Hybrid polymer networks were then prepared by copolymerization of AA-chitosan with thermosensitive poly(N-isopropylacrylamide) (NIPAm) and AA-chitosan with and poly(2-hydroxyethyl methacrylate) (HEMA). *In vitro* and *in vivo* studies involving broad-spectrum antibacterials (chloramphenicol, norfloxacin) and anticholinergics (pilocarpine, atropine) demonstrated that HEMA containing hybrid polymer networks appeared to be the most promising [143].

Chitosan (CS) has the ability to form thermosensitive hydrogels when crosslinked with β-glycerophosphate (β-GP). β-GP neutralizes the positive charge of chitosan, weakens the charge repulsion among CS chains making possible the formation of hydrogen bonds, that provides to chitosan aqueous solution a thermally sensitive neutral character (within the pH range of 6.8-7.2) [144, 145]. Gelation rate is dependent on temperature and chitosan deacetylation degree. These hydrogels have shown to sustain the *in vitro* release of many compounds depending on their molecular weight and the presence of lysozyme in the release media over a period of several hours to a month [55, 146, 147]. Also, they have been successfully used to deliver ellagic acid to treat brain cancer [148], and sustained release of paclitaxel at tumor resection sites, namely one intratumoral injection of the paclitaxel containing CS-β-GP hydrogel was as efficacious as four intravenous injections of Taxolw in inhibiting the growth of EMT-6 cancer cells in mice [55].

As pure chitosan only dissolves in acidic environment [149], chitosan has been substituted in some formulations by water-soluble derivatives. Research on quaternization of chitosan to obtain water-soluble chitosan derivatives has shown for instance that N-[(2-hydroxy-3-trimethylammonium) propyl] chitosan chloride (or HTCC) shows better hygroscopic property, better antibacterial [150] and antioxidant activity [151] comparing to chitosan. Getting inspired by the thermosensitive CS/ β-GP formulation and the better properties of HTCC, Wu *et al.* modified chitosan by a quaternization reaction and combined it with β-GP to produce a thermosensitive but also pH sensitive new hydrogel (HTTC/ β-GP). The resulting formulation showed to be transparent when forming the gel at 37°C, oppositely to CS/ β-GP formulation, making it an attractive candidate for ocular drug formulations. Moreover, doxorubicin hydrochloride (DX), incorporated as model drug, was released quickly in acidic condition and quite slowly with low initial release in basic condition [152].

Gelatin molecules have been recently added to the CS/ β-GP formulation to form chitosan/gelatin/β-GP systems. When introducing 1% of gelatin, the gelation time was shortened from about 10 min to 1 min at 31.1-33.8°C and the gel strength improved at physiological temperature [153, 154]. This particular chitosan/ gelatin/β-GP system was loaded with 200 μM of ferulic acid (FA), a natural phenolic antioxidant, to treat oxidative damage to cornea. The FA containing hydrogels showed potential to decrease the inflammation-level and apoptosis of the corneal epithelial cells under oxidative stress and to promote the corneal wound healing in a rabbit corneal alkali burn model [155]. The same group has also demonstrated its potential to deliver latanoprost *via* a subconjunctival route for the treatment of glaucoma. Prolonged sustained release was observed *in vitro* for at least 28 days, *i.e.* 70% of latanoprost was released after that period with no initial burst effect and no cytotoxic effect was observed on human corneal epithelial cells. *In vivo* as well, a steady drug concentration was observed in the aqueous humor and the formulation showed biocompatibility towards human corneal cells, no haemolytic effect and no signs of inflammation at the end of the study (60 days). A rabbit model of glaucoma was established by intravitreal injection of triamcinolone acetonide. After a single subconjunctival injection, latanoprost-loaded C/G/GP hydrogel showed to successfully decrease the ocular hypertension [154].

Another promising latanoprost loaded formulation for glaucoma treatment relied on a gel composed of assembled colloids. Basically, nanocapsules of amphiphilically modified chitosan (carboxymethylhexanoyl chitosan (CHC)) are mixed with β-GP and form thermo-gelling solutions which, upon increased temperature, aggregate into a continuous colloidal network. The gels, forming in 4 min at 37 °C and in 32 min at 4 °C, are composed of a polymer-rich CHC nanocapsule network phase and an aqueous inter-nanocapsule phase, both being continuous throughout the colloidal gel. Interestingly, these gels present also a shear-reversible gelation which means they can actually be injected from the gel state. The therapeutic effect of a single administration of this latanoprost loaded CHC/β-GP gel was proven in a rabbit glaucoma model wherein intraocular pressure was lowered for up to 40 days [156]. These developed colloidal chitosan based gels demonstrated further excellent biocompatibility [157] and low toxicity toward human retinal pigmented epithelium cells [158].

Chitosan has also been combined with other polymers to improve hydrogel's features such as other polysaccharides [159] or synthetic thermosensitive polymers [160 - 162]. Chitosan, in combination with Poloxamers, have been developed for the delivery of ciprofloxacin for treating ocular infections caused by Pseudomona aeruginosa [160] and fluconazole against fungal keratitis [161, 162]. Poloxamers are thermosensitive amphiphilic block copolymers that comprise various blocks of hydrophilic poly(ethylene oxide) (PEO) and hydrophobic poly(propylene oxide) (PPO) that are organized in a tri-block polymer structure (PEO)/(PPO)/(PEO). In the aqueous solution, various segments aggregate into micelles. At high concentrations and temperatures, the micelles self-organize and form a gel [163, 164]. Gratieri *et al.* have optimized a particular formulation of chitosan (DD of 75-85%) and Poloxamer 407, wherein chitosan was found to improve the mechanical strength and texture properties of poloxamer formulations and also confer mucoadhesive properties in a concentration-dependent manner. After a 10-min instillation of the poloxamer/chitosan 16:1 formulation in human eyes, 50-60% of the gel was still in contact with the cornea surface, which represents a fourfold increased retention in comparison with a conventional solution [161]. This particular hydrogel formulation further showed its potential to enhance ocular bioavailability in the treatment of fungal keratitis. *In vitro* release experiments showed that the combination of chitosan and Poloxamer showed a superior capacity for sustaining fluconazole release. *Ex vivo* permeation studies across porcine cornea confirmed that the chitosan/Poloxamer have a permeation-enhancing effect that is independent of chitosan concentration in the range from 0.5 to 1.5% w/w, *i.e.* about 3.5-fold greater total amount of fluconazole permeated when compared to chitosan solutions alone [162].

Blending chitosan with oxidized alginate to form composite polysaccharide hydrogels has shown that gelation time (from 10 s to 5 min) and degradation rate of these hydrogels could be tuned by varying the oxidized alginate content. *In vitro* release study showed that the loaded avastin had an initial burst release at early stage (within 4 h) followed by a sustained release manner in the period of 3 days. With the increase of oxidized alginate concentration, the release rate of avastin from hydrogels decreased [159].

In a different approach, chitosan hydrogel was used as mucoadhesive matrix for levofloxacin encapsulated PLGA nanoparticles. As PLGA is an anionic polymer, hence non-mucoadhesive, drug loaded PLGA nanoparticles are incorporated in chitosan in order to enhance the nanoparticles retention in the eye. These PLGA nanoparticles did not affect the gelation pH or the viscosity of the final formulation. Preliminar work indicated the levofloxacin-loaded PLGA nanoparticles/chitosan hydrogel remained at the rabbit corneal surface for longer time than the chitosan hydrogel alone, the suspension of nanoparticles and a marketed formulation [165].

Besides the possibility of tuning the properties of chitosan formulations, cross-linkers need to be fully removed and preferentially non-cytotoxic. For instance, among the available chemical cross-linkers for chitosan [134], genipin (GP), that is obtained from the fruits of *Gardenia jasminoides Ellis* [166], provides chitosan membranes with increased tensile strength but decreased strain-at-fracture, swelling ratio and degradation rate [76]. Genipin has been characterized as about 10000 times less cytotoxic than glutaraldehyde [76, 167]. So the biocompatibility of GP cross-linked chitosan formulations was tested in view of ocular applications. Genipin was shown to promote a higher compatibility than other cross-linkers towards retinal pigment epithelial cells [168]. *In vivo*, rabbits bearing GP cross-linked chitosan exhibited no signs of ocular inflammation and improved preservation of corneal endothelial cell density after 24 weeks compared to glutaraldehyde treated or non-treated chitosan [169].

It is worthwhile to note that although the rabbit has been used for toxicity assessment due to some anatomical similarities to the human eye such as size and aqueous humor volume [170], it also exhibits significant differences, *i.e.* corneal thickness around 0.35 mm against 0.5 to 0.7 mm in the human eye [171], slower blink reflex and a nictitating membrane (absent in humans) [170]. So, careful investigation should be undertaken to understand the cellular fate when exposing the human tissues to chitosan based drug delivery systems. Although rabbits are the traditional model for ophthalmic formulations evaluation, human volunteers were preferred for the study of the ophthalmic retention time of a new chitosan/Poloxamer (16/1) in-situ forming hydrogel. After instillation, no adverse or irritant effects such as blinking, conjunctival redness or discharge were observed in the short or long term effect of using this formulation [161].

Nasal Delivery

Nasal route offers some advantages that include a low dose requirement due to the large surface area available for drug absorption owing to the presence of microvilli on the epithelial surface, a highly vascularized subepithelial layer, the avoidance of hepatic first pass metabolism, a rapid achievement of therapeutic blood level and a good patient acceptability [172]. Similarly to what occurs with the ocular route, it is expected that the chitosan based formulations transform *in situ* into a hydrogel at body temperature after being administered into nasal cavity, decreasing the nasal mucociliary clearance and a controlled drug release.

Chitosan solutions for insulin intranasal formulations have shown a good nasal tolerance and an increased permeability of insulin across the nasal mucosa both *in vitro* and *in vivo* in rats [173]. Based on this evidence, many formulations based on chitosan hydrogel have been investigated as a nasal route to deliver insulin. A formulation based on quaternized chitosan HTCC based hydrogel was prepared for the delivery of insulin. HTCC was mixed to poly(ethylene glycol) (PEG) with a small amount of α-β-GP. The obtained formulation (HTCC-PEG-GP) underwent sol-gel transition around 37 °C in several minutes. In a rat model, no apparent cytoxicity was found after application, an enhanced absorption of insulin was observed in nasal cavity and decreased glucose levels (40-50% of initial blood glucose concentration) were measured in blood for at least 4-5h after administration [174].

To avoid the use of glycerophosphate, a phosphate free formulation for insulin delivery was proposed by Agrawal *et al.* using blends of chitosan and poly vinyl alcohol (PVA). All the formulations formed a gel at 37 °C within 12 min and release insulin up to 6h suggesting that 1% of chitosan is enough to widen the tight junctions. Comparing to an insulin subcutaneous injection that resulted in maximum reduction within the first hour after administration, the maximum reduction of blood glucose levels was observed for four hours after administrating the formulation. This slow release phenomenon through the CS/PVA gel was suggested to be useful to control the hypoglycaemic shock which may be seen with injectable insulin [175].

An alternative quaternized chitosan derivative obtained by the reductive methylation of chitosan, the N-trimethyl chitosan chloride (TMC), has been recognized as the HTCC to promote enhanced absorption *via* the opening of the tight junctions located between epithelial cells over a wide pH range especially in neutral environments where chitosan is ineffective [176]. In rats, its degree of quaternization plays an important role in the absorption enhancement properties across the nasal epithelia in a neutral environment. For instance, maximum absorption of mannitol was observed at a degree of quaternization of 48% [177]. Also, it presents enhanced mucoadhesive properties and improved solubility profiles [176]. TMC, as HTTC, was used to optimize coformulations with PEG and GP, wherein the quaternization degree and molecular weight of TMC was varied. Rheological evaluations demonstrated that TMC with a low degree of quaternization and a medium average molecular weight provide the TMC/PEG/GP hydrogels with good water-holding capacity, good affinity for mucosal surfaces, a sol-gel transition at 32.5 °C within 7 min (temperature range relevant for the nasal environment (between 32 and 35 °C) [178], strong mucoadhesive potential, and a rheological synergy with mucus [179]. This TMC/PEG/GP formulation was further developed with a fixed degree of quaternization (33%) and tested *in vivo* in a rat model. It showed to reduce blood glucose level over ca. 24 hours, highlighting a potential to be used as once-a-day dosage form for the delivery of insulin through nasal route [65].

Another formulation has demonstrated a highly prolonged hyperglycaemic effect with a sustained insulin release of up to 20h in diabetic rats. This hydrogel was prepared by crosslinking chitosan (CS) with glutaraldehyde (GA), interpenetrate Poloxamer gels and glycine (Gly). Insulin release was sustained for about six time longer than when release from Poloxamer gels. Regarding the nasal cavity of the rats, no morphological difference was observed beyond two weeks after the tests [180].

More recently, the thermosensitive system HTCC/β-GP was also investigated as an intranasal vaccine delivery for the adenovirus-based Ebola antigen. HTCC/β-GP was prepared with different degrees of quaternization (41%, 59%, 79.5%, and 99%). Intranasal immunization is a very effective route as the immune response following intranasal administration can provide protection at the administration site and at various effector sites as part of the common mucosal immune system [181]. Moderate quarternized HTCC (60% and 79.5%) hydrogel/antigen formulations induced highest immune response in serum, as well as in the mucosal response in lung wash, which was attributed to the prolonged antigen residence time due to the thermal-sensitivity of this hydrogel. Moreover, this formulation exhibited a low toxicity to nasal tissue and epithelial cells even by frequently intranasal dosing of hydrogel [182].

Oral Delivery

Oral drug delivery route is the most comfortable route of self-administering drugs. In chitosan hydrogels based oral drug delivery, formulations have been developed for buccal delivery, gastrointestinal delivery and colon-specific drug delivery. Since pH variations through the gastrointestinal tract vary from 1 to 7.5 [183]. pH sensitivity is a pivotal characteristic when preparing orally administrable hydrogels.

Colon-specific Delivery

Clinically, the largest niche within oral delivery is the targeting of the colon [184]. Colon has a unique metabolic activity wherein drug absorption is highly enhanced comparing to the small intestine. For instance, protein drugs can be absorbed better due to the reduced proteolytic activity in the colon. Moreover, some drugs have serious deleterious effects on the gastric mucosa [184]. There are a wide range of chitosan based formulations that have been developed to exhibit specific colon delivery. Namely, chitosan based polyelectrolyte complex based on chitosan cross-linked with γ-poly(glutamic acid) (γ-PGA) form hydrogels with increased swelling at neutral pH [185]. Polyelectrolyte complexes (PECs) are the association formed between oppositely charged compounds. These are formed due to electrostatic interaction between oppositely charged polyions avoiding the use of chemical cross linking agents and thereby reducing the possible toxicity and other undesirable effects of these reagents [186].

Carboxymethyl chitosan (CMC) is one of the water soluble derivative of chitosan, characterized by carboxymethyl substituent on some or both the amino and primary hydroxyl sites of the glucosamine units of the chitosan structure. By introducing carboxymethyl groups, the solubility of chitosan at neutral and alkaline pH is drastically increased. It has also demonstrated pH sensitivity in aqueous solutions because of its COOH and NH_2 groups [187]. Glutaraldehyde cross-linked CMC hydrogels have shown that with increasing degree of deacetylation (DD) and substitution (DS), the hydrogel changed from polyampholyte into polycations or polyanions, respectively. Osmotic forces and intermolecular interactions among CMC chains regulate the gel swelling behavior. Bovine serum albumin-loaded gels exhibited a slow release in the gastric simulated fluid and a rapid one in the intestinal one [188]. A similar formulation based on glutaraldehyde cross-linked CMC for colon targeted ornidazole delivery showed *in vitro* a pH responsive swelling characteristic with very less swelling at pH 1.2 and quick swelling at pH 6.8 followed by linear swelling at pH 7.4 [189].

Because the ionization degree of carboxylic groups in CMC decreases in acidic conditions, methoxy poly (ethylene glycol) (mPEG) has been grafted to CMC to increase its aqueous solubility over a wide pH range [190]. mPEG grafted CMC (mPEG-g-CMC) was combined with polyanionic alginate to construct the matrix of a pH sensitive interpenetrating network. Alginate is pH-sensitive, non-toxic and biodegradable at high pH value when administered orally. However, alginate can rapidly dissolve at intestinal pH, leading to burst drug release. To overcome this limitation, mPEG-g-CMC was co-gelled with alginate. The effect of grafting mPEG to CMC against just simply mixing mPEG and CMC in the properties of the resultant hydrogels was investigated. Both formulations maintained good pH sensitivity. The loading capacity of bovine serum albumin, as a protein drug model, in the mPEG-g-CMC/alginate hydrogel was enhanced in comparison with the hydrogel prepared by physically mixing mPEG. The burst release of the protein was slightly decreased at pH 1.2, while the release at pH 7.4 was improved, suggesting that the mPEG-g-CMC/alginate pH-sensitive hydrogel are suitable for site-specific protein drug delivery in the intestine [31].

Islam *et al.* developed a pH sensitive blend based on chitosan and PVA. The hydrogels swelling properties against different media could be tuned by varying the amount of PVA in CS/PVA blend, wherein an increasing amount of PVA leads to a decreased swelling degree and an increased thermal stability. The hydrogels were tested for dexamethasone release in simulated gastric fluid (SGF, pH 1.2) and in simulated intestinal fluid (SIF pH 6.8). A release about 9.37% over a period of 2 h was observed in SGF and its transfer to SIF showed consistent release of remaining drug up to 7 h. This release profile of dexamethasone is of interest as it fulfills the US Pharmacopeia requirement (USP XXIV) [191]. Another formulation based on CMC/PVA blend was developed with added alginate microspheres to control and sustain the delivery of BSA, wherein the incorporation of alginate microspheres improved the swelling ratio and thermal stability of the final hydrogels [192].

With the intent of producing a formulation that would reach the colon as intact as possible for the delivery of diphenhydramine HCl (DPH), a highly water-soluble antihistamine, Bawa *et al.* prepared lyophilized polyelectrolyte matrices constituted by a combination of natural polysaccharide polymers such as pectin and chitosan (CS) and a partially hydrolyzed synthetic thermosensitive polymer (HPAAM). By modifying the concentrations of the CS, pectin, and HPAAm and their combinations, the applicability of these matrices could be tuned for effective site-specific and controlled oral drug delivery. After 2h, no DPH was released from CS-HPAAm complexes, whereas 28.2% and 82.2% was released from pectin-HPAAm and pectin-CS-HPAAm ones, respectively. After 4h, complete DPH release was achieved from pectin-CS-HPAAm matrices in contrast to only 35% from CS-HPAAm matrices. Mechanically, CS-HPAAm matrices were most resilient and pectin-CS-HPAAm were the least [193].

Chitosan has also been graft copolymerized with pH sensitive polymers. These polymers that exhibit pH sensitivity contain pendant acid (*e.g.* carboxylic or sulfonic acids) or basic (*e.g.* ammonium salts) groups that either accept or release protons in response to changes in environmental pH [194]. Within this strategy, chitosan was graft copolymerized to poly(acrylic acid) (PAA). After incubating the resulting hydrogels with cecal and colonic enzymatic media from rats, it was found that largest degree of swelling and larger rate of degradation were found at pH 7.4. The degree of swelling was shown to be predominantly affected by the molar ratio of hydrophilic groups/hydrophobic groups and the degree of cross-linking [195]. Following a similar approach, chitosan (CS) was combined with itaconic acid (IA) and methacrylic acid (MMA). MMA content was varied in these hydrogel formulations and it was found to improve the mechanical properties. The ionic bonds were shown to disintegrate at pH above 6, where chitosan amino groups are not dissociated leading to a high swelling of the hydrogel at pH = 6-8 [196].

A particular kind of polymeric network, the interpenetrating polymer network (IPN), has attracted attention to produce hydrogels due to their water imbibing property. A chitosan based IPN occurs when a cross-linked chitosan network is allowed to swell in an aqueous solution of polymer monomers. These monomers are then polymerized, forming a physically entangled polymer mesh called an interpenetrating network. There are also semi-IPNs where only one of the polymer networks is cross-linked, while the second polymer remains in its linear state [197]. Chitosan derivatives such as carboxymethyl chitosan (CMC) and N,O-carboxymethyl chitosan, (NOCC) have been studied for the formation of IPN to endow the final hydrogel with biocompatibility and temperature sensitivity [31, 198, 199]. Semi-IPN polyampholyte hydrogels were prepared by using CMC and poly(N-isopropylacrylamide) (PNIPAm) [200]. In another study, CMC was combined with poly(N-isopropylacrylamide-co-methacrylic acid), P(NIPA-co-MAA), to form pH and temperature sensitive IPNs for site-specific riboflavin delivery. The phase transition temperature was adjusted to 37 °C by changing the MAA content. The riboflavin release was only about 6% at pH 1.8 for 2 h (due to the polymer-polymer interaction (H-bond)) and about 13% before the drug-loaded hydrogel reached the position with pH 6.8, where after sustained release was achieved [198].

Another IPN was developed by mixing the water soluble chitosan derivative *N,O*-carboxymethyl chitosan, (NOCC) and alginate blended with genipin. At pH 1.2, the swelling ratio of the genipin-cross-linked NOCC/alginate hydrogel was limited due to formation of hydrogen bonds between NOCC and alginate. At pH 7.4, the carboxylic acid groups became progressively ionized creating electrostatic repulsion and resulting in a more significant swelling. Bovine serum albumin release at pH 1.2 was relatively low (20%), while increased to 80% at pH 7.4 [199].

Das *et al.* presented an interesting new system based on a pH-sensitive chitosan and poly(vinyl alcohol) (PVA) IPN with solid inclusion complexes of drug and cyclodextrins (CDs). CDs are cyclic oligosaccharides with lipophilic inner cavities and hydrophilic outer surfaces that have the ability to form noncovalent complexes with many guest molecules [201, 202]. The addition of the preformed drug-cyclodextrin inclusion complexes in the hydrogel matrix significantly prolonged the delivery in comparison to the free drug. *In vitro* tests performed with naproxen [203] and diclofenac sodium as drug included in the cyclodextrins revealed that the presence of β-cyclodextrin influences the relaxation rate of the polymer chains, leading to slower release of the drug. Negligible drug release was observed in the simulated gastric fluid and a sustained release was measured in the intestinal fluid with the two tested drugs: naproxen [203] and diclofenac sodium [204]. This cyclodextrins containing IPN formulation appears as a valuable option for colon-specific delivery, particularly to deliver drugs that have poor water solubility or cause deleterious effect on the epithelium of the gastric mucosa such as non-steroidal anti-inflammatory drugs.

IPNs have also been prepared as superporous hydrogels for the delivery of insulin [205 - 207]. Superporous hydrogels (SPHs) are porous hydrophilic crosslinked structures with the ability of absorbing aqueous fluids up to a few hundred times their own weight. They are also characterized by a fast swelling, *i.e.* generally reached in a fraction of a minute and a high porosity, *i.e.* final average pore size is about of 200 μm [208]. Insulin delivery by the oral route is still challenging because it is degraded by the acidic stomach environment and its absorption through the gastrointestinal mucosa into the blood stream has been questioned. To achieve a successful colonic delivery *via* the oral route, drug carriers need to be designed to protect the drug from the harsher environments and then be released into the proximal colon [209]. Yin *et al.* developed a pH-sensitive superporous hydrogel of interpenetrating network (SPH IPN) composed of poly(acrylic acid-co-acrylamide)/o-carboxymethyl chitosan (O-CMCS). Due to the cross-linked *O*-CMCS network, *in vitro* mucoadhesive force and mechanical properties, including compression and tensile modulus, of the SPH-IPN were greatly improved when compared with chitosan SPHs. Insulin release from the SPH-IPNs exhibited sensitivity towards pH and ionic strength [205]. Through their abilities to bind Ca^{2+} and to entrap the enzymes, swollen SPH-IPNs could attach mechanically and mucoadhere to the intestinal wall, thus achieving improved retentive properties compared to commonly used mucoadhesive excipient Carbopol® 934. Transport of insulin across rat intestine and colon *ex vivo* was enhanced around 2 to 3-fold after application of the SPH-IPN [206]. The effect of polymer integrity on insulin absorption mechanisms was further investigated by comparing integral SPH-IPN (I-SPH-IPN) and powdered SPH-IPN (P-SPH-IPN). I-SPH-IPN showed marked superiority to P-SPH-IPN in *in vivo* enzymatic inhibition, paracellular permeability of insulin, and intestinal retention (orally delivered I-SPH-IPN was retained for more than 8h while P-SPH-IPN was rapidly eliminated). Moreover, oral administration of insulin-loaded I-SPH-IPN yielded notable insulin absorption and hypoglycemic effect, while P-SPH-IPN was ineffective [207]. The biological safety of these superporous hydrogels was evaluated through an extensive cytotoxicity and genotoxicity testing. SPH-IPN caused minimal damage towards cell viability, lysosomal activity, and metabolic activity following both direct and indirect treatment on RBL-2H3 and Caco-2 cells. Moreover, they did not induce cell apoptosis or DNA breakage. They did not increase micronucleus (MN) incidence in mouse bone marrow [210]. Oral acute and sub-acute toxicity study in mice confirmed the good tolerance of SPH-IPN [207].

Other systems are being investigated for insulin oral delivery. A carboxymethyl starch derivative was produced by cross-linking chitosan with carboxymethyl starch (SCC) in the presence of citric acid. This derivative was further copolymerized with methacrylic acid (MMA) and poly(ethylene glycol) monomethyl ether methacrylate (PEGMA). All the prepared hydrogels demonstrated a protective effect on insulin and maintained its biological activity even after being treated with gastric fluid. Insulin molecules have the tendency to attach to polar groups due to hydrogen bonding. So, carboxylic acids present in the SCC were needed to increase insulin loading to the hydrogel but also for the pH-sensitive character. Increases of chitosan and MMA content increased the bioadhesivity. By regulating the cross-linking percentage of the MAA copolymers, pH-sensitive hydrogels with improved optimal hydrolysis rates were obtained. The hydrolysis of the drug-polymer conjugates were performed at pH 1 and 7.4 at 37 °C. The MAA in the feed monomer mixtures resulted in less swelling in simulated gastric fluid but greater swelling in intestinal fluids [211].

Recently, oil in water microemulsions were introduced into a chitosan hydrogel for the creation of a hydrophobic drug delivery system. The design is based on chitosan crosslinked with genipin with embedded biocompatible non-ionic microemulsion (ME). A prolonged release period of 48 h in water, and of 24 h in phosphate buffer saline (PBS) of pH 7.4 was demonstrated for Nile red and curcumin [212].

For the particular case of delivering the antitumor doxorubicin drug for liver cancer, nano-Hap particles have been added to a sodium alginate/chitosan hydrogels since *in vitro* experiments have shown that nano-HAP have inhibition effects on liver cancer cells. The sodium alginate/chitosan/hydroxyapatite

(SA/CS/HAP) nanocomposite hydrogel were prepared with different amounts, of nano-HAP (from 0.6 up to 5.0%). The gelation percentage increased with the increase of HAP content. *In vitro* studies showed that doxorubicin release was pH dependent, with a release up to 95% at pH 5, but 60% at pH 7.4 [213].

Buccal Delivery

An optimal buccal delivery system should stay in the oral cavity for few hours and release the drug toward the mucosa in a sustained manner. Besides avoiding the hepatic first-pass metabolism, buccal delivery also avoids the acidity and proteolytic activity of the rest of the gastrointestinal tract. It can be used to treat several diseases such as periodontal disease, stomatitis, mucositis, fungal and viral infections, and oral cancers [214]. Chitosan is advantageous for buccal application due to its mucoadhesivity and absorption enhancer properties. Chitosan hydrogel was found to exert a marked permeabilizing effect on buccal mucosa [215]. Chitosan has also an inherent antimicrobial character [216] and has shown to inhibit the adhesion of *Candida albicans* to human buccal cells, with or without the use of antifungal agents such as chlorhexidine gluconate [217].

Oral mucositis is a disease that results from the toxic effects of chemotherapeutic agents and irradiation and has serious consequences. On the other hand, the pineal hormone melatonin inhibits the production of free radicals that mediate the toxicity of chemotherapy. Melatonin-loaded chitosan hydrogels were developed for the prophylaxis and/or treatment of oral mucositis [218]. Nystatin loaded gelatin/chitosan hydrogels also demonstrated to have the palliative effects of an occlusive dressing against oral mucositis, the potential for delivering nystatin concentrations above its minimum inhibition concentration and for extending the retention time in the oral cavity due to bioadhesive properties [219].

Although bioadhesive, chitosan hydrogels may still have limited mucoadhesion in wet conditions. To surpass this and inspired by the excellent wet adhesion of marine mussel adhesive protein [220], a catechol-functionalized chitosan (Cat-CS) hydrogel with two degrees of catechol conjugation (9% and 19%) were developed in a recent work [221]. Interestingly, while chitosan is normally soluble only at acidic pH, catechol functionalization converts it into a highly water soluble polymer at physiological pH [222]. Also, catechol functionalization does not alter the gelation time and mechanical properties of the chitosan hydrogels but significantly enhanced the *in vitro* mucoadhesion to porcine mucosal membrane, wherein Cat-CS maintained contact after 6 h, whereas all of the chitosan hydrogels lost contact after 1.5 h. These new hydrogel systems sustained the release of lidocaine for about 3 h. *In vivo*, lidocaine was detected in the rabbit's serum at concentration about 1 ng/ml only from the cathecol-functionalized chitosan. The authors attributed that observation to the higher mucoadhesivity of Cat-CS comparing to chitosan allowing the drug to be absorbed through the mucosa [221].

Gastrointestinal Delivery

Gastroretentive drug delivery systems can remain in the gastric region for several hours and hence significantly prolong the gastric residence time of drugs. Prolonged gastric retention improves bioavailability, and is specially useful for drugs that are poorly soluble in intestinal medium and readily soluble in acidic medium but also drugs that are intended for local action on the gastroduodenal wall [223]. A semi-IPN hydrogel was developed for antibiotics delivery in acidic conditions by crosslinking chitosan and polyvinyl pyrrolidone (PVP) with glutaraldehyde. *In vitro* release demonstrated that formulation containing chitosan (2% w/v) and PVP (4% w/v) released antibiotics such as repaglinide and clarithromycin after 12h in acidic conditions. Under acidic conditions, these hydrogels exhibited high mucoadhesion and high swelling due to the protonation of a primary amino group on chitosan [224, 225]. Another semi-IPN was formulated by crosslinking chitosan and PVP blend with glutaraldehyde. Hydrogels were prepared as porous freeze-dried and as air-dried. Porous freeze-dried hydrogels exhibited superior pH-dependent swelling properties and drug-release ability over non-porous air-dried hydrogels [226].

Oppositely to the formulations specific for colon, buccal and gastroretentive applications, other formulations show an equal drug release in different pH. This equal stability in different media can be an advantage if the intent is to release the drug at a constant rate regardless pH changes during its passage through gastro-intestinal tract. One example of these formulations is the combination of chitosan (DD of 79.7%) with xanthan to form polyionic complex (Xa-CS) hydrogels to be used as capsules for the delivery of theophylline (THP), one of the most common bronchodilator drugs in chronic obstructive pulmonary disease [227]. Molecular weight of CS, the pH of CS solutions and the initial Xa concentration were shown to influence the ability of forming the polyionic complex network [228]. Various Xa-CS mass ratios (1.7, 2.2 and 2.7) were studied. Complex with a mass ratio of 2.2 demonstrated both stability under acidic conditions and sufficient swelling degree for drug entrapment. THP *in vitro* release profiles at pH = 1.5 and pH = 7.4 did not exhibit significant differences due to the fact that the matrix is equally stable in both dissolution media. *In vivo* administration of THP-loaded capsules in rats showed that the complex increases the THP bioavailability [229].

Transdermal Delivery

Transdermal drug delivery systems possess also benefits such as increasing bioavailability and therapeutic efficacy, bypassing hepatic first-pass metabolism. Chitosan and its derivatives such as N-trimethyl chitosan and mono-carboxylmethyl chitosan have also demonstrate the ability to destroy the tight junctions of the keratin cells resulting in a decreased skin barrier and an improved drug percutaneous permeation property. This property was related to significant changes in the secondary structure of the keratin induced by chitosan and its derivatives and also on cell membrane potential and fluidity [230 - 232]. The particular transdermal permeation enhancement of TMC for testosterone was shown to increase with the increase of quaternization degree [233].

Fang *et al.* have demonstrated the feasibility of using chitosan hydrogels for the delivery of buprenorphine under the application of iontophoresis or electroporation, separately or together. Both iontophoresis and electroporation enhanced the skin permeation of buprenorphine from the chitosan hydrogels [234]. Iontophoresis has also been used to increase skin permeation and drug retention through chitosan hydrogels loaded with doxorubicin (DOX) for the treatment of melanoma [235].

Following a different approach, a transdermal patch was developed by combining a rate controlling chitosan membrane and a chitosan hydrogel as a lidocaine hydrochloride reservoir. Lidocaine hydrochloride release was found slower through chitosan membranes with a higher degree of deacetylation (% DD) and with a larger thickness. The most prolonged release *in vitro* was obtained with a 95% DD chitosan rate controlling membrane. On the forearm of human volunteers, patches with 70% and 95% DD membranes delayed the anesthetic effect [236].

In order to improve the adhesiveness of chitosan to the skin and subsequently the contact time and drug permeation, chitosan has been combined with glycolic acid to produce hydrogels for progesterone delivery. Glycolic acid had a significant positive impact on the adhesive properties of the chitosan-glycolic acid hydrogel. Hydrogels performance was compared to the one of chitosan-EDTA and carrageenan. Skin diffusion tests using dermatomed porcine skin demonstrated that the highest cumulative permeation of progesterone after 48 hours was obtained from chitosan-glycolic acid hydrogels due to their higher adhesiveness [237].

Insulin delivery is also being investigated for the transdermal route by nano-insulin loaded chitosan/PVA blends crosslinked with glutaraldehyde [238] and CS/β-GP hydrogel with biodegradable poly(3-hydroxybutyrate-co-3-hydroxy-hexanoate) nanoparticles (NP) containing insulin phospholipid [239]. The performance of the insulin-containing NP loaded hydrogels (NP-CS/GP) was compared with free-insulin loaded hydrogels (INS-CS/GP). *In vitro* release studies showed that 96.41% of total insulin was released from the INS-CS/GP within 16 days while only 19.11% of total insulin was released from the NP-CS/GP hydrogel within 31 days as the insulin release is dually impeded by both NP and gel matrix. In diabetic rats, the hypoglycemic effect of NP-CS/GP following subcutaneous injection lasted for >5 days, much longer than the effect caused by INS-CS/GP or other long-acting insulin formulations [239].

FILMS AS DRUG CARRIERS

Chitosan is soluble in acid solutions and can easily form films and membranes by casting/solvent evaporation technique. This method is the most widely found in the literature mainly due to the low cost that the system setup incurs at the laboratory scale [14, 18, 240, 241]. In a typical procedure, a chitosan solution is prepared by dissolving the chitosan in dilute acid solutions with pH below 6.0. The solution is filtered and then cast on a mold. After the solvent evaporation, the film is easily separated from the mold surface. To prepare a medicated film the drug is introduced into a polymeric matrix dispersed in the form of inclusions to promote a controlled release by diffusion.

Chitosan membranes/films delivery systems are mostly studied for encapsulation and release of drugs for the application of wound healing and tissue regeneration [5, 30, 242 - 244]. The efficacy of chitosan in promoting wound healing was first reported in 1978 [17, 245]. Chitosan films could serve as the basis for tissue growth, activate macrophages, and promote cell proliferation and multilayer tissue organization [244, 246]. This polymer has a haemostatic effect, facilitating the natural blood coagulation, inducing fibroblast proliferation and stimulating the activation of macrophages, which promotes more rapid healing of the wound and cicatrices [3, 247, 248].

Polymer films for drug delivery are one of the most prospective medical applications of chitosan. Today, the research of chitosan films with drugs is focused on the study of behavior of drug release. Some parameters that determine the efficacy of polymeric films in delivery of therapeutic compounds are thickness, surface morphology, dissolution medium, degree of swelling and degradation behavior [30, 244].

The focus of this review is to provide an insight into the drug carriers based on chitosan and its derivatives, through the various anatomical routes by which the drug may be introduced in human body, such as transdermal, buccal, ocular and local administration.

Transdermal Delivery

The drug administration through skin have been recognized as an interesting way to provide controlled delivery of drugs to the systemic circulation, which maintains the constant medication levels in the blood, decreases the side effects and improves the bioavailability by circumvention of hepatic first pass metabolism [249 - 251]. Many approaches have been investigated to enhance the drug permeation through the barrier of skin for the use in transdermal drug delivery. In view of this, chitosan/hyaluronan films were developed to improve bioavailability of thiocolchicoside, a muscle relaxant with anti-inflammatory and analgesic effects. The lower the hydration of the polymeric network, the lower is the drug diffusion through the films and its permeation through the skin, confirming that the selection of a suitable polymeric weight ratio and appropriate preparative conditions allows the modulation of film functional properties to achieve effective release of drug [252]. Chitosan has also been combined with other oppositely charged polymer like polyacrylic acid (PAA) to prepare polyelectrolyte complexes (PEC), that may be applied on skin without the risk of a significant interference in the barrier function or cause sensitization. The high ionic interactions density between the polymers in PEC provides less pH-dependent swelling-eroding behavior and exhibited a suitable controlled drug release profile [251]. Another approach has shown that microneedle-mediated transdermal transport increases the permeability of human skin to many drugs, by several orders of magnitude [250, 253].

Propranolol, a beta-blocker used to treat tremors, angina, hypertension and other heart or circulatory conditions, has short biological half-life, is highly lipophilic and undergoes high first pass metabolism by the liver, thus on average only about 25% of drug reaches the systemic circulation [254]. Chitosan film containing propranolol hydrochloride and cineol is one potential alternative to overcome undesirable characteristics of oral administration, in order to improve the penetration of medication. This film is nonirritant and achieved high permeation through rat skin, making it adaptable for accomplishing transdermal drug delivery [255]. Chitosan membranes cross-linking with glutaraldehyde has also shown the ability to deliver the drug on rabbit pinna skin when tested by conducting *in vitro* studies in modified Franz diffusion cells [256]. The influence of glycyrrhizin and chitosan on rat epidermis and its correlation with percutaneous permeation characteristics of carvedilol were investigated [257]. The application of patches containing glycyrrhizin-chitosan mixture resulted in sustained release of carvedilol, which was able to control the hypertension in deoxycorticosterone acetate induced hypertensive rats through 28 hours.

In order to minimize the side effects (abdominal pain, nausea or headache) of some antibiotic when orally administrated, a dual antimicrobial release system was developed by using chitosan films as a topical drug delivery vehicle [14]. The combination of silver nanoparticles (AgNPs) with antibiotics, such as ciprofloxacin hydrochloride (CFX), has a synergic effect against several bacteria, becoming a powerful tool to treat microbial infections. The release of silver nanoparticles showed excellent inhibition against Streptococcus and Escherichia coli species [258]. CFX loaded bi-layer composite membranes based on alginate and chitosan inhibited the growth of microorganisms for longer than 7 days. More importantly, the pharmacodynamics studies showed that there was an effective improvement in the wound healing process and no edema or erythema after administration of membranes on the rabbit skin after 14 days [259].

Glimepiride is an antidiabetic sulphonylurea drug which often causes bioavailability problems, dizziness and gastric disturbances. Permeation studies through rat abdominal skin revealed that high flux values were obtained from chitosan films containing glimepiride-β-cyclodextrin complex. Furthermore, these formulations demonstrated a therapeutic efficacy sustained for about 48 hours when tested *in vivo* on diabetic rats [260]. Other antidiabetic drugs with severe side effect, and low oral bioavailability have also been investigated for transdermal route [261].

Buccal Delivery

The buccal mucosa is a barrier, providing protection to underlying tissue, but is more permeable than other alternative routes such as the skin [262]. The excellent blood supply, ability to recover from stress of damage and permeability profile make the buccal cavity an attractive site for local and systemic drug delivery [263, 264]. The structure of chitosan is important in determining its mucoadhesive properties since it binds to the mucosal epithelium *via* ionic bonds between its amino group and sialic acid residues on mucosa [265, 266].

Mucoadhesive buccal films are preferred in terms of flexibility, comfort, good adaptation to the mucosal surface, as well as improved patient compliance due to their small size and reduced thickness, compared to lozenges and tablets [241]. Moreover they can ensure an accurate drug dosing with respect to liquid formulations (gels and ointment), which can be easily washed away by saliva [267]. Thus, great attention has been focused on this dosage form due to possibility to deliver various therapeutic drug classes such as peptides [268, 269], antiemetic agent [270], antibiotic [271], anti-inflammatory [264] and vaccines as an alternative to currently used administration.

The incorporation of drug/Cyclodextrin complexes into chitosan films demonstrated to be a suitable strategy to obtain an effective formulation for buccal delivery of drugs with low aqueous solubility, such as bupivacaine hydrochloride and triclosan [272]. Supercritical solution impregnation could also be amenable for processing poorly soluble drugs and/or those exhibiting extensive first-pass metabolisms. In this technique, the binary mixture of drug and supercritical fluid (usually CO_2) could facilitate the mass transfer in the medication load process. This approach has been successfully implemented in preparation of chitosan film loaded with ibuprofen [264]. The *ex vivo* release profiles showed that the films could deliver the drug across the rabbit buccal mucosa, and up to 70% of the ibuprofen was released from the matrix in 460 minutes.

Gelatin has been recently added to the chitosan to form innovative mucoadhesive system for buccal delivery of propranolol hydrochloride [267]. The film showed limited drug diffusion, possibly due to strong interactions between pharmacologic agent and polyelectrolyte complexes. Thus the addition of mannitol, a hydrophilic absorbing material, was tested to overcome this drawback, since entries of water favor the polymeric chain mobility and the drug diffusion through the hydrated films. In this way the desirable film characteristics were attained ensuring the necessary dose for treatment of hypertension and atrial fibrillation, suggesting the possible use of this formulation for treatment of chronic diseases. The film prepared by crosslinking chitosan with poly(N-vinyl-2-pyrrolidone) and genipin has also proved satisfactory release of this medication [273].

With the intent of producing a formulation for buccal cancer's treatment, Costa *et al.* [240] developed a chitosan mucoadhesive film containing 5-Aminolevulinic acid (5-ALA), a biological precursor for photodynamic therapy (PDT), which can be activated intracellularly by red light at 630 nm, forming reactive oxygen species that accumulate in cancerous tissue, leading to cell destruction. In this case is necessary enhance retention of the drug in the tissue for a sufficient time for carrying out the PDT. The *in vitro* studies of drug retention and mucoadhesion in porcine mucosa showed a possible synergistic effect of 5-ALA associated with the chitosan film, favoring increased retention at the application site and hence can be exploited for the treatment of tumors in the oral mucosa.

The peptide-based drugs for treating diabetes can rapidly denature or degrade before reach their targets, when are administered orally [79]. In addition, due to the relatively large molecular size their bioavailability through the buccal mucosa is quite low [268]. This account demonstrates the feasibility to use nanoparticles for oral insulin delivery incorporated into carrier vehicles, such as mucoadhesive films. These films adhere to the mucosal surface and transiently open the tight junctions between contiguous epithelial cells, allowing that the drug permeates through the opened paracellular pathway and moves to the systemic circulation [63, 79, 274]. In fact, several works considering chitosan films impregnated with peptides-loaded nanoparticles have been reported in literature [79, 269, 275 - 278].

Drug delivery in the oral cavity may be particularly useful in the treatment of periodontal diseases that require a sustained drug delivery, as demonstrated by Mazzarino *et al.* [279] through the model of curcumin release and also to alleviate drawback in the treatment of periodontitis [271].

Ocular Delivery

Drug delivery system in the ocular cavity faces many challenges due to the anatomic structure and protective mechanisms of the eye. Low corneal drug absorption, less than 5% of the drug applied penetrates the cornea and reaches the intraocular tissue, requires a frequent dosing regimen through the instillations of drops to achieve the therapeutic effects [18, 280]. Thus new approaches have been developed to enhance the bioavailability of ophthalmic drugs by prolonging the contact time between the formulations and the corneal/conjunctival epithelium and reducing systemic side effects

Rodrigues *et al.* [281] have compared and characterized mono and bilayer chitosan films as a carrier for dexamethasone. This medication is a corticosteroid used in eye drops to treat inflammation after cataract surgery, corneal operations and others applications. A sustained *in vitro* drug release was obtained in which the release from bilayer chitosan films was much slower (weeks) than that from monolayer films (few hours). The maximum release rate was observed in the system containing one layer of chitosan film where 39.3% drug was released in the first 2h and progressively to 89.6% after 8h.

Mucoadhesive chitosan films of single application to be placed in the conjunctival sac as delivery system for cyclosporine were investigated [18], in order to evaluate their cytotoxicity and anti-inflammatory activity. Glycerol and hydroxypropyl-β-cyclodextrin were added to the formulations, providing flexible films and enhancing drug absorption *in vivo*. Lower drug release was measured from chitosan films containing a higher amount of plasticizer. Moreover, the results revealed absence of cytotoxic effects and significant suppression of interleukin-2 secretion in concanavalin A stimulated Jurkat T cells.

Chitosan film has also been combined with timolol maleate to produce formulations of single administration, equally effective in reducing intraocular pressure in rabbits as twice a day instillation of commercial ophthalmic timolol maleate solution. Developed films represent alternative methods aimed at improving conventional ocular therapy. TM-loaded chitosan films decreased the intraocular pressure levels over a 10-week period up to the 35th measurement (each measure lasting for 48h), being safe and efficient and presents a promising future as an ocular drug delivery system in the treatment and prevention of glaucoma [19].

More recently, crosslinked Chitosan/PVA (polyvinyl alcohol) film loaded 5-Fluorouracil was investigated to determine the drug release properties [282]. This therapeutic agent is known for its widespread use as potentially effective drugs against postsurgical proliferative vitreoretinopathy, being one of the most often used antimetabolites in ophthalmology. Evidences have demonstrated that this drug could also harness the host immune system to prevent cancer progression. According to the results, almost all the drug was diffused from the crosslinked Ch/PVA films after 60h and no toxic action was observed on eye structure of chinchilla rabbits, leading to lack of inflammatory reaction.

SCAFFOLDS AS DRUG CARRIERS

In tissue engineering, scaffolds refer to biodegradable structures that provide a spatial and temporal environment for cellular activity and tissue regeneration [283]. Biomaterials used as scaffolds in tissue engineering applications must meet certain requirements in order to successfully perform their functions. During the regeneration process, the scaffold must offer structural and mechanical restoration of the damaged tissues, gradually degrade into biocompatible products and also present an interconnected porous structure - to support cell attachment, subsequent cell proliferation and differentiation and extracellular matrix (ECM) synthesis. Ideally, the scaffold should mimic in some extent the native ECM and degrade in a controlled way such that it is completely resorbed by the time the tissue is regenerated. Therefore, the success of a 3D scaffold in tissue engineering applications is largely dependent on its nature, composition, and associated properties [284, 285].

Chitosan and its derivatives, owing to their chemical versatility and recognized biological performance, have been identified as an important class of biomaterials [286]. A variety of fabrication techniques have been developed to produce chitosan-based 3D porous scaffolds for application in tissue-engineering and regenerative medicine [287]. The most frequently used is the phase separation and freeze-drying (lyophilization) technique [288 - 292]. In this procedure, a chitosan solution is prepared in dilute acetic acid, introduced into a mold and subjected to control freezing. Upon freezing, ice crystals form in the chitosan solution, and are phase-separated from the chitosan salt. During the subsequent freeze-drying step, the ice crystals sublimate to yield a porous 3D structure. Finally, the scaffold is neutralized and hydrated using sodium hydroxide and/or ethanol. The scaffold pore structure, in terms of pore diameter, orientation and interconnectivity, varies depending on multiple processing parameters including freezing temperature, thermal gradients, and polymer concentration [293]. With this method, uniform scaffold porosity can be difficult to achieve because freezing conditions can vary throughout the scaffold volume.

Another technique used to fabricate chitosan-based scaffolds is solvent-evaporation/particulate leaching method [294]. It is technically simple and easy to carry out and it is often combined with phase separation and lyophilization. In this procedure a sacrificial porogen is mixed with the chitosan solution and then leached out into a solvent following the lyophilization step [295]. This results in two levels of porosity generated by the multi-step process [296, 297]. Commonly-used porogen include salts, sugars, paraffin and gelatin. Pore size and extent of porosity can be controlled by the type, size and concentration of porogen [298]. For example, sodium chloride, a cubic-shaped porogen, generates higher scaffold porosity, thinner pore walls and enhanced pore interconnectivity at higher porogen concentrations. Given a constant porogen concentration, a larger diameter salt particle leads to larger diameter pores, thinner pore walls and thus lower scaffold mechanical strength [299]. In particular, it has been demonstrated that spherical particles result in more uniform pore morphology and pore interconnectivity than cubic particles [300, 301]. Common porogens are generally inexpensive and easy to handle, but one drawback of particulate leaching is related to its time consuming nature. In addition, there can be a lack of control of pore interconnectivity during processing because porogens can be heterogeneously dispersed within the polymer phase [299].

As an alternative to the use of a solid phase to induce porosity, gas foaming exploits the nucleation and growth of gas bubbles within a polymeric matrix to yield a porous structure [302]. The gas bubbles can either be formed by foaming/blowing agent *via* chemical reaction [303] or be released from a pre-saturated gas-polymer mixture at a high pressure [304]. A foaming/blowing agent is a substance that is mixed into the pre-polymer and generates a gas when it chemically decomposes. Gas foaming does not necessitate the use of organic solvents, but pore formation and porosity depend on the rate of nucleation and gas diffusion, which can be difficult to control [299]. Carbon dioxide is often used for foaming of biomaterials because of its moderate critical temperature (31 °C) and pressure (73 bar) and low toxicity [305]. Gas foaming with supercritical carbon dioxide has been used for fabrication of porous chitosan scaffolds [306]. More recently, rapid prototyping technologies have allowed for solid free-form fabrication directly from a computer-aided design model. Rapid prototyping utilizes the computer-aided design data (CAD) to produce 3D scaffolds with precise control over its architecture [307]. Highly reproducible geometries with enhanced mechanical properties are possible with this technique [293].

With respect to scaffolds intended for delivery, they have been developed as vehicle for growth factors, genes and cells. This approach has incorporated the drug delivery concept into tissue engineering in order to achieve accelerated tissue regeneration. For example, single or multiple growth factors can be loaded onto various tissue-engineered constructs for the repair of skin, blood vessel, nerve, cartilage, and bone [308 - 310]. The other approach is to develop surface-modified biopharmaceuticals scaffolds for tissue engineering and regenerative medicine [311, 312].

For bone tissue engineering, porous 3D chitosan-based scaffolds have been extensively studied and beneficial biological activity of chitosan on bone regeneration has been demonstrated in many reports [290, 313]. However, chitosan has some limitations in inducing rapid bone regeneration at initial stages. Bone formation after implanting these matrices occurs over a long period (several months or years) [314]. Additional functions for chitosan materials are necessary to shorten bone forming period and improve their efficacy. Incorporation of active biomolecules such as growth factors, serum proteins, small peptide sequences, and other bioactive molecules has been used as a strategy which is highly beneficial for obtaining improved bone regeneration [315]. Gümüderelioğlu and Aday [316] investigated the effects of heparin-functionalized chitosan scaffolds prepared by a freeze-drying method on the activity of preosteoblasts. Two different methods for immobilization of heparin were successfully performed. In the first method, functionalization of the scaffolds was achieved by means of electrostatic interactions between negatively charged heparin and positively charged chitosan. Second immobilization method consisted in covalently immobilize heparin using 1-(3-dimethylaminopropyl)-3-ethylcarbodiimide (EDAC) and N-hydroxysuccinimide (NHS). Both immobilization methods affected cellular activity through different mechanisms of action. Cell-culture studies demonstrated that, while covalently immobilized heparin-containing scaffolds induced both cell proliferation and differentiation, electrostatically-immobilized heparin-containing scaffolds supported cellular differentiation.

In another study Park *et al.* [317] developed composite chitosan scaffolds containing anionic carbohydrate, such as chondroitin 4-sulfate (CS) or alginate (AG), with biomimetic apatite layer on their surfaces, and investigate their capacity to deliver progenitor cells (bone marrow stromal cells, BMSC) and model proteins with net-positive (histone) and net-negative charge (bovine serum albumin, BSA). The incorporation of CS or AG in chitosan scaffolds increased compressive modulus of the materials and enhanced apatite formation. Regarding the delivery character, initial burst release of histone was significantly higher than that of BSA from chitosan scaffold, while the addition of CS or AG in the scaffolds significantly reduced this initial burst release indicating strong electrostatic interaction between histone and negatively charged CS or AG. Also, the apatite layer created on scaffold surfaces significantly reduced the initial burst release of both BSA and histone.

Chitosan based scaffolds have been combined with chitosan microspheres for reducing initial burst release of tetracycline hydrochloride (TC) - an antibiotic used to treat osteomyelitis [57]. In this study [318], drug loaded chitosan microspheres were incorporated into calcium polyphosphate/chitosan/aldehyde alginate composite scaffolds. Drug-loaded chitosan microspheres were prepared by crosslinking-emulsion method with glutaraldehyde and aldehyde alginate was used here as a biological crosslinking agent to crosslink CS/calcium poly-phosphate scaffolds. The results indicated that encapsulating the drug-loaded chitosan microspheres into the biopolymer/ceramic scaffolds reduced initial drug burst release effectively and prolongs drug release time up to 30 days.

In cartilage repair, chitosan-based scaffolds are one of the most important biomaterials. They can promote the proliferation of chondrocytes and deliver growth factors in a controlled fashion. When compared to soft hydrogels, the 3D porous scaffolds exhibit higher mechanical strength for repairing cartilage. Mattioli-Belmonte *et al.* [319] investigated the effect of BMP-7 associated with N,N-dicarboxymethyl chitosan to induce repair of femoral articular cartilage lesion in rabbit, hypothesizing a synergism of their respective biological effects. Indeed, BMP-7 enhanced the *in vivo* proliferation of chondrocytes, leading to partial healing of the articular surface of the lesions. Whereas, chitosan was linked to increased proliferation of fibrovascular tissue, the initial process of ossification; and was crucial as a BMP-7 carrier. Lee *et al.* [320] reported porous collagen/ chitosan/ glycosaminoglycan scaffolds loaded with TGF-β1. The scaffold exhibited controlled release of TGF-β1 and promoted cartilage regeneration. Moreover, addition of chitosan to the collagen scaffold was seen to improve the mechanical properties and the stability of the collagen network by inhibiting the action of collagenases [321]. In another study [322], a porous freeze-dried chitosan scaffold incorporating TGF-β1-containing microspheres was used for the treatment of cartilage defects. TGF-β1 was released in a sustained fashion, and promoted chondrocyte proliferation and matrix synthesis.

Chitosan/collagen scaffolds prepared through the freeze-drying techniques and combined with pEGFP-TGF-β1 [323] have also potential for cartilage tissue engineering. Chitosan/collagen scaffolds were loaded with plasmid encoding human transforming growth factor-β1 (TGF-β1). Afterwards human bone marrow stem cells were seeded in the scaffold, and gene transfection was traced by enhanced green fluorescent protein (EGFP). The results demonstrated the feasibility of using gene-activated chitosan/collagen scaffolds for primary MSCs' culture and expression of TGF-β1 gene *in vitro*. The gene-activated scaffolds showed better cytocompatibility than the pure chitosan/collagen scaffold. The scaffold containing pEGFP-TGF-β1 exhibited the highest proliferation rate, and the gene expression of type II collagen and aggrecan was upregulated in the pEGFP-TGF-β1 scaffold.

Silk fibroin has also been introduced in chitosan-based scaffolds to repair cartilage defects in a rabbit knee [324]. Bone marrow-derived mesenchymal stem cells (BMSCs) were seeded in the silk fibroin/chitosan scaffolds and allowed almost a fully repair comparing to the silk fibroin/chitosan scaffold alone and the control group.

More recently, regeneration of complex tissues such as osteochondral tissue, involving simultaneous regeneration of bone and cartilage is of great interest [325]. Chen *et al.* [326] formulated a multi-phasic chitosan-based scaffold designed to differentiate a single stem cell population by localized gene delivery. The formulation is based on a bilayer gene-activated osteochondral scaffold consisting of plasmid TGF-β1-activated chitosan-gelatin scaffold for the chondrogenic part and plasmid BMP-2-activated hydroxyapatite/chitosan-gelatin scaffold for the osteogenic one. The results showed that spatially controlled and localized gene delivery system in the bilayer integrated scaffolds could induce the mesenchymal stem cells in different layers to differentiate into chondrocytes and osteoblasts *in vitro*, respectively, and simultaneously support the articular cartilage and subchondral bone regeneration in a rabbit knee osteochondral defect model.

Chitosan-based scaffolds were also proposed as promising candidates to improve skin regeneration and to promote vascularization. Biodegradable chitosan-gelatin microspheres containing bFGF were fabricated and incorporated into a porous chitosan/gelatin scaffold [327]. The release kinetics of bFGF showed a fast release (23.7%) at the initial phase in the first 2 days, and the ultimate accumulated release was approximately 71.8% by day 14, indicating an extended time course for complete release. Human fibroblasts seeded on chitosan/gelatin scaffolds with and without bFGF-loaded chitosan-gelatin microspheres (bFGF-MS) were incubated *in vitro* for 2 weeks and showed that, compared to chitosan-gelatin scaffolds alone, the scaffolds with bFGF-MS significantly augmented the proliferation and glycosaminoglycan (GAG) synthesis of human fibroblasts.

In order to develop effective treatment of wounds, a series of norfloxacin-loaded scaffolds were synthesized by combining collagen with two different types of chitosan using a freeze-drying technique [328]. All scaffolds showed almost 100% drug release within 24 h. After 28 days of wound dressing Albino rats with selected norfloxacin-loaded or unloaded collagen/chitosan scaffolds, it was found that the tissue regeneration time was faster compared to non-treated wounds. The wound treatment with the drug-loaded scaffolds led to a normal structure of an intact epidermal layer as well as the underlying dermis, suggesting that the investigated norfloxacin-loaded collagen/chitosan scaffold is a potential candidate for skin regeneration application.

Ciprofloxacin delivery is also being investigated for the transdermal route by chitosan based scaffolds. A thiolated chitosan with poly(N-isopropyl acrylamide) loaded with ciprofloxacin showed antibacterial properties to the virulent bacteria E. coli and was found to modulate the release of the incorporated ciprofloxacin in a sustained fashion reflecting its suitability to protect a wound for a prolonged period. This material exhibited adequate mechanical strength and was easily removable from a substrate by lowering the temperature without causing any pain or discomfort [329, 330].

For neural tissue engineering applications, chitosan based composite scaffolds can also offer new options [289]. Novel chitosan/gelatin /hyaluronic acid (HA)/ heparin sulfate (HS) composite scaffolds demonstrated that the presence of hyaluronic acid and heparin sulfate in the scaffolds significantly promoted initial neural stem and progenitor cells adhesion and supported long-time growth in three-dimensional environment. Moreover, neural stem and progenitor cells also maintained multilineage differentiation potentials with enhanced neuronal differentiation upon induction in these scaffolds.

Chitosan-based composites scaffolds also demonstrated suitability for cardiac tissue engineering applications. Since chitosan is nonconductive, its electrical properties should be improved by adding, for instance, conductive materials. Carbon nanofibers were used as doping material to develop a highly conductive chitosan-based composite [331]. Chitosan/carbon scaffolds presented an elastic modulus of 28.1±3.3 kPa, similar to rat myocardium, and excellent electrical properties, with a conductivity of 0.25±0.09 S/m. The metabolic activity of neonatal rat heart cells in chitosan/carbon constructs was significantly higher as compared to cells in chitosan scaffolds, presumably through enhanced transmission of electrical signals between the cells as carbon nanofibers led to increased expression of cardiac-specific genes involved in muscle contraction and electrical coupling.

Interestingly 3D chitosan based scaffolds were identified being capable of better replicate the response to targeted gene delivery *in vivo*. The work was performed by Kui Wang *et al* [332] who investigated if prostate cancer (PCa) cells cultured in 3D chitosan-alginate porous scaffolds could model cationic nanoparticles (NP)-mediated gene targeted delivery to tumors. The efficacy of NP-targeted gene delivery was evaluated in PCa cells in 2D cultures *vs* 3D scaffolds. The results revealed that PCa cells cultured in the 3D scaffolds grew into tumor spheroids and displayed characteristics of higher malignancy as compared to those in 2D cultures. Moreover, targeted gene delivery was only significantly observed in cells cultured in scaffolds, whereas cells cultured on 2D plates showed no difference in gene delivery between targeted and non-target control NPs. *In vivo* NP evaluation confirmed targeted gene delivery, indicating that only 3D chitosan-alginate scaffolds correctly modeled NP-mediated targeted delivery *in vivo*.

In bone tissue engineering, polymeric scaffolds such as chitosan scaffolds are commonly combined with bioactive glass to improve biological performance through improved mechanical properties and bone-bonding ability. Wers *et al.* have prepared chitosan/bioactive glass 46S6 to serve as gentamicin sulfate delivery for bone applications *via* freeze-drying method and investigated the effect of bioactive glass presence on the gentamicin delivery. An increase of bioactive glass content induced a slower release of gentamicin but did not affect its relaxation time [333]. In orthopaedics procedures, bacterial infection remains a major problem [334]. Local delivery of antibiotics offers a better alternative to systemic administration, since it allows to reach high antibiotic concentration at the site of action. This is particularly important in the initial post-surgery time for preventing the formation of biofilm onto prosthesis surface. Doratti *et al.* have investigated a formulation that can be used either as an injectable *in situ* gelifying device or as moldable scaffold to be used as filler at prosthesis implantation site with the ability of antibiotic delivery. This formulation is based on chitosan chloride crosslinked with glycerol 2-phosphate disodium salt and laden with bovine bone substitute granules and gentamicin loaded polylactide-coglycolide-co-poly ethylene glycol (PLGA PEG) microparticles. This formulation demonstrated a high bactericidal action in the first 4 h due to superimposable effect of gentamicin release and bacteriostatic chitosan properties. A controlled gentamicin release was achieved and measured up to 14 days [335].

In addition to synthetic drugs such as gentamicin, chitosan scaffolds are also being investigated to deliver naturally occurring drugs such as curcumin. Curcumin, particularly found in saffron, has given recent evidences of improving wound healing and having anti-inflammatory, antimicrobial, antioxidant, and especially anticancer properties. Curcumin containing chitosan scaffolds were successfully prepared by particle aggregation method [336].

For lung cancer therapy, chitosan has been combined with polyethylene oxide (PEO) and graphene oxide (GO) and processed in the form of electrospun nanofibers to deliver doxorubicin (DOX), which is usually delivered by intravenous administration and is associated to side effects. Fibers with a minimum diameter of about 85 nm could be produced. In this system, graphene oxide is used to induce a strong bond with doxorubicin through π-π stacking interaction leading to a high DOX loading (98%) and a controlled release. DOX release was found to be pH dependent. Lowering pH from 7 to 5.3 induced a faster release due to the reduced interaction between DOX and GO. This formulation exhibited further a stronger cytotoxicity towards human lung epithelial carcinoma A549 cells in comparison to free DOX [337].

Collectively, the studies described in this section show that in the last decade, significant progresses have been made toward the use of 3D chitosan-based scaffolds for the delivery of biopharmaceuticals in tissue engineering applications.

CONCLUDING REMARKS

The development and characterization of chitosan based drug delivery system with different morphology summarized in this chapter demonstrate that it is possible to prepare chitosan drug carriers in almost any form. Furthermore, this therapeutic system can be used for the delivery of proteins, antibiotics, anti-inflammatory, growth factors, as well as in gene therapy and bioimaging applications. Chitosan represents an ample resource with great medical potential and despite the recognized success, in the scientific scope, currently there are no products in the market using the capacity of this polymer for therapeutic application. Novel formulations need to consider *"in vivo"* models to assess the drug carriers and their toxicity, thus, longer term effect on cells and tissues should be the subject of thorough research. In conclusion, we believe that with the current investigations and clinical trials, the future holds promise for chitosan based drug delivery systems to be available to patients, since this polymer will provide health benefits over current formulations.

CONFLICT OF INTEREST

The authors confirm that they have no conflict of interest to declare for this publication.

ACKNOWLEDGEMENTS

This work was developed within the scope of the project CICECO-Aveiro Institute of Materials, POCI-01-0145-FEDER-007679 (FCT Ref. UID /CTM /50011/2013), financed by national funds through the FCT/MEC and when appropriate co-financed by FEDER under the PT2020 Partnership Agreement. A.L.D.S. acknowledges FCT/MCTES for the research contract and funding of an exploratory project under the Program 'Investigador FCT' 2014. The authors would like to express their thanks to the Capes, CNPq, Fapergs and Federal University of Santa Maria for supporting this work. Bárbara J. Leite Ferreira thanks Portuguese Foundation for Science and Technology (FCT) her post-doctoral grant - Ref. SFRH/BPD/81113/2011.

REFERENCES

[1] Ravi Kumar MN. A review of chitin and chitosan applications. React Funct Polym 2000; 46: 1-27. [http://dx.doi.org/10.1016/S1381-5148(00)00038-9]

[2] Prabaharan M. Review paper: chitosan derivatives as promising materials for controlled drug delivery. J Biomater Appl 2008; 23(1): 5-36. [http://dx.doi.org/10.1177/0885328208091562] [PMID: 18593819]

[3] Baldrick P. The safety of chitosan as a pharmaceutical excipient. Regul Toxicol Pharmacol 2010; 56(3): 290-9. [http://dx.doi.org/10.1016/j.yrtph.2009.09.015] [PMID: 19788905]

[4] Dash M, Chiellini F, Ottenbrite RM, Chiellini E. Chitosan - A versatile semi-synthetic polymer in biomedical applications. Prog Polym Sci 2011; 36: 981-1014. [http://dx.doi.org/10.1016/j.progpolymsci.2011.02.001]

[5] Croisier F, Jérôme C. Chitosan-based biomaterials for tissue engineering. Eur Polym J 2013; 49: 780-92. [http://dx.doi.org/10.1016/j.eurpolymj.2012.12.009]

[6] Rouget C. Des substances amylacees dans le tissue des animux, specialement les articules (Chitine). Compt Rend 1859; 48: 792.

[7] Teng D. From chitin to chitosan. Chitosan-based hydrogels functions and applications. CRC Press . 2011; pp. 1-37. [http://dx.doi.org/10.1201/b11048-2]

[8] Mima S, Miya M, Iwamoto R, Yoshikawa S. Highly deacetylated chitosan and its properties. J Appl Polym Sci 1983; 28: 1909-17. [http://dx.doi.org/10.1002/app.1983.070280607]

[9] Bhattarai N, Gunn J, Zhang M. Chitosan-based hydrogels for controlled, localized drug delivery. Adv Drug Deliv Rev 2010; 62(1): 83-99. [http://dx.doi.org/10.1016/j.addr.2009.07.019] [PMID: 19799949]

[10] Bernkop-Schnürch A, Dünnhaupt S. Chitosan-based drug delivery systems. Eur J Pharm Biopharm 2012; 81(3): 463-9. [http://dx.doi.org/10.1016/j.ejpb.2012.04.007] [PMID: 22561955]

[11] Hu L, Sun Y, Wu Y. Advances in chitosan-based drug delivery vehicles. Nanoscale 2013; 5(8): 3103-11. [http://dx.doi.org/10.1039/c3nr00338h] [PMID: 23515527]

[12] Yang Y, Wang S, Wang Y, Wang X, Wang Q, Chen M. Advances in self-assembled chitosan nanomaterials for drug delivery. Biotechnol Adv 2014; 32(7): 1301-16. [http://dx.doi.org/10.1016/j.biotechadv.2014.07.007] [PMID: 25109677]

[13] Kean T, Thanou M. Biodegradation, biodistribution and toxicity of chitosan. Adv Drug Deliv Rev 2010; 62(1): 3-11. [http://dx.doi.org/10.1016/j.addr.2009.09.004] [PMID: 19800377]

[14] Aranaz I, Harris R, Navarro-García F, Heras A, Acosta N. Chitosan based films as supports for dual antimicrobial release. Carbohydr Polym 2016; 146: 402-10. [http://dx.doi.org/10.1016/j.carbpol.2016.03.064] [PMID: 27112890]

[15] Mura S, Nicolas J, Couvreur P. Stimuli-responsive nanocarriers for drug delivery. Nat Mater 2013; 12(11): 991-1003. [http://dx.doi.org/10.1038/nmat3776] [PMID: 24150417]

[16] Luo Y, Wang Q. Recent development of chitosan-based polyelectrolyte complexes with natural polysaccharides for drug delivery. Int J Biol Macromol 2014; 64: 353-67. [http://dx.doi.org/10.1016/j.ijbiomac.2013.12.017] [PMID: 24360899]

[17] Denkbas EB. Perspectives on: chitosan drug delivery systems based on their geometries. J Bioact Compat Polym 2006; 21: 351-68. [http://dx.doi.org/10.1177/0883911506066930]

[18] Hermans K, Van den Plas D, Kerimova S, *et al.* Development and characterization of mucoadhesive chitosan films for ophthalmic delivery of cyclosporine A. Int J Pharm 2014; 472(1-2): 10-9. [http://dx.doi.org/10.1016/j.ijpharm.2014.06.017] [PMID: 24929014]

[19] Fulgêncio GdeO, Viana FA, Ribeiro RR, Yoshida MI, Faraco AG, Cunha-Júnior AdaS. New mucoadhesive chitosan film for ophthalmic drug delivery of timolol maleate: *in vivo* evaluation. J Ocul Pharmacol Ther 2012; 28(4): 350-8. [http://dx.doi.org/10.1089/jop.2011.0174] [PMID: 22320419]

[20] Schipper NG, Olsson S, Hoogstraate JA, deBoer AG, Vårum KM, Artursson P. Chitosans as absorption enhancers for poorly absorbable drugs 2: mechanism of absorption enhancement. Pharm Res 1997; 14(7): 923-9. [http://dx.doi.org/10.1023/A:1012160102740] [PMID: 9244151]

[21] Smith J, Wood E, Dornish M. Effect of chitosan on epithelial cell tight junctions. Pharm Res 2004; 21(1): 43-9. [http://dx.doi.org/10.1023/B:PHAM.0000012150.60180.e3] [PMID: 14984256]

[22] Okamoto Y, Kawakami K, Miyatake K, Morimoto M, Shigemasa Y, Minami S. Analgesic effects of chitin and chitosan. Carbohydr Polym 2002; 49: 249-52. [http://dx.doi.org/10.1016/S0144-8617(01)00316-2]

[23] Aranaz I, Mengíbar M, Harris R, *et al.* Functional characterization of chitin and chitosan. Curr Chem Biol 2009; 3: 203-30. [http://dx.doi.org/10.2174/187231309788166415]

[24] Yang J, Tian F, Wang Z, Wang Q, Zeng YJ, Chen SQ. Effect of chitosan molecular weight and deacetylation degree on hemostasis. J Biomed Mater Res B Appl Biomater 2008; 84(1): 131-7. [http://dx.doi.org/10.1002/jbm.b.30853] [PMID: 17514656]

[25] Ong S-Y, Wu J, Moochhala SM, Tan M-H, Lu J. Development of a chitosan-based wound dressing with improved hemostatic and antimicrobial properties. Biomaterials 2008; 29(32): 4323-32. [http://dx.doi.org/10.1016/j.biomaterials.2008.07.034] [PMID: 18708251]

[26] Mak AF, Sun S. Intelligent chitosan-based hydrogels as multifunctional materials. In: Shahinpoor M, Schneider H-J, Eds. Intell Mater. Cambridge: Royal Society of Chemistry 2007; pp. 447-63. [http://dx.doi.org/10.1039/9781847558008-00447]

[27] Dünnhaupt S, Barthelmes J, Iqbal J, *et al. In vivo* evaluation of an oral drug delivery system for peptides based on S-protected thiolated chitosan. J Control Release 2012; 160(3): 477-85. [http://dx.doi.org/10.1016/j.jconrel.2012.04.020] [PMID: 22542699]

[28] Morimoto M, Saimoto H, Shigemasa Y. Control of functions of chitin and chitosan by chemical modification. Trends Glycosci Glycotechnol 2002; 14: 205-22. [http://dx.doi.org/10.4052/tigg.14.205]

[29] Zhu X, Su M, Tang S, *et al.* Synthesis of thiolated chitosan and preparation nanoparticles with sodium alginate for ocular drug delivery. Mol Vis 2012; 18: 1973-82.
[PMID: 22876124]

[30] Upadhyaya L, Singh J, Agarwal V, Tewari RP. The implications of recent advances in carboxymethyl chitosan based targeted drug delivery and tissue engineering applications. J Control Release 2014; 186: 54-87.
[http://dx.doi.org/10.1016/j.jconrel.2014.04.043] [PMID: 24806482]

[31] Yang J, Chen J, Pan D, Wan Y, Wang Z. pH-sensitive interpenetrating network hydrogels based on chitosan derivatives and alginate for oral drug delivery. Carbohydr Polym 2013; 92(1): 719-25.
[http://dx.doi.org/10.1016/j.carbpol.2012.09.036] [PMID: 23218359]

[32] Wang LC, Chen XG, Zhong DY, Xu QC. Study on poly(vinyl alcohol)/carboxymethyl-chitosan blend film as local drug delivery system. J Mater Sci Mater Med 2007; 18(6): 1125-33.
[http://dx.doi.org/10.1007/s10856-007-0159-5] [PMID: 17268861]

[33] Vårum KM, Myhr MM, Hjerde RJ, Smidsrød O. *In vitro* degradation rates of partially N-acetylated chitosans in human serum. Carbohydr Res 1997; 299(1-2): 99-101.
[http://dx.doi.org/10.1016/S0008-6215(96)00332-1] [PMID: 9129298]

[34] Muzzarelli RA. Human enzymatic activities related to the therapeutic administration of chitin derivatives. Cell Mol Life Sci 1997; 53(2): 131-40.
[http://dx.doi.org/10.1007/PL00000584] [PMID: 9118001]

[35] Ren D, Yi H, Wang W, Ma X. The enzymatic degradation and swelling properties of chitosan matrices with different degrees of N-acetylation. Carbohydr Res 2005; 340(15): 2403-10.
[http://dx.doi.org/10.1016/j.carres.2005.07.022] [PMID: 16109386]

[36] Zhang H, Neau SH. *In vitro* degradation of chitosan by a commercial enzyme preparation: effect of molecular weight and degree of deacetylation. Biomaterials 2001; 22(12): 1653-8.
[http://dx.doi.org/10.1016/S0142-9612(00)00326-4] [PMID: 11374467]

[37] Freier T, Koh HS, Kazazian K, Shoichet MS. Controlling cell adhesion and degradation of chitosan films by N-acetylation. Biomaterials 2005; 26(29): 5872-8.
[http://dx.doi.org/10.1016/j.biomaterials.2005.02.033] [PMID: 15949553]

[38] Lu G, Sheng B, Wang G, *et al.* Controlling the degradation of covalently cross-linked carboxymethyl chitosan utilizing bimodal molecular weight distribution. J Biomater Appl 2009; 23(5): 435-51.
[http://dx.doi.org/10.1177/0885328208091661] [PMID: 18697877]

[39] Kafedjiiski K, Föger F, Hoyer H, Bernkop-Schnürch A, Werle M. Evaluation of in vitro enzymatic degradation of various thiomers and cross-linked thiomers. Drug Dev Ind Pharm 2007; 33(2): 199-208.
[http://dx.doi.org/10.1080/03639040600762651] [PMID: 17454052]

[40] McConnell EL, Murdan S, Basit AW. An investigation into the digestion of chitosan (noncrosslinked and crosslinked) by human colonic bacteria. J Pharm Sci 2008; 97(9): 3820-9.
[http://dx.doi.org/10.1002/jps.21271] [PMID: 18186460]

[41] Hirano S, Seino H, Akiyama Y, Nonaka I. Chitosan: A biocompatible material for oral and intravenous administrations. Prog Biomed Polym 1990; 283-90.
[http://dx.doi.org/10.1007/978-1-4899-0768-4_28]

[42] Park K. Controlled drug delivery systems: past forward and future back. J Control Release 2014; 190: 3-8.
[http://dx.doi.org/10.1016/j.jconrel.2014.03.054] [PMID: 24794901]

[43] Mezei M, Gulasekharam V. Liposomesa selective drug delivery system for the topical route of administration. Lotion dosage form. Life Sci 1980; 26(18): 1473-7.
[http://dx.doi.org/10.1016/0024-3205(80)90268-4] [PMID: 6893068]

[44] Hoffman AS. The origins and evolution of controlled drug delivery systems. J Control Release 2008; 132(3): 153-63.
[http://dx.doi.org/10.1016/j.jconrel.2008.08.012] [PMID: 18817820]

[45] Jain KK. Drug delivery systems - an overview. In: Jain KK, Ed. Drug Deliv Syst. New York: Humana Press 2008; pp. 1-50.
[http://dx.doi.org/10.1007/978-1-59745-210-6_1]

[46] Zhang Y, Chan HF, Leong KW. Advanced materials and processing for drug delivery: the past and the future. Adv Drug Deliv Rev 2013; 65(1): 104-20.
[http://dx.doi.org/10.1016/j.addr.2012.10.003] [PMID: 23088863]

[47] Thanou M, Verhoef JC, Junginger HE. Oral drug absorption enhancement by chitosan and its derivatives. Adv Drug Deliv Rev 2001; 52(2): 117-26.
[http://dx.doi.org/10.1016/S0169-409X(01)00231-9] [PMID: 11718935]

[48] Mao S, Sun W, Kissel T. Chitosan-based formulations for delivery of DNA and siRNA. Adv Drug Deliv Rev 2010; 62(1): 12-27.
[http://dx.doi.org/10.1016/j.addr.2009.08.004] [PMID: 19796660]

[49] Park JH, Saravanakumar G, Kim K, Kwon IC. Targeted delivery of low molecular drugs using chitosan and its derivatives. Adv Drug Deliv Rev 2010; 62(1): 28-41.
[http://dx.doi.org/10.1016/j.addr.2009.10.003] [PMID: 19874862]

[50] Vashist AA, Vashist AA, Gupta YK, Ahmad S. Recent advances in hydrogel based drug delivery systems for the human body. J Mater Chem B Mater Biol Med 2014; 2: 147.
[http://dx.doi.org/10.1039/C3TB21016B]

[51] Zhou HY, Zhang YP, Zhang WF, Chen XG. Biocompatibility and characteristics of injectable chitosan-based thermosensitive hydrogel for drug delivery. Carbohydr Polym 2011; 83: 1643-51.
[http://dx.doi.org/10.1016/j.carbpol.2010.10.022]

[52] Das RK, Kasoju N, Bora U. Encapsulation of curcumin in alginate-chitosan-pluronic composite nanoparticles for delivery to cancer cells. Nanomedicine Nanotechnology. Biol Med 2010; 6: 153-60.
[http://dx.doi.org/10.1016/j.nano.2009.05.009]

[53] Riva R, Ragelle H, des Rieux A, *et al.* Chitosan and chitosan derivatives in drug delivery and tissue engineering Chitosan Biomater II. Advances in Polymer Science. Springer 2011; pp. 19-44.
[http://dx.doi.org/10.1007/12_2011_137]

[54] Wolinsky JB, Colson YL, Grinstaff MW. Local drug delivery strategies for cancer treatment: gels, nanoparticles, polymeric films, rods, and wafers. J Control Release 2012; 159(1): 14-26.
[http://dx.doi.org/10.1016/j.jconrel.2011.11.031] [PMID: 22154931]

[55] Ruel-Gariépy E, Shive M, Bichara A, *et al.* A thermosensitive chitosan-based hydrogel for the local delivery of paclitaxel. Eur J Pharm Biopharm 2004; 57(1): 53-63.
[http://dx.doi.org/10.1016/S0939-6411(03)00095-X] [PMID: 14729080]

[56] Trapani A, Di Gioia S, Ditaranto N, *et al.* Systemic heparin delivery by the pulmonary route using chitosan and glycol chitosan nanoparticles. Int J Pharm 2013; 447(1-2): 115-23.
[http://dx.doi.org/10.1016/j.ijpharm.2013.02.035] [PMID: 23454518]

[57] Fraimow HS. Systemic antimicrobial therapy in osteomyelitis. Semin Plast Surg 2009; 23(2): 90-9.
[http://dx.doi.org/10.1055/s-0029-1214161] [PMID: 20567731]

[58] Mackichan JJ, Doshi DH. Pharmacokinetic considerations for drug delivery. In: Gibaldi M, Lee M, Desai A, Eds. Gibaldi's drug Deliv Syst Pharm care. ASHP 2007.

[59] Mignani S, El Kazzouli S, Bousmina M, Majoral J-P. Expand classical drug administration ways by emerging routes using dendrimer drug delivery systems: a concise overview. Adv Drug Deliv Rev 2013; 65(10): 1316-30.
[http://dx.doi.org/10.1016/j.addr.2013.01.001] [PMID: 23415951]

[60] Loftsson T. Drug pharmacokinetics after alternative routes of administration. Essent Pharmacokinet 2015; pp. 105-18.
[http://dx.doi.org/10.1016/B978-0-12-801411-0.00004-4]

[61] Jones D. Applications of polymers for controlled drug delivery. Pharm. Appl. Polym. Drug Deliv., Rapra Technology 2004; pp. 16-39.

[62] Gao N, Li XJ. Controlled drug delivery using microfluidic devices Microfluid Devices Biomed Appl. Woodhead Publishing 2013; pp. 167-84.
[http://dx.doi.org/10.1533/9780857097040.2.167]

[63] Chen M-C, Mi F-L, Liao Z-X, *et al.* Recent advances in chitosan-based nanoparticles for oral delivery of macromolecules. Adv Drug Deliv Rev 2013; 65(6): 865-79.
[http://dx.doi.org/10.1016/j.addr.2012.10.010] [PMID: 23159541]

[64] Illum L. Nasal drug deliverypossibilities, problems and solutions. J Control Release 2003; 87(1-3): 187-98.
[http://dx.doi.org/10.1016/S0168-3659(02)00363-2] [PMID: 12618035]

[65] Nazar H, Caliceti P, Carpenter B, *et al.* A once-a-day dosage form for the delivery of insulin through the nasal route: *in vitro* assessment and *in vivo* evaluation. Biomater Sci 2013; 1: 306-14.
[http://dx.doi.org/10.1039/C2BM00132B]

[66] Hejazi R, Amiji M. Chitosan-based gastrointestinal delivery systems. J Control Release 2003; 89(2): 151-65.
[http://dx.doi.org/10.1016/S0168-3659(03)00126-3] [PMID: 12711440]

[67] Du H, Liu M, Yang X, Zhai G. The design of pH-sensitive chitosan-based formulations for gastrointestinal delivery. Drug Discov Today 2015; 20(8): 1004-11.
[http://dx.doi.org/10.1016/j.drudis.2015.03.002] [PMID: 25769687]

[68] Kumar A, Vimal A, Kumar A. Why Chitosan? From properties to perspective of mucosal drug delivery. Int J Biol Macromol 2016; 91: 615-22.
[http://dx.doi.org/10.1016/j.ijbiomac.2016.05.054] [PMID: 27196368]

[69] Hye Kim J, Il Kim S, Kwon IB, Hyun Kim M, Ik Lim J. Simple fabrication of silver hybridized porous chitosan-based patch for transdermal drug-delivery system. Mater Lett 2013; 95: 48-51.
[http://dx.doi.org/10.1016/j.matlet.2012.12.078]

[70] Agnihotri SA, Mallikarjuna NN, Aminabhavi TM. Recent advances on chitosan-based micro- and nanoparticles in drug delivery. J Control Release 2004; 100(1): 5-28.
[http://dx.doi.org/10.1016/j.jconrel.2004.08.010] [PMID: 15491807]

[71] Liu Z, Jiao Y, Wang Y, Zhou C, Zhang Z. Polysaccharides-based nanoparticles as drug delivery systems. Adv Drug Deliv Rev 2008; 60(15): 1650-62.
[http://dx.doi.org/10.1016/j.addr.2008.09.001] [PMID: 18848591]

[72] Mizrahy S, Peer D. Polysaccharides as building blocks for nanotherapeutics. Chem Soc Rev 2012; 41(7): 2623-40.
[http://dx.doi.org/10.1039/C1CS15239D] [PMID: 22085917]

[73] Ohya Y, Shiratani M, Kobayashi H, Ouchi T. Release behavior of 5-fluorouracil from chitosan-gel nanospheres immobilizing 5-fluorouracil coated with polysaccharides and their cell specific cytotoxicity. J Macromol Sci Part A 1994; 31(5): 629-42.
[http://dx.doi.org/10.1080/10601329409349743]

[74] Manickam B, Sreedharan R, Elumalai M. Genipin - the natural water soluble cross-linking agent and its importance in the modified drug delivery systems: an overview. Curr Drug Deliv 2014; 11(1): 139-45.
[http://dx.doi.org/10.2174/15672018113106660059] [PMID: 24041312]

[75] Kawadkar J, Chauhan MK. Intra-articular delivery of genipin cross-linked chitosan microspheres of flurbiprofen: preparation, characterization, *in vitro* and *in vivo* studies. Eur J Pharm Biopharm 2012; 81(3): 563-72.
[http://dx.doi.org/10.1016/j.ejpb.2012.04.018] [PMID: 22595132]

[76] Mi F-L, Tan Y-C, Liang H-C, Huang R-N, Sung H-W. *In vitro* evaluation of a chitosan membrane cross-linked with genipin. J Biomater Sci Polym Ed 2001; 12(8): 835-50.
[http://dx.doi.org/10.1163/156856201753113051] [PMID: 11718480]

[77] Motwani SK, Chopra S, Talegaonkar S, Kohli K, Ahmad FJ, Khar RK. Chitosan-sodium alginate nanoparticles as submicroscopic reservoirs for ocular delivery: formulation, optimisation and *in vitro* characterisation. Eur J Pharm Biopharm 2008; 68(3): 513-25.
[http://dx.doi.org/10.1016/j.ejpb.2007.09.009] [PMID: 17983737]

[78] Tavakol M, Vasheghani-Farahani E, Dolatabadi-Farahani T, Hashemi-Najafabadi S. Sulfasalazine release from alginate-N,O-carboxymethyl chitosan gel beads coated by chitosan. Carbohydr Polym 2009; 77(2): 326-30.
[http://dx.doi.org/10.1016/j.carbpol.2009.01.005]

[79] Sung HW, Sonaje K, Liao ZX, Hsu L-W, Chuang E-Y. pH-responsive nanoparticles shelled with chitosan for oral delivery of insulin: from mechanism to therapeutic applications. Acc Chem Res 2012; 45(4): 619-29.
[http://dx.doi.org/10.1021/ar200234q] [PMID: 22236133]

[80] Huo M, Zhang Y, Zhou J, *et al.* Synthesis and characterization of low-toxic amphiphilic chitosan derivatives and their application as micelle carrier for antitumor drug. Int J Pharm 2010; 394(1-2): 162-73.
[http://dx.doi.org/10.1016/j.ijpharm.2010.05.001] [PMID: 20457237]

[81] Liu K-H, Chen B-R, Chen S-Y, Liu D-M. Self-assembly behavior and doxorubicin-loading capacity of acylated carboxymethyl chitosans. J Phys Chem B 2009; 113(35): 11800-7.
[http://dx.doi.org/10.1021/jp902103p] [PMID: 19678632]

[82] Wang YS, Jiang Q, Li R-S, *et al.* Self-assembled nanoparticles of cholesterol-modified O-carboxymethyl chitosan as a novel carrier for paclitaxel. Nanotechnology 2008; 19(14): 145101. [http://dx.doi.org/10.1088/0957-4484/19/14/145101] [PMID: 21817751]

[83] Gaspar VM, Sousa F, Queiroz JA, Correia IJ. Formulation of chitosan-TPP-pDNA nanocapsules for gene therapy applications. Nanotechnology 2011; 22(1): 015101.
[http://dx.doi.org/10.1088/0957-4484/22/1/015101] [PMID: 21135452]

[84] Katas H, Alpar HO. Development and characterisation of chitosan nanoparticles for siRNA delivery. J Control Release 2006; 115(2): 216-25.
[http://dx.doi.org/10.1016/j.jconrel.2006.07.021] [PMID: 16959358]

[85] Gan Q, Wang T, Cochrane C, McCarron P. Modulation of surface charge, particle size and morphological properties of chitosan-TPP nanoparticles intended for gene delivery. Colloids Surf B Biointerfaces 2005; 44(2-3): 65-73.
[http://dx.doi.org/10.1016/j.colsurfb.2005.06.001] [PMID: 16024239]

[86] Shu XZ, Zhu KJ. Chitosan/gelatin microspheres prepared by modified emulsification and ionotropic gelation. J Microencapsul 2001; 18(2): 237-45.
[http://dx.doi.org/10.1080/02652040010000415] [PMID: 11253940]

[87] Peniche C, Argüelles-Monal W, Peniche H, Acosta N. Chitosan: an attractive biocompatible polymer for microencapsulation. Macromol Biosci 2003; 3(10): 511-20.
[http://dx.doi.org/10.1002/mabi.200300019]

[88] Rodrigues S, da Costa AM, Grenha A. Chitosan/carrageenan nanoparticles: effect of cross-linking with tripolyphosphate and charge ratios. Carbohydr Polym 2012; 89(1): 282-9.
[http://dx.doi.org/10.1016/j.carbpol.2012.03.010] [PMID: 24750635]

[89] Liu T-Y, Hu S-H, Liu D-M, Chen S-Y, Chen I-W. Biomedical nanoparticle carriers with combined thermal and magnetic responses. Nano Today 2009; 4(1): 52-65. [http://dx.doi.org/10.1016/j.nantod.2008.10.011]

[90] Wang JJ, Zeng ZW, Xiao RZ, *et al.* Recent advances of chitosan nanoparticles as drug carriers. Int J Nanomedicine 2011; 6: 765-74. [http://dx.doi.org/10.2147/ijn.s17296] [PMID: 21589644]

[91] Tan ML, Choong PF, Dass CR. Review: doxorubicin delivery systems based on chitosan for cancer therapy. J Pharm Pharmacol 2009; 61(2): 131-42. [http://dx.doi.org/10.1211/jpp.61.02.0001] [PMID: 19178759]

[92] Son YJ, Jang JS, Cho YW, *et al.* Biodistribution and anti-tumor efficacy of doxorubicin loaded glycol-chitosan nanoaggregates by EPR effect. J Control Release 2003; 91(1-2): 135-45. [http://dx.doi.org/10.1016/S0168-3659(03)00231-1] [PMID: 12932645]

[93] Hyung Park J, Kwon S, Lee M, *et al.* Self-assembled nanoparticles based on glycol chitosan bearing hydrophobic moieties as carriers for doxorubicin: *in vivo* biodistribution and anti-tumor activity. Biomaterials 2006; 27(1): 119-26. [http://dx.doi.org/10.1016/j.biomaterials.2005.05.028] [PMID: 16023198]

[94] Yoon HY, Son S, Lee SJ, *et al.* Glycol chitosan nanoparticles as specialized cancer therapeutic vehicles: sequential delivery of doxorubicin and Bcl-2 siRNA. Sci Rep 2014; 4: 6878. [http://dx.doi.org/10.1038/srep06878] [PMID: 25363213]

[95] Min HS, You DG, Son S, *et al.* Echogenic glycol chitosan nanoparticles for ultrasound-triggered cancer theranostics. Theranostics 2015; 5(12): 1402-18. [http://dx.doi.org/10.7150/thno.13099] [PMID: 26681985]

[96] Na JH, Koo H, Lee S, *et al.* Real-time and non-invasive optical imaging of tumor-targeting glycol chitosan nanoparticles in various tumor models. Biomaterials 2011; 32(22): 5252-61. [http://dx.doi.org/10.1016/j.biomaterials.2011.03.076] [PMID: 21513975]

[97] Kim J-Y, Ryu JH, Schellingerhout D, *et al.* Direct imaging of cerebral thromboemboli using computed tomography and fibrin-targeted gold nanoparticles. Theranostics 2015; 5(10): 1098-114. [http://dx.doi.org/10.7150/thno.11679] [PMID: 26199648]

[98] Lee SJ, Koo H, Lee D-E, *et al.* Tumor-homing photosensitizer-conjugated glycol chitosan nanoparticles for synchronous photodynamic imaging and therapy based on cellular on/off system. Biomaterials 2011; 32(16): 4021-9. [http://dx.doi.org/10.1016/j.biomaterials.2011.02.009] [PMID: 21376388]

[99] Mukhopadhyay P, Mishra R, Rana D, Kundu PP. Strategies for effective oral insulin delivery with modified chitosan nanoparticles: A review. Prog Polym Sci 2012; 37(11): 1457-75. [http://dx.doi.org/10.1016/j.progpolymsci.2012.04.004]

[100] Werle M, Bernkop-Schnürch A. Thiolated chitosans: useful excipients for oral drug delivery. J Pharm Pharmacol 2008; 60(3): 273-81. [http://dx.doi.org/10.1211/jpp.60.3.3001] [PMID: 18284806]

[101] Thanou M, Florea BI, Langemeÿer MW, Verhoef JC, Junginger HE. N-trimethylated chitosan chloride (TMC) improves the intestinal permeation of the peptide drug buserelin *in vitro* (Caco-2 cells) and *in vivo* (rats). Pharm Res 2000; 17(1): 27-31. [http://dx.doi.org/10.1023/A:1007558206506] [PMID: 10714604]

[102] Lopes M, Shrestha N, Correia A, *et al.* Dual chitosan/albumin-coated alginate/dextran sulfate nanoparticles for enhanced oral delivery of insulin. J Control Release 2016; 232: 29-41. [http://dx.doi.org/10.1016/j.jconrel.2016.04.012] [PMID: 27074369]

[103] Sarett SM, Nelson CE, Duvall CL. Technologies for controlled, local delivery of siRNA. J Control Release 2015; 218: 94-113. [http://dx.doi.org/10.1016/j.jconrel.2015.09.066] [PMID: 26476177]

[104] Gao S, Hein S, Dagnæs-Hansen F, *et al.* Megalin-mediated specific uptake of chitosan/siRNA nanoparticles in mouse kidney proximal tubule epithelial cells enables AQP1 gene silencing. Theranostics 2014; 4(10): 1039-51.
[http://dx.doi.org/10.7150/thno.7866] [PMID: 25157280]

[105] Ragelle H, Riva R, Vandermeulen G, *et al.* Chitosan nanoparticles for siRNA delivery: optimizing formulation to increase stability and efficiency. J Control Release 2014; 176: 54-63.
[http://dx.doi.org/10.1016/j.jconrel.2013.12.026] [PMID: 24389132]

[106] Buschmann MD, Merzouki A, Lavertu M, Thibault M, Jean M, Darras V. Chitosans for delivery of nucleic acids. Adv Drug Deliv Rev 2013; 65(9): 1234-70.
[http://dx.doi.org/10.1016/j.addr.2013.07.005] [PMID: 23872012]

[107] Liao ZX, Peng SF, Chiu YL, *et al.* Enhancement of efficiency of chitosan-based complexes for gene transfection with poly(γ-glutamic acid) by augmenting their cellular uptake and intracellular unpackage. J Control Release 2014; 193: 304-15.
[http://dx.doi.org/10.1016/j.jconrel.2014.04.024] [PMID: 24780267]

[108] Fleige E, Quadir MA, Haag R. Stimuli-responsive polymeric nanocarriers for the controlled transport of active compounds: concepts and applications. Adv Drug Deliv Rev 2012; 64(9): 866-84.
[http://dx.doi.org/10.1016/j.addr.2012.01.020] [PMID: 22349241]

[109] Schmaljohann D. Thermo- and pH-responsive polymers in drug delivery. Adv Drug Deliv Rev 2006; 58(15): 1655-70.
[http://dx.doi.org/10.1016/j.addr.2006.09.020] [PMID: 17125884]

[110] Taghizadeh B, Taranejoo S, Monemian SA, *et al.* Classification of stimuli-responsive polymers as anticancer drug delivery systems. Drug Deliv 2015; 22(2): 145-55.
[http://dx.doi.org/10.3109/10717544.2014.887157] [PMID: 24547737]

[111] Karimi M, Eslami M, Sahandi-Zangabad P, *et al.* pH-Sensitive stimulus-responsive nanocarriers for targeted delivery of therapeutic agents. Wiley Interdiscip Rev Nanomed Nanobiotechnol 2016; 8(5): 696-716.
[http://dx.doi.org/10.1002/wnan.1389] [PMID: 26762467]

[112] Rao W, Wang H, Han J, *et al.* Chitosan-decorated doxorubicin-encapsulated nanoparticle targets and eliminates tumor reinitiating cancer stem-like cells. ACS Nano 2015; 9(6): 5725-40.
[http://dx.doi.org/10.1021/nn506928p] [PMID: 26004286]

[113] Stephen ZR, Kievit FM, Veiseh O, *et al.* Redox-responsive magnetic nanoparticle for targeted convection-enhanced delivery of O^6-benzylguanine to brain tumors. ACS Nano 2014; 8(10): 10383-95.
[http://dx.doi.org/10.1021/nn503735w] [PMID: 25247850]

[114] Wang Y, Wang J, Xu H, Ge L, Zhu J. Investigation of dual-sensitive nanogels based on chitosan and N-isopropylacrylamide and its intelligent drug delivery of 10-hydroxycamptothecine. Drug Deliv 2015; 22(6): 803-13.
[http://dx.doi.org/10.3109/10717544.2014.883219] [PMID: 24512347]

[115] Zhang W, Gilstrap K, Wu L, *et al.* Synthesis and characterization of thermally responsive Pluronic F127-chitosan nanocapsules for controlled release and intracellular delivery of small molecules. ACS Nano 2010; 4(11): 6747-59.
[http://dx.doi.org/10.1021/nn101617n] [PMID: 21038924]

[116] Wang H, Di J, Sun Y, *et al.* Biocompatible PEG-Chitosan@Carbon dots hybrid nanogels for two-photon fluorescence imaging, near-infrared light/pH dual-responsive drug carrier, and synergistic therapy. Adv Funct Mater 2015; 25(34): 5537-47.
[http://dx.doi.org/10.1002/adfm.201501524]

[117] Wang C, Mallela J, Garapati US, *et al.* A chitosan-modified graphene nanogel for noninvasive controlled drug release. Nanomedicine (Lond) 2013; 9(7): 903-11.

[http://dx.doi.org/10.1016/j.nano.2013.01.003] [PMID: 23352802]

[118] Hernández R, Sacristán J, Asín L, *et al.* Magnetic hydrogels derived from polysaccharides with improved specific power absorption: potential devices for remotely triggered drug delivery. J Phys Chem B 2010; 114(37): 12002-7.
[http://dx.doi.org/10.1021/jp105556e] [PMID: 20806925]

[119] Mahdavinia GR, Etemadi H, Soleymani F. Magnetic/pH-responsive beads based on caboxymethyl chitosan and κ-carrageenan and controlled drug release. Carbohydr Polym 2015; 128: 112-21.
[http://dx.doi.org/10.1016/j.carbpol.2015.04.022] [PMID: 26005146]

[120] Zhao P, Liu H, Deng H, *et al.* A study of chitosan hydrogel with embedded mesoporous silica nanoparticles loaded by ibuprofen as a dual stimuli-responsive drug release system for surface coating of titanium implants. Colloids Surf B Biointerfaces 2014; 123: 657-63.
[http://dx.doi.org/10.1016/j.colsurfb.2014.10.013] [PMID: 25456989]

[121] Jain T, Kumar S, Dutta P. Theranostics: A way of modern medical diagnostics and the role of chitosan. J Mol Genet Med 2015; 09: 1-5.
[http://dx.doi.org/10.4172/1747-0862.1000159]

[122] Wang C, Ravi S, Garapati US, *et al.* Multifunctional chitosan magnetic-graphene (CMG) nanoparticles: a theranostic platform for tumor-targeted co-delivery of drugs, genes and MRI contrast agents. J Mater Chem B 2013; 1(35): 4396-405.
[http://dx.doi.org/10.1039/c3tb20452a] [PMID: 24883188]

[123] Sahoo AK, Banerjee S, Ghosh SS, Chattopadhyay A. Simultaneous RGB emitting Au nanoclusters in chitosan nanoparticles for anticancer gene theranostics. ACS Appl Mater Interfaces 2014; 6(1): 712-24.
[http://dx.doi.org/10.1021/am4051266] [PMID: 24281656]

[124] Arias JL, Reddy LH, Couvreur P. Fe_3O_4/chitosan nanocomposite for magnetic drug targeting to cancer. J Mater Chem 2012; 22: 7622.
[http://dx.doi.org/10.1039/c2jm15339d]

[125] Chen JH, Xing HT, Guo HX, Li GP, Weng W, Hu SR. Preparation, characterization and adsorption properties of a novel 3-aminopropyltriethoxysilane functionalized sodium alginate porous membrane adsorbent for Cr(III) ions. J Hazard Mater 2013; 248-249: 285-94.
[http://dx.doi.org/10.1016/j.jhazmat.2013.01.042] [PMID: 23399906]

[126] Chen R, Wang X, Yao X, Zheng X, Wang J, Jiang X. Near-IR-triggered photothermal/photodynamic dual-modality therapy system *via* chitosan hybrid nanospheres. Biomaterials 2013; 34(33): 8314-22.
[http://dx.doi.org/10.1016/j.biomaterials.2013.07.034] [PMID: 23896004]

[127] Key J, Cooper C, Kim AY, *et al. In vivo* NIRF and MR dual-modality imaging using glycol chitosan nanoparticles. J Control Release 2012; 163(2): 249-55.
[http://dx.doi.org/10.1016/j.jconrel.2012.07.038] [PMID: 22902594]

[128] Xi D, Dong S, Meng X, Lu Q, Meng L, Ye J. Gold nanoparticles as computerized tomography (CT) contrast agents. RSC Advances 2012; 2(33): 12515-24.
[http://dx.doi.org/10.1039/c2ra21263c]

[129] Chen R, Zheng X, Qian H, Wang X, Wang J, Jiang X. Combined near-IR photothermal therapy and chemotherapy using gold-nanorod/chitosan hybrid nanospheres to enhance the antitumor effect. Biomater Sci 2013; 1(3): 285-93.
[http://dx.doi.org/10.1039/C2BM00138A]

[130] Hoffman AS. Hydrogels for biomedical applications. Adv Drug Deliv Rev 2012; 64: 18-23.
[http://dx.doi.org/10.1016/j.addr.2012.09.010] [PMID: 11755703]

[131] Berger J, Reist M, Mayer JM, Felt O, Peppas NA, Gurny R. Structure and interactions in covalently and ionically crosslinked chitosan hydrogels for biomedical applications. Eur J Pharm Biopharm 2004; 57(1): 19-34.

[http://dx.doi.org/10.1016/S0939-6411(03)00161-9] [PMID: 14729078]

[132] Giri TK, Thakur A, Alexander A. Ajazuddin, Badwaik H, Tripathi DK. Modified chitosan hydrogels as drug delivery and tissue engineering systems: present status and applications. Acta Pharm Sin B 2012; 2(5): 439-49. [http://dx.doi.org/10.1016/j.apsb.2012.07.004]

[133] Rufaihah AJ, Seliktar D. Hydrogels for therapeutic cardiovascular angiogenesis. Adv Drug Deliv Rev 2016; 96: 31-9. [http://dx.doi.org/10.1016/j.addr.2015.07.003] [PMID: 26212158]

[134] Shen X, Shamshina JL, Berton P, Gurau G, Rogers RD. Hydrogels based on cellulose and chitin: fabrication, properties, and applications. Green Chem 2015; 18(1): 53-75. [http://dx.doi.org/10.1039/c5gc02396c]

[135] Maurice D. Kinetics of topically applied ophthalmic drugs. In: Deliv OD, Ed. Marco Fabrizio Saettone, Massimo Bucci P, Speiser. New York: Springer 1987; pp. 19-26.

[136] Urtti A. Challenges and obstacles of ocular pharmacokinetics and drug delivery. Adv Drug Deliv Rev 2006; 58(11): 1131-5. [http://dx.doi.org/10.1016/j.addr.2006.07.027] [PMID: 17097758]

[137] Almeida H, Amaral MH, Lobão P, Lobo JM. *In situ* gelling systems: a strategy to improve the bioavailability of ophthalmic pharmaceutical formulations. Drug Discov Today 2014; 19(4): 400-12. [http://dx.doi.org/10.1016/j.drudis.2013.10.001] [PMID: 24120893]

[138] Kute PR, Gondkar SB, Saudagar RB. Ophthalmic in-situ gel: an overview. World J Pharm Pharm Sci 2015; 4: 549-68.

[139] Felt O, Furrer P, Mayer JM, Plazonnet B, Buri P, Gurny R. Topical use of chitosan in ophthalmology: tolerance assessment and evaluation of precorneal retention. Int J Pharm 1999; 180(2): 185-93. [http://dx.doi.org/10.1016/S0378-5173(99)00003-4] [PMID: 10370189]

[140] Wang W, Xu D. Viscosity and flow properties of concentrated solutions of chitosan with different degrees of deacetylation. Int J Biol Macromol 1994; 16(3): 149-52. [http://dx.doi.org/10.1016/0141-8130(94)90042-6] [PMID: 7981162]

[141] Lehr CM, Bouwstra JA, Schacht EH, Junginger HE. *In vitro* evaluation of mucoadhesive properties of chitosan and some other natural polymers. Int J Pharm 1992; 78: 43-8. [http://dx.doi.org/10.1016/0378-5173(92)90353-4]

[142] He P, Davis SS, Illum L. *In vitro* evaluation of the mucoadhesive properties of chitosan microspheres. Int J Pharm 1998; 166: 75-88. [http://dx.doi.org/10.1016/S0378-5173(98)00027-1]

[143] Verestiuc L, Nastasescu O, Barbu E, Sarvaiya I, Green KL, Tsibouklis J. Functionalized chitosan/NIPAM (HEMA) hybrid polymer networks as inserts for ocular drug delivery: synthesis, *in vitro* assessment, and *in vivo* evaluation. J Biomed Mater Res A 2006; 77(4): 726-35. [http://dx.doi.org/10.1002/jbm.a.30668] [PMID: 16555266]

[144] Chenite A, Chaput C, Wang D, *et al.* Novel injectable neutral solutions of chitosan form biodegradable gels *in situ*. Biomaterials 2000; 21(21): 2155-61. [http://dx.doi.org/10.1016/S0142-9612(00)00116-2] [PMID: 10985488]

[145] Cho J, Heuzey MC, Bégin A, Carreau PJ. Physical gelation of chitosan in the presence of β-glycerophosphate: the effect of temperature. Biomacromolecules 2005; 6(6): 3267-75. [http://dx.doi.org/10.1021/bm050313s] [PMID: 16283755]

[146] Ruel-Gariépy E, Chenite A, Chaput C, Guirguis S, Leroux J. Characterization of thermosensitive chitosan gels for the sustained delivery of drugs. Int J Pharm 2000; 203(1-2): 89-98. [http://dx.doi.org/10.1016/S0378-5173(00)00428-2] [PMID: 10967431]

[147] Rahmati M, Samadikuchaksaraei A, Mozafari M. Insight into the interactive effects of β-glycerophosphate molecules on thermosensitive chitosan-based hydrogels. Bioinspired. Biomim Nanobiomaterials 2016; 5: 67-73.
[http://dx.doi.org/10.1680/jbibn.15.00022]

[148] Kim S, Nishimoto SK, Bumgardner JD, Haggard WO, Gaber MW, Yang Y. A chitosan/beta-glycerophosphate thermo-sensitive gel for the delivery of ellagic acid for the treatment of brain cancer. Biomaterials 2010; 31(14): 4157-66.
[http://dx.doi.org/10.1016/j.biomaterials.2010.01.139] [PMID: 20185170]

[149] Pillai CK, Paul W, Sharma CP. Chitin and chitosan polymers: Chemistry, solubility and fiber formation. Prog Polym Sci 2009; 34: 641-78.
[http://dx.doi.org/10.1016/j.progpolymsci.2009.04.001]

[150] Lim SH, Hudson SM. Synthesis and antimicrobial activity of a water-soluble chitosan derivative with a fiber-reactive group. Carbohydr Res 2004; 339(2): 313-9.
[http://dx.doi.org/10.1016/j.carres.2003.10.024] [PMID: 14698889]

[151] Zhang X, Geng X, Jiang H, Li J, Huang J. Synthesis and characteristics of chitin and chitosan with the (2-hydroxy-3-trimethylammonium)propyl functionality, and evaluation of their antioxidant activity *in vitro*. Carbohydr Polym 2012; 89(2): 486-91.
[http://dx.doi.org/10.1016/j.carbpol.2012.03.032] [PMID: 24750748]

[152] Wu J, Su ZG, Ma GH. A thermo- and pH-sensitive hydrogel composed of quaternized chitosan/glycerophosphate. Int J Pharm 2006; 315(1-2): 1-11.
[http://dx.doi.org/10.1016/j.ijpharm.2006.01.045] [PMID: 16616819]

[153] Cheng Y-H, Yang S-H, Su W-Y, *et al.* Thermosensitive chitosan-gelatin-glycerol phosphate hydrogels as a cell carrier for nucleus pulposus regeneration: an *in vitro* study. Tissue Eng Part A 2010; 16(2): 695-703.
[http://dx.doi.org/10.1089/ten.tea.2009.0229] [PMID: 19769528]

[154] Cheng YH, Hung KH, Tsai TH, *et al.* Sustained delivery of latanoprost by thermosensitive chitosan-gelatin-based hydrogel for controlling ocular hypertension. Acta Biomater 2014; 10(10): 4360-6.
[http://dx.doi.org/10.1016/j.actbio.2014.05.031] [PMID: 24914827]

[155] Tsai CY, Woung LC, Yen JC, *et al.* Thermosensitive chitosan-based hydrogels for sustained release of ferulic acid on corneal wound healing. Carbohydr Polym 2016; 135: 308-15.
[http://dx.doi.org/10.1016/j.carbpol.2015.08.098] [PMID: 26453882]

[156] Hsiao MH, Chiou SH, Larsson M, *et al.* A temperature-induced and shear-reversible assembly of latanoprost-loaded amphiphilic chitosan colloids: characterization and *in vivo* glaucoma treatment. Acta Biomater 2014; 10(7): 3188-96.
[http://dx.doi.org/10.1016/j.actbio.2014.03.016] [PMID: 24681374]

[157] Chien Y, Liao YW, Liu DM, *et al.* Corneal repair by human corneal keratocyte-reprogrammed iPSCs and amphiphatic carboxymethyl-hexanoyl chitosan hydrogel. Biomaterials 2012; 33(32): 8003-16.
[http://dx.doi.org/10.1016/j.biomaterials.2012.07.029] [PMID: 22858046]

[158] Hsiao MH, Larsson M, Larsson A, *et al.* Design and characterization of a novel amphiphilic chitosan nanocapsule-based thermo-gelling biogel with sustained *in vivo* release of the hydrophilic anti-epilepsy drug ethosuximide. J Control Release 2012; 161(3): 942-8.
[http://dx.doi.org/10.1016/j.jconrel.2012.05.038] [PMID: 22652548]

[159] Xu X, Weng Y, Xu L, Chen H. Sustained release of Avastin® from polysaccharides cross-linked hydrogels for ocular drug delivery. Int J Biol Macromol 2013; 60: 272-6.
[http://dx.doi.org/10.1016/j.ijbiomac.2013.05.034] [PMID: 23748006]

[160] Varshosaz J, Tabbakhian M, Salmani Z. Designing of a thermosensitive chitosan/poloxamer *in situ* gel for ocular delivery of ciprofloxacin. Open Drug Deliv J 2008; 2: 61-70.
[http://dx.doi.org/10.2174/1874126600802010061]

[161] Gratieri T, Gelfuso GM, Rocha EM, Sarmento VH, de Freitas O, Lopez RF. A poloxamer/chitosan *in situ* forming gel with prolonged retention time for ocular delivery. Eur J Pharm Biopharm 2010; 75(2): 186-93.
[http://dx.doi.org/10.1016/j.ejpb.2010.02.011] [PMID: 20188828]

[162] Gratieri T, Gelfuso GM, de Freitas O, Rocha EM, Lopez RF. Enhancing and sustaining the topical ocular delivery of fluconazole using chitosan solution and poloxamer/chitosan *in situ* forming gel. Eur J Pharm Biopharm 2011; 79(2): 320-7.
[http://dx.doi.org/10.1016/j.ejpb.2011.05.006] [PMID: 21641994]

[163] Edsman K, Carlfors J, Petersson R. Rheological evaluation of poloxamer as an *in situ* gel for ophthalmic use. Eur J Pharm Sci 1998; 6(2): 105-12.
[http://dx.doi.org/10.1016/S0928-0987(97)00075-4] [PMID: 9795025]

[164] Dumortier G, Grossiord JL, Agnely F, Chaumeil JC. A review of poloxamer 407 pharmaceutical and pharmacological characteristics. Pharm Res 2006; 23(12): 2709-28.
[http://dx.doi.org/10.1007/s11095-006-9104-4] [PMID: 17096184]

[165] Gupta H, Aqil M, Khar RK, Ali A, Bhatnagar A, Mittal G. An alternative *in situ* gel-formulation of levofloxacin eye drops for prolong ocular retention. J Pharm Bioallied Sci 2015; 7(1): 9-14.
[http://dx.doi.org/10.4103/0975-7406.149810] [PMID: 25709330]

[166] Ramos-de-la-Peña AM, Renard CM, Wicker L, Montañez JC, García-Cerda LA, Contreras-Esquivel JC. Environmental friendly cold-mechanical/sonic enzymatic assisted extraction of genipin from genipap (Genipa americana). Ultrason Sonochem 2014; 21(1): 43-9.
[http://dx.doi.org/10.1016/j.ultsonch.2013.06.008] [PMID: 23871416]

[167] Sung H-W, Huang R-N, Huang LL, Tsai C-C. *In vitro* evaluation of cytotoxicity of a naturally occurring cross-linking reagent for biological tissue fixation. J Biomater Sci Polym Ed 1999; 10(1): 63-78.
[http://dx.doi.org/10.1163/156856299X00289] [PMID: 10091923]

[168] Lai J-Y, Li Y-T, Wang T-P. *In vitro* response of retinal pigment epithelial cells exposed to chitosan materials prepared with different cross-linkers. Int J Mol Sci 2010; 11(12): 5256-72.
[http://dx.doi.org/10.3390/ijms11125256] [PMID: 21614206]

[169] Lai JY. Biocompatibility of genipin and glutaraldehyde cross-linked chitosan materials in the anterior chamber of the eye. Int J Mol Sci 2012; 13(9): 10970-85.
[http://dx.doi.org/10.3390/ijms130910970] [PMID: 23109832]

[170] Rittenhouse KD, Pollack GM. Microdialysis and drug delivery to the eye. Adv Drug Deliv Rev 2000; 45(2-3): 229-41.
[http://dx.doi.org/10.1016/S0169-409X(00)00112-5] [PMID: 11108976]

[171] Wadhwa S, Paliwal R, Paliwal SR, Vyas SP. Nanocarriers in ocular drug delivery: an update review. Curr Pharm Des 2009; 15(23): 2724-50.
[http://dx.doi.org/10.2174/138161209788923886] [PMID: 19689343]

[172] Ugwoke MI, Agu RU, Verbeke N, Kinget R. Nasal mucoadhesive drug delivery: background, applications, trends and future perspectives. Adv Drug Deliv Rev 2005; 57(11): 1640-65.
[http://dx.doi.org/10.1016/j.addr.2005.07.009] [PMID: 16182408]

[173] Yu S, Zhao Y, Wu F, *et al.* Nasal insulin delivery in the chitosan solution: *in vitro* and *in vivo* studies. Int J Pharm 2004; 281(1-2): 11-23.
[http://dx.doi.org/10.1016/j.ijpharm.2004.05.007] [PMID: 15288339]

[174] Wu J, Wei W, Wang LY, Su ZG, Ma GH. A thermosensitive hydrogel based on quaternized chitosan and poly(ethylene glycol) for nasal drug delivery system. Biomaterials 2007; 28(13): 2220-32.
[http://dx.doi.org/10.1016/j.biomaterials.2006.12.024] [PMID: 17291582]

[175] Agrawal AK, Gupta PN, Khanna A, *et al.* Development and characterization of in situ gel system for nasal insulin delivery. Pharmazie 2010; 65(3): 188-93.

[PMID: 20383938]

[176] van der Merwe SM, Verhoef JC, Verheijden JH, Kotzé AF, Junginger HE. Trimethylated chitosan as polymeric absorption enhancer for improved peroral delivery of peptide drugs. Eur J Pharm Biopharm 2004; 58(2): 225-35.
[http://dx.doi.org/10.1016/j.ejpb.2004.03.023] [PMID: 15296951]

[177] Hamman JH, Stander M, Kotzé AF. Effect of the degree of quaternisation of N-trimethyl chitosan chloride on absorption enhancement: *in vivo* evaluation in rat nasal epithelia. Int J Pharm 2002; 232(1-2): 235-42.
[http://dx.doi.org/10.1016/S0378-5173(01)00914-0] [PMID: 11790507]

[178] Dahl R, Mygind N. Anatomy, physiology and function of the nasal cavities in health and disease. Adv Drug Deliv Rev 1998; 29(1-2): 3-12.
[http://dx.doi.org/10.1016/S0169-409X(97)00058-6] [PMID: 10837577]

[179] Nazar H, Fatouros DG, van der Merwe SM, *et al.* Thermosensitive hydrogels for nasal drug delivery: the formulation and characterisation of systems based on N-trimethyl chitosan chloride. Eur J Pharm Biopharm 2011; 77(2): 225-32.
[http://dx.doi.org/10.1016/j.ejpb.2010.11.022] [PMID: 21130876]

[180] Chung T, Liu D, Yang J. Effects of interpenetration of thermo-sensitive gels by crosslinking of chitosan on nasal delivery of insulin : *In vitro* characterization and *in vivo* study. Carbohydr Polym 2010; 82: 316-22.
[http://dx.doi.org/10.1016/j.carbpol.2010.04.068]

[181] Davis SS. Nasal vaccines. Adv Drug Deliv Rev 2001; 51(1-3): 21-42.
[http://dx.doi.org/10.1016/S0169-409X(01)00162-4] [PMID: 11516777]

[182] Wu Y, Wu S, Hou L, *et al.* Novel thermal-sensitive hydrogel enhances both humoral and cell-mediated immune responses by intranasal vaccine delivery. Eur J Pharm Biopharm 2012; 81(3): 486-97.
[http://dx.doi.org/10.1016/j.ejpb.2012.03.021] [PMID: 22507968]

[183] Evans DF, Pye G, Bramley R, Clark AG, Dyson TJ, Hardcastle JD. Measurement of gastrointestinal pH profiles in normal ambulant human subjects. Gut 1988; 29(8): 1035-41.
[http://dx.doi.org/10.1136/gut.29.8.1035] [PMID: 3410329]

[184] Rubinstein A. Colonic drug delivery. Drug Discov Today Technol 2005; 2(1): 33-7.
[http://dx.doi.org/10.1016/j.ddtec.2005.05.021] [PMID: 24981753]

[185] Park BG, Kang H-S, Lee W, Kim JS, Son T-I. Reinforcement of pH-responsive γ-poly(glutamic acid)/chitosan hydrogel for orally administrable colon-targeted drug delivery. J Appl Polym Sci 2013; 127: 832-6.
[http://dx.doi.org/10.1002/app.37565]

[186] Lankalapalli S, Kolapalli VR. Polyelectrolyte complexes: a review of their applicability in drug delivery technology. Indian J Pharm Sci 2009; 71(5): 481-7.
[http://dx.doi.org/10.4103/0250-474X.58165] [PMID: 20502564]

[187] Upadhyaya L, Singh J, Agarwal V, Tewari RP. Biomedical applications of carboxymethyl chitosans. Carbohydr Polym 2013; 91(1): 452-66.
[http://dx.doi.org/10.1016/j.carbpol.2012.07.076] [PMID: 23044156]

[188] Chen L, Tian Z, Du Y. Synthesis and pH sensitivity of carboxymethyl chitosan-based polyampholyte hydrogels for protein carrier matrices. Biomaterials 2004; 25(17): 3725-32.
[http://dx.doi.org/10.1016/j.biomaterials.2003.09.100] [PMID: 15020148]

[189] Vaghani SS, Patel MM, Satish CS. Synthesis and characterization of pH-sensitive hydrogel composed of carboxymethyl chitosan for colon targeted delivery of ornidazole. Carbohydr Res 2012; 347(1): 76-82.
[http://dx.doi.org/10.1016/j.carres.2011.04.048] [PMID: 22099382]

[190] Casettari L, Vllasaliu D, Castagnino E, Stolnik S, Howdle S, Illum L. PEGylated chitosan derivatives: Synthesis, characterizations and pharmaceutical applications. Prog Polym Sci 2012; 37: 659-85. [http://dx.doi.org/10.1016/j.progpolymsci.2011.10.001]

[191] Islam A, Yasin T. Controlled delivery of drug from pH sensitive chitosan/poly (vinyl alcohol) blend. Carbohydr Polym 2012; 88: 1055-60. [http://dx.doi.org/10.1016/j.carbpol.2012.01.070]

[192] Liu Q, Zuo Q, Guo R, Hong A, Li C, Zhang Y, *et al.* Fabrication and characterization of carboxymethyl chitosan/poly(vinyl alcohol) hydrogels containing alginate microspheres for protein delivery. J Bioact Compat Polym 2015; 30: 397-411. [http://dx.doi.org/10.1177/0883911515578761]

[193] Bawa P, Pillay V, Choonara YE, du Toit LC, Ndesendo VM, Kumar P. A composite polyelectrolytic matrix for controlled oral drug delivery. AAPS PharmSciTech 2011; 12(1): 227-38. [http://dx.doi.org/10.1208/s12249-010-9576-8] [PMID: 21225384]

[194] Qiu Y, Park K. Environment-sensitive hydrogels for drug delivery. Adv Drug Deliv Rev 2001; 53(3): 321-39. [http://dx.doi.org/10.1016/S0169-409X(01)00203-4] [PMID: 11744175]

[195] Gong S, Tu H, Zheng H, Xu H, Yin Y. Chitosan-g-PAA hydrogels for colon-specific drug delivery: Preparation, swelling behavior and *in vitro* degradability. J Wuhan Univ Technol Mater Sci Ed 2010; 25: 248-51. [http://dx.doi.org/10.1007/s11595-010-2248-4]

[196] Milosavljevic NB, Milasinovic NZ, Popovic IG, Filipovic JM, Krusic MK. Preparation and characterization of pH-sensitive hydrogels based on chitosan, itaconic acid and methacrylic acid. Polym Int 2011; 60: 443-52. [http://dx.doi.org/10.1002/pi.2967]

[197] Shivashankar M, Mandal BK. A review on interpenetrating polymer network. Int J Pharm Pharm Sci 2012; 4: 1-7.

[198] Chen S, Liu M, Jin S, Chen YP. PH-/temperature-sensitive carboxymethyl chitosan/poly(N-isopropylacrylamide-co-methacrylic acid) IPN: Preparation, characterization and sustained release of riboflavin. Polym Bull 2014; 71: 719-34. [http://dx.doi.org/10.1007/s00289-013-1088-8]

[199] Chen S-C, Wu Y-C, Mi F-L, Lin Y-H, Yu L-C, Sung H-W. A novel pH-sensitive hydrogel composed of N,O-carboxymethyl chitosan and alginate cross-linked by genipin for protein drug delivery. J Control Release 2004; 96(2): 285-300. [http://dx.doi.org/10.1016/j.jconrel.2004.02.002] [PMID: 15081219]

[200] Guo BL, Gao QY. Preparation and properties of a pH/temperature-responsive carboxymethyl chitosan/poly(N-isopropylacrylamide)semi-IPN hydrogel for oral delivery of drugs. Carbohydr Res 2007; 342(16): 2416-22. [http://dx.doi.org/10.1016/j.carres.2007.07.007] [PMID: 17669378]

[201] Bibby DC, Davies NM, Tucker IG. Mechanisms by which cyclodextrins modify drug release from polymeric drug delivery systems. Int J Pharm 2000; 197(1-2): 1-11. [http://dx.doi.org/10.1016/S0378-5173(00)00335-5] [PMID: 10704788]

[202] Challa R, Ahuja A, Ali J, Khar RK. Cyclodextrins in drug delivery: an updated review. AAPS PharmSciTech 2005; 6(2): E329-57. [http://dx.doi.org/10.1208/pt060243] [PMID: 16353992]

[203] Das S, Subuddhi U. Cyclodextrin mediated controlled release of naproxen from pH-sensitive chitosan/poly(vinyl alcohol) hydrogels for colon targeted delivery. Ind Eng Chem Res 2013; 52: 14192-200. [http://dx.doi.org/10.1021/ie402121f]

[204] Das S, Subuddhi U. Controlled and targeted delivery of diclofenac sodium to the intestine from pH-Responsive chitosan/poly(vinyl alcohol) interpenetrating polymeric network hydrogels. Polym Sci Ser A 2016; 58: 154-66.
[http://dx.doi.org/10.1134/S0965545X16020048]

[205] Yin L, Fei L, Cui F, Tang C, Yin C. Superporous hydrogels containing poly(acrylic acid-c--acrylamide)/O-carboxymethyl chitosan interpenetrating polymer networks. Biomaterials 2007; 28(6): 1258-66.
[http://dx.doi.org/10.1016/j.biomaterials.2006.11.008] [PMID: 17118443]

[206] Yin L, Ding J, Fei L, *et al.* Beneficial properties for insulin absorption using superporous hydrogel containing interpenetrating polymer network as oral delivery vehicles. Int J Pharm 2008; 350(1-2): 220-9.
[http://dx.doi.org/10.1016/j.ijpharm.2007.08.051] [PMID: 17976932]

[207] Yin L, Ding J, Zhang J, He C, Tang C, Yin C. Polymer integrity related absorption mechanism of superporous hydrogel containing interpenetrating polymer networks for oral delivery of insulin. Biomaterials 2010; 31(12): 3347-56.
[http://dx.doi.org/10.1016/j.biomaterials.2010.01.045] [PMID: 20116843]

[208] Omidian H, Park K, Rocca JG. Recent developments in superporous hydrogels. J Pharm Pharmacol 2007; 59(3): 317-27.
[http://dx.doi.org/10.1211/jpp.59.3.0001] [PMID: 17331335]

[209] Chaturvedi K, Ganguly K, Nadagouda MN, Aminabhavi TM. Polymeric hydrogels for oral insulin delivery. J Control Release 2013; 165(2): 129-38.
[http://dx.doi.org/10.1016/j.jconrel.2012.11.005] [PMID: 23159827]

[210] Yin L, Zhao X, Cui L, *et al.* Cytotoxicity and genotoxicity of superporous hydrogel containing interpenetrating polymer networks. Food Chem Toxicol 2009; 47(6): 1139-45.
[http://dx.doi.org/10.1016/j.fct.2009.01.043] [PMID: 19425232]

[211] Mahkam M. Modified chitosan cross-linked starch polymers for oral insulin delivery. J Bioact Compat Polym 2010; 25: 406-18.
[http://dx.doi.org/10.1177/0883911510369038]

[212] Delmar K, Bianco-Peled H. Composite chitosan hydrogels for extended release of hydrophobic drugs. Carbohydr Polym 2016; 136: 570-80.
[http://dx.doi.org/10.1016/j.carbpol.2015.09.072] [PMID: 26572389]

[213] Abou Taleb MF, Alkahtani A, Mohamed SK. Radiation synthesis and characterization of sodium alginate/chitosan/hydroxyapatite nanocomposite hydrogels: a drug delivery system for liver cancer. Polym Bull 2015; 72: 725-42.
[http://dx.doi.org/10.1007/s00289-015-1301-z]

[214] Rossi S, Sandri G, Caramella CM. Buccal drug delivery: A challenge already won? Drug Discov Today Technol 2005; 2(1): 59-65.
[http://dx.doi.org/10.1016/j.ddtec.2005.05.018] [PMID: 24981756]

[215] Senel S, Kremer MJ, Kaş S, Wertz PW, Hincal AA, Squier CA. Enhancing effect of chitosan on peptide drug delivery across buccal mucosa. Biomaterials 2000; 21(20): 2067-71.
[http://dx.doi.org/10.1016/S0142-9612(00)00134-4] [PMID: 10966016]

[216] Kong M, Chen XG, Xing K, Park HJ. Antimicrobial properties of chitosan and mode of action: a state of the art review. Int J Food Microbiol 2010; 144(1): 51-63.
[http://dx.doi.org/10.1016/j.ijfoodmicro.2010.09.012] [PMID: 20951455]

[217] Şenel S, İkinci G, Kaş S, Yousefi-Rad A, Sargon MF, Hincal AA. Chitosan films and hydrogels of chlorhexidine gluconate for oral mucosal delivery. Int J Pharm 2000; 193(2): 197-203.
[http://dx.doi.org/10.1016/S0378-5173(99)00334-8] [PMID: 10606782]

[218] Perchyonok T, Zhang S, Oberholzer T. Novel melatonin-chitosan hydrogels as suitable oral bio-drug delivery systems to fight oral mucositis : synergy of antioxidants and bioactives in action. Curr Org Chem 2012; 16: 2430-6.
[http://dx.doi.org/10.2174/138527212803520173]

[219] Perchyonok VT, Zhang S, Oberholzer T. Chitosan and gelatin based prototype delivery systems for the treatment of oral mucositis: from material to performance *in vitro*. Curr Drug Deliv 2013; 10(1): 144-50.
[http://dx.doi.org/10.2174/1567201811310010020] [PMID: 23017090]

[220] Xu J, Soliman GM, Barralet J, Cerruti M. Mollusk glue inspired mucoadhesives for biomedical applications. Langmuir 2012; 28(39): 14010-7.
[http://dx.doi.org/10.1021/la3025414] [PMID: 22950962]

[221] Xu J, Strandman S, Zhu JX, Barralet J, Cerruti M. Genipin-crosslinked catechol-chitosan mucoadhesive hydrogels for buccal drug delivery. Biomaterials 2015; 37: 395-404.
[http://dx.doi.org/10.1016/j.biomaterials.2014.10.024] [PMID: 25453967]

[222] Kim K, Ryu JH, Lee DY, Lee H. Bio-inspired catechol conjugation converts water-insoluble chitosan into a highly water-soluble, adhesive chitosan derivative for hydrogels and LbL assembly. Biomater Sci 2013; 1: 783-90.
[http://dx.doi.org/10.1039/c3bm00004d]

[223] Bagul US, Patil R V, Shirsath YA, Nikam AJ. Stomach specific drug delivery systems : a review. 2012; 4: 1527-45.

[224] Vaghani SS, Patel MM. Hydrogels based on interpenetrating network of chitosan and polyvinyl pyrrolidone for pH-sensitive delivery of repaglinide. Curr Drug Discov Technol 2011; 8(2): 126-35.
[http://dx.doi.org/10.2174/157016311795563848] [PMID: 21513486]

[225] Vaghani SS, Patel MM. pH-sensitive hydrogels based on semi-interpenetrating network (semi-IPN) of chitosan and polyvinyl pyrrolidone for clarithromycin release. Drug Dev Ind Pharm 2011; 37(10): 1160-9.
[http://dx.doi.org/10.3109/03639045.2011.563422] [PMID: 21417603]

[226] Risbud MV, Hardikar AA, Bhat SV, Bhonde RR. pH-sensitive freeze-dried chitosan-polyvinyl pyrrolidone hydrogels as controlled release system for antibiotic delivery. J Control Release 2000; 68(1): 23-30.
[http://dx.doi.org/10.1016/S0168-3659(00)00208-X] [PMID: 10884576]

[227] Dumitriu S, Magny P, Montané D, Vidal PF, Chornet E. Polyionic hydrogels obtained by complexation between xanthan and chitosan: their properties as supports for enzyme immobilization. J Bioact Compat Polym 1994; 9: 184-209.
[http://dx.doi.org/10.1177/088391159400900205]

[228] Argin-Soysal S, Kofinas P, Lo YM. Effect of complexation conditions on xanthan-chitosan polyelectrolyte complex gels. Food Hydrocoll 2009; 23: 202-9.
[http://dx.doi.org/10.1016/j.foodhyd.2007.12.011]

[229] Popa N, Novac O, Profire L, Lupusoru CE, Popa MI. Hydrogels based on chitosan-xanthan for controlled release of theophylline. J Mater Sci Mater Med 2010; 21(4): 1241-8.
[http://dx.doi.org/10.1007/s10856-009-3937-4] [PMID: 19924518]

[230] Cerchiara T, Luppi B, Bigucci F, Orienti I, Zecchi V. Physically cross-linked chitosan hydrogels as topical vehicles for hydrophilic drugs. J Pharm Pharmacol 2002; 54(11): 1453-9.
[http://dx.doi.org/10.1211/00223570281] [PMID: 12495547]

[231] He W, Guo X, Xiao L, Feng M. Study on the mechanisms of chitosan and its derivatives used as transdermal penetration enhancers. Int J Pharm 2009; 382(1-2): 234-43.
[http://dx.doi.org/10.1016/j.ijpharm.2009.07.038] [PMID: 19686826]

[232] Zhou X, You G, Liu D, Yao K. *In vitro* investigation on the interactions of oligo-chitosan with keratin. Acta Polym Sin 2009; 781-5.

[233] He W, Guo X, Zhang M. Transdermal permeation enhancement of N-trimethyl chitosan for testosterone. Int J Pharm 2008; 356(1-2): 82-7. [http://dx.doi.org/10.1016/j.ijpharm.2007.12.050] [PMID: 18337030]

[234] Fang J-Y, Sung KC, Wang J-J, Chu C-C, Chen K-T. The effects of iontophoresis and electroporation on transdermal delivery of buprenorphine from solutions and hydrogels. J Pharm Pharmacol 2002; 54(10): 1329-37. [http://dx.doi.org/10.1211/002235702760345392] [PMID: 12396293]

[235] Taveira SF, Nomizo A, Lopez RF. Effect of the iontophoresis of a chitosan gel on doxorubicin skin penetration and cytotoxicity. J Control Release 2009; 134(1): 35-40. [http://dx.doi.org/10.1016/j.jconrel.2008.11.002] [PMID: 19047006]

[236] Thein-Han WW, Stevens WF. Transdermal delivery controlled by a chitosan membrane. Drug Dev Ind Pharm 2004; 30(4): 397-404. [http://dx.doi.org/10.1081/DDC-120030934] [PMID: 15132182]

[237] Kählig H, Hasanovic A, Biruss B, Höller S, Grim J, Valenta C. Chitosan-glycolic acid: a possible matrix for progesterone delivery into skin. Drug Dev Ind Pharm 2009; 35(8): 997-1002. [http://dx.doi.org/10.1080/03639040902755189] [PMID: 19365777]

[238] Zu Y, Zhang Y, Zhao X, *et al.* Preparation and characterization of chitosan-polyvinyl alcohol blend hydrogels for the controlled release of nano-insulin. Int J Biol Macromol 2012; 50(1): 82-7. [http://dx.doi.org/10.1016/j.ijbiomac.2011.10.006] [PMID: 22020189]

[239] Peng Q, Sun X, Gong T, *et al.* Injectable and biodegradable thermosensitive hydrogels loaded with PHBHHx nanoparticles for the sustained and controlled release of insulin. Acta Biomater 2013; 9(2): 5063-9. [http://dx.doi.org/10.1016/j.actbio.2012.09.034] [PMID: 23036950]

[240] Costa IdosS, Abranches RP, Garcia MT, Pierre MB. Chitosan-based mucoadhesive films containing 5-aminolevulinic acid for buccal cancers treatment. J Photochem Photobiol B 2014; 140: 266-75. [http://dx.doi.org/10.1016/j.jphotobiol.2014.08.005] [PMID: 25190225]

[241] Morales JO, McConville JT. Manufacture and characterization of mucoadhesive buccal films. Eur J Pharm Biopharm 2011; 77(2): 187-99. [http://dx.doi.org/10.1016/j.ejpb.2010.11.023] [PMID: 21130875]

[242] Altiok D, Altiok E, Tihminlioglu F. Physical, antibacterial and antioxidant properties of chitosan films incorporated with thyme oil for potential wound healing applications. J Mater Sci Mater Med 2010; 21(7): 2227-36. [http://dx.doi.org/10.1007/s10856-010-4065-x] [PMID: 20372985]

[243] Levi-Polyachenko N, Jacob R, Day C, Kuthirummal N. Chitosan wound dressing with hexagonal silver nanoparticles for hyperthermia and enhanced delivery of small molecules. Colloids Surf B Biointerfaces 2016; 142: 315-24. [http://dx.doi.org/10.1016/j.colsurfb.2016.02.038] [PMID: 26970819]

[244] Fedotov AY, Sevost'yanov MA, Sergienko KV, *et al.* Chitosan-based films with medicines. Inorg Mater: Appl Res 2014; 5: 330-3. [http://dx.doi.org/10.1134/S2075113314040212]

[245] Balassa LL, Prudden JF. Applications of chitin and chitosan in wound healing acceleration. In: Zikakis JP, Ed. Chitin, chitosan and related enzymes. 296-305.

[246] Muzzarelli R, Baldassarre V, Conti F, *et al.* Biological activity of chitosan: ultrastructural study. Biomaterials 1988; 9(3): 247-52. [http://dx.doi.org/10.1016/0142-9612(88)90092-0] [PMID: 3408796]

[247] Kojima K, Okamoto Y, Kojima K, *et al.* Effects of chitin and chitosan on collagen synthesis in wound healing. J Vet Med Sci 2004; 66(12): 1595-8. [http://dx.doi.org/10.1292/jvms.66.1595] [PMID: 15644615]

[248] Muzzarelli RA, Mattioli-Belmonte M, Pugnaloni A, Biagini G. Biochemistry, histology and clinical uses of chitins and chitosans in wound healing. EXS 1999; 87: 251-64. [PMID: 10906965]

[249] Lim JI, Lee W-K. Preparation of macroporous chitosan patch using co-solvent as a transdermal drug-delivery system. J Porous Mater 2014; 21: 783-8. [http://dx.doi.org/10.1007/s10934-014-9825-4]

[250] Xie Y, Xu B, Gao Y. Controlled transdermal delivery of model drug compounds by MEMS microneedle array. Nanomedicine Nanotechnology. Biol Med 2005; 1: 184-90. [http://dx.doi.org/10.1016/j.nano.2005.03.001]

[251] Silva CL, Pereira JC, Ramalho A, Pais AA, Sousa JJ. Films based on chitosan polyelectrolyte complexes for skin drug delivery: Development and characterization. J Membr Sci 2008; 320: 268-79. [http://dx.doi.org/10.1016/j.memsci.2008.04.011]

[252] Bigucci F, Abruzzo A, Saladini B, Gallucci MC, Cerchiara T, Luppi B. Development and characterization of chitosan/hyaluronan film for transdermal delivery of thiocolchicoside. Carbohydr Polym 2015; 130: 32-40. [http://dx.doi.org/10.1016/j.carbpol.2015.04.067] [PMID: 26076598]

[253] Larrañeta E, Lutton RE, Woolfson AD, Donnelly RF. Microneedle arrays as transdermal and intradermal drug delivery systems: Materials science, manufacture and commercial development. Mater Sci Eng Rep 2016; 104: 1-32. [http://dx.doi.org/10.1016/j.mser.2016.03.001]

[254] Ahad A, Al-Jenoobi FI, Al-Mohizea AM, Akhtar N, Raish M, Aqil M. Systemic delivery of β-blockers *via* transdermal route for hypertension. Saudi Pharm J 2015; 23(6): 587-602. [http://dx.doi.org/10.1016/j.jsps.2013.12.019] [PMID: 26702253]

[255] Rasool BK, Aziz US, Sarheed O, Rasool AA. Design and evaluation of a bioadhesive film for transdermal delivery of propranolol hydrochloride. Acta Pharm 2011; 61(3): 271-82. [http://dx.doi.org/10.2478/v10007-011-0025-3] [PMID: 21945906]

[256] Thacharodi D, Rao KP. Development and *in vitro* evaluation of chitosan-based transdermal drug delivery systems for the controlled delivery of propranolol hydrochloride. Biomaterials 1995; 16(2): 145-8. [http://dx.doi.org/10.1016/0142-9612(95)98278-M] [PMID: 7734649]

[257] Sapra B, Jain S, Tiwary AK. Transdermal delivery of carvedilol containing glycyrrhizin and chitosan as permeation enhancers: biochemical, biophysical, microscopic and pharmacodynamic evaluation. Drug Deliv 2008; 15(7): 443-54. [http://dx.doi.org/10.1080/10717540802327047] [PMID: 18712622]

[258] Paul S, Jayan A, Sasikumar CS. Physical, chemical and biological studies of gelatin/chitosan based transdermal films with embedded silver nanoparticles. Asian Pacific J Trop Dis 2015; 5: 975-86. [http://dx.doi.org/10.1016/S2222-1808(15)60968-9]

[259] Han F, Dong Y, Song A, Yin R, Li S. Alginate/chitosan based bi-layer composite membrane as potential sustained-release wound dressing containing ciprofloxacin hydrochloride. Appl Surf Sci 2014; 311: 626-34. [http://dx.doi.org/10.1016/j.apsusc.2014.05.125]

[260] Ammar HO, Salama HA, El-Nahhas SA, Elmotasem H. Design and evaluation of chitosan films for transdermal delivery of glimepiride. Curr Drug Deliv 2008; 5(4): 290-8. [http://dx.doi.org/10.2174/156720108785915005] [PMID: 18855598]

[261] Ahad A, Al-Saleh AA, Akhtar N, Al-Mohizea AM, Al-Jenoobi FI. Transdermal delivery of antidiabetic drugs: formulation and delivery strategies. Drug Discov Today 2015; 20(10): 1217-27. [http://dx.doi.org/10.1016/j.drudis.2015.06.002] [PMID: 26100737]

[262] Montenegro-Nicolini M, Morales JO. Overview and future potential of buccal mucoadhesive films as drug delivery systems for biologics. AAPS PharmSciTech 2016; 1-12. [http://dx.doi.org/10.1208/s12249-016-0525-z] [PMID: 27084567]

[263] Hearnden V, Sankar V, Hull K, *et al.* New developments and opportunities in oral mucosal drug delivery for local and systemic disease. Adv Drug Deliv Rev 2012; 64(1): 16-28. [http://dx.doi.org/10.1016/j.addr.2011.02.008] [PMID: 21371513]

[264] Tang C, Guan Y-X, Yao S-J, Zhu Z-Q. Preparation of ibuprofen-loaded chitosan films for oral mucosal drug delivery using supercritical solution impregnation. Int J Pharm 2014; 473(1-2): 434-41. [http://dx.doi.org/10.1016/j.ijpharm.2014.07.039] [PMID: 25079432]

[265] Mansuri S, Kesharwani P, Jain K, Tekade RK, Jain NK. Mucoadhesion: A promising approach in drug delivery system. React Funct Polym 2016; 100: 151-72. [http://dx.doi.org/10.1016/j.reactfunctpolym.2016.01.011]

[266] Russo E, Selmin F, Baldassari S, *et al.* A focus on mucoadhesive polymers and their application in buccal dosage forms. J Drug Deliv Sci Technol 2016; 32: 113-25. [http://dx.doi.org/10.1016/j.jddst.2015.06.016]

[267] Abruzzo A, Bigucci F, Cerchiara T, Cruciani F, Vitali B, Luppi B. Mucoadhesive chitosan/gelatin films for buccal delivery of propranolol hydrochloride. Carbohydr Polym 2012; 87: 581-8. [http://dx.doi.org/10.1016/j.carbpol.2011.08.024]

[268] Giovino C, Ayensu I, Tetteh J, Boateng JS. An integrated buccal delivery system combining chitosan films impregnated with peptide loaded PEG-b-PLA nanoparticles. Colloids Surf B Biointerfaces 2013; 112: 9-15. [http://dx.doi.org/10.1016/j.colsurfb.2013.07.019] [PMID: 23928054]

[269] Mortazavian E, Dorkoosh FA, Rafiee-Tehrani M. Design, characterization and *ex vivo* evaluation of chitosan film integrating of insulin nanoparticles composed of thiolated chitosan derivative for buccal delivery of insulin. Drug Dev Ind Pharm 2014; 40(5): 691-8. [http://dx.doi.org/10.3109/03639045.2014.886590] [PMID: 24524272]

[270] Park D-M, Song Y-K, Jee J-P, Kim HT, Kim C-K. Development of chitosan-based ondansetron buccal delivery system for the treatment of emesis. Drug Dev Ind Pharm 2012; 38(9): 1077-83. [http://dx.doi.org/10.3109/03639045.2011.639076] [PMID: 22182584]

[271] Khan G, Yadav SK, Patel RR, Nath G, Bansal M, Mishra B. Development and evaluation of biodegradable chitosan films of metronidazole and levofloxacin for the management of periodontitis. AAPS PharmSciTech 2015; 1-14. [http://dx.doi.org/10.1208/s12249-015-0466-y] [PMID: 26689408]

[272] Jug M, Maestrelli F, Mura P. Native and polymeric β-cyclodextrins in performance improvement of chitosan films aimed for buccal delivery of poorly soluble drugs. J Incl Phenom Macrocycl Chem 2012; 74: 87-97. [http://dx.doi.org/10.1007/s10847-011-0086-4]

[273] Aldana AA, González A, Strumia MC, Martinelli M. Preparation and characterization of chitosan/genipin/poly(N-vinyl-2-pyrrolidone) films for controlled release drugs. Mater Chem Phys 2012; 134: 317-24. [http://dx.doi.org/10.1016/j.matchemphys.2012.02.071]

[274] Montero-Padilla S, Velaga S, Morales JO. Buccal dosage forms: general considerations for pediatric patients. AAPS PharmSciTech 2016; 1-10. [http://dx.doi.org/10.1208/s12249-016-0567-2] [PMID: 27301872]

[275] Rachmawati H, Haryadi BM, Anggadiredja K, Suendo V. Intraoral film containing insulin-phospholipid microemulsion: formulation and *in vivo* hypoglycemic activity study. AAPS PharmSciTech 2015; 16(3): 692-703.
[http://dx.doi.org/10.1208/s12249-014-0258-9] [PMID: 25511810]

[276] Colonna C, Genta I, Perugini P, *et al.* 5-methyl-pyrrolidinone chitosan films as carriers for buccal administration of proteins. AAPS PharmSciTech 2006; 7(3): 70.
[http://dx.doi.org/10.1208/pt070370] [PMID: 17025236]

[277] Morales JO, Huang S, Williams RO III, McConville JT. Films loaded with insulin-coated nanoparticles (ICNP) as potential platforms for peptide buccal delivery. Colloids Surf B Biointerfaces 2014; 122: 38-45.
[http://dx.doi.org/10.1016/j.colsurfb.2014.05.025] [PMID: 25016543]

[278] He C, Cui F, Yin L, Qian F, Tang C, Yin C. A polymeric composite carrier for oral delivery of peptide drugs: bilaminated hydrogel film loaded with nanoparticles. Eur Polym J 2009; 45: 368-76.
[http://dx.doi.org/10.1016/j.eurpolymj.2008.11.004]

[279] Mazzarino L, Borsali R, Lemos-Senna E. Mucoadhesive films containing chitosan-coated nanoparticles: a new strategy for buccal curcumin release. J Pharm Sci 2014; 103(11): 3764-71.
[http://dx.doi.org/10.1002/jps.24142] [PMID: 25187001]

[280] Lin HR, Chang PC. Novel pluronic-chitosan micelle as an ocular delivery system. J Biomed Mater Res - Part B Appl Biomater 2013; 101 B: 689-99.
[http://dx.doi.org/10.1002/jbm.b.32871]

[281] Rodrigues LB, Leite HF, Yoshida MI, Saliba JB, Cunha AS Jr, Faraco AA. *In vitro* release and characterization of chitosan films as dexamethasone carrier. Int J Pharm 2009; 368(1-2): 1-6.
[http://dx.doi.org/10.1016/j.ijpharm.2008.09.047] [PMID: 18955123]

[282] Baiyrkhanova A, Ismailova A, Botabekova T, Enin E, Semenova Y. Crosslinked Chitosan/PVA film, suturated with 5- fluorouracil for the prevention of proliferative vitreoretinopathy. Int J Drug Deliv Technol 2016; 6: 47-51.

[283] Rodríguez-Vázquez M, Vega-Ruiz B, Ramos-Zúñiga R, Saldaña-Koppel DA, Quiñones-Olvera LF. Chitosan and its potential use as a scaffold for tissue engineering in regenerative medicine. BioMed Res Int 2015; 2015: 821279.
[http://dx.doi.org/10.1155/2015/821279] [PMID: 26504833]

[284] Jiang T, Deng M, Fattah WI. Chitosan-based biopharmaceutical scaffolds in tissue engineering and regenerative medicine. In: Sarmento B, Neves J das, Eds. Chitosan-Based Syst Biopharm Deliv Target Polym Ther. Chichester, UK: John Wiley & Sons, Ltd 2012; pp. 393-427.

[285] Dhandayuthapani B, Yoshida Y, Maekawa T, Kumar DS. Polymeric scaffolds in tissue engineering application : a review. Int J Polym Sci 2011; 1-19.
[http://dx.doi.org/10.1155/2011/290602]

[286] Jiang T, Kumbar SG, Nair LS, Laurencin CT. Biologically active chitosan systems for tissue engineering and regenerative medicine. Curr Top Med Chem 2008; 8(4): 354-64.
[http://dx.doi.org/10.2174/156802608783790974] [PMID: 18393897]

[287] Wang S, Lu L, Wang C, Gao C, Wang X. Polymeric biomaterials for tissue engineering applications. Int J Polym Sci 2010; 1-2.

[288] Pattnaik S, Nethala S, Tripathi A, Saravanan S, Moorthi A, Selvamurugan N. Chitosan scaffolds containing silicon dioxide and zirconia nano particles for bone tissue engineering. Int J Biol Macromol 2011; 49(5): 1167-72.
[http://dx.doi.org/10.1016/j.ijbiomac.2011.09.016] [PMID: 21968009]

[289] Guan S, Zhang X-L, Lin X-M, Liu T-Q, Ma X-H, Cui Z-F. Chitosan/gelatin porous scaffolds containing hyaluronic acid and heparan sulfate for neural tissue engineering. J Biomater Sci Polym Ed 2013; 24(8): 999-1014.

[http://dx.doi.org/10.1080/09205063.2012.731374] [PMID: 23647254]

[290] Sainitya R, Sriram M, Kalyanaraman V, *et al.* Scaffolds containing chitosan/carboxymethyl cellulose/mesoporous wollastonite for bone tissue engineering. Int J Biol Macromol 2015; 80: 481-8. [http://dx.doi.org/10.1016/j.ijbiomac.2015.07.016] [PMID: 26188305]

[291] Shavandi A, Bekhit Ael-D, Ali MA, Sun Z. Bio-mimetic composite scaffold from mussel shells, squid pen and crab chitosan for bone tissue engineering. Int J Biol Macromol 2015; 80: 445-54. [http://dx.doi.org/10.1016/j.ijbiomac.2015.07.012] [PMID: 26187191]

[292] Wen P, Gao J, Zhang Y, *et al.* Fabrication of chitosan scaffolds with tunable porous orientation structure for tissue engineering. J Biomater Sci Polym Ed 2010; 5063: 37-41. [PMID: 20557692]

[293] Saravanan S, Leena RS, Selvamurugan N. Chitosan based biocomposite scaffolds for bone tissue engineering. Int J Biol Macromol 2016; 93(Pt B): 1354-65. [http://dx.doi.org/10.1016/j.ijbiomac.2016.01.112] [PMID: 26845481]

[294] Lim JI, Lee Y-K, Shin J-S, Lim K-J. Preparation of interconnected porous chitosan scaffolds by sodium acetate particulate leaching. J Biomater Sci Polym Ed 2011; 22(10): 1319-29. [http://dx.doi.org/10.1163/092050610X504783] [PMID: 20594410]

[295] Chen VJ, Ma PX. Nano-fibrous poly(L-lactic acid) scaffolds with interconnected spherical macropores. Biomaterials 2004; 25(11): 2065-73. [http://dx.doi.org/10.1016/j.biomaterials.2003.08.058] [PMID: 14741621]

[296] Ma J, Wang H, He B, Chen J. A preliminary *in vitro* study on the fabrication and tissue engineering applications of a novel chitosan bilayer material as a scaffold of human neofetal dermal fibroblasts. Biomaterials 2001; 22(4): 331-6. [http://dx.doi.org/10.1016/S0142-9612(00)00188-5] [PMID: 11205436]

[297] Karageorgiou V, Kaplan D. Porosity of 3D biomaterial scaffolds and osteogenesis. Biomaterials 2005; 26(27): 5474-91. [http://dx.doi.org/10.1016/j.biomaterials.2005.02.002] [PMID: 15860204]

[298] Yu H, Matthew HW, Wooley PH, Yang SY. Effect of porosity and pore size on microstructures and mechanical properties of poly-epsilon-caprolactone- hydroxyapatite composites. J Biomed Mater Res B Appl Biomater 2008; 86(2): 541-7. [http://dx.doi.org/10.1002/jbm.b.31054] [PMID: 18335434]

[299] Levengood SL, Zhang M. Chitosan-based scaffolds for bone tissue engineering. J Mater Chem B Mater Biol Med 2014; 2(21): 3161-84. [http://dx.doi.org/10.1039/c4tb00027g] [PMID: 24999429]

[300] Draghi L, Resta S, Pirozzolo MG, Tanzi MC. Microspheres leaching for scaffold porosity control. J Mater Sci Mater Med 2005; 16(12): 1093-7. [http://dx.doi.org/10.1007/s10856-005-4711-x] [PMID: 16362206]

[301] Zhang J, Wu L, Jing D, Ding J. A comparative study of porous scaffolds with cubic and spherical macropores. Polymer (Guildf) 2005; 46: 4979-85. [http://dx.doi.org/10.1016/j.polymer.2005.02.120]

[302] Annabi N, Nichol JW, Zhong X, *et al.* Controlling the porosity and microarchitecture of hydrogels for tissue engineering. Tissue Eng Part B Rev 2010; 16(4): 371-83. [http://dx.doi.org/10.1089/ten.teb.2009.0639] [PMID: 20121414]

[303] Tuncer Caykara, Sıla Küçüktepe, Eylem Turan. Swelling characteristics of thermo- sensitive poly[(2-diethylaminoethyl methacrylate)-co-(N,N-dimethylacrylamide)] porous hydrogels. Polym Int 2007; 56: 532-7. [http://dx.doi.org/10.1002/pi.2166]

[304] Lips PA, Velthoen IW, Dijkstra PJ, Wessling M, Feijen J. Gas foaming of segmented poly(ester amide) films. Polymer (Guildf) 2005; 46: 9396-403.

[http://dx.doi.org/10.1016/j.polymer.2005.07.052]

[305] Şenel S, McClure SJ. Potential applications of chitosan in veterinary medicine. Adv Drug Deliv Rev 2004; 56(10): 1467-80. [http://dx.doi.org/10.1016/j.addr.2004.02.007] [PMID: 15191793]

[306] Ji C, Annabi N, Khademhosseini A, Dehghani F. Fabrication of porous chitosan scaffolds for soft tissue engineering using dense gas CO2. Acta Biomater 2011; 7(4): 1653-64. [http://dx.doi.org/10.1016/j.actbio.2010.11.043] [PMID: 21130905]

[307] Barbetta A, Colosi C, Costantini M, *et al.* Rapid prototyping of chitosan-coated alginate scaffolds through the use of a 3D fiber deposition technique. J Mater Chem B Mater Biol Med 2014; 2: 6779-91. [http://dx.doi.org/10.1039/C4TB00732H]

[308] Richardson TP, Peters MC, Ennett AB, Mooney DJ. Polymeric system for dual growth factor delivery. Nat Biotechnol 2001; 19(11): 1029-34. [http://dx.doi.org/10.1038/nbt1101-1029] [PMID: 11689847]

[309] Simmons CA, Alsberg E, Hsiong S, Kim WJ, Mooney DJ. Dual growth factor delivery and controlled scaffold degradation enhance *in vivo* bone formation by transplanted bone marrow stromal cells. Bone 2004; 35(2): 562-9. [http://dx.doi.org/10.1016/j.bone.2004.02.027] [PMID: 15268909]

[310] Zhang Y, Cheng X, Wang J, *et al.* Novel chitosan/collagen scaffold containing transforming growth factor-β1 DNA for periodontal tissue engineering. Biochem Biophys Res Commun 2006; 344(1): 362-9. [http://dx.doi.org/10.1016/j.bbrc.2006.03.106] [PMID: 16600180]

[311] Hubbell JA. Bioactive biomaterials. Curr Opin Biotechnol 1999; 10(2): 123-9. [http://dx.doi.org/10.1016/S0958-1669(99)80021-4] [PMID: 10209141]

[312] Ratner BD. The engineering of biomaterials exhibiting recognition and specificity. J Mol Recognit 1996; 9(5-6): 617-25. [http://dx.doi.org/10.1002/(SICI)1099-1352(199634/12)9:5/6<617::AID-JMR310>3.0.CO;2-D] [PMID: 9174947]

[313] Di Martino A, Sittinger M, Risbud MV. Chitosan: a versatile biopolymer for orthopaedic tissue-engineering. Biomaterials 2005; 26(30): 5983-90. [http://dx.doi.org/10.1016/j.biomaterials.2005.03.016] [PMID: 15894370]

[314] Khor E, Lim LY. Implantable applications of chitin and chitosan. Biomaterials 2003; 24(13): 2339-49. [http://dx.doi.org/10.1016/S0142-9612(03)00026-7] [PMID: 12699672]

[315] Malafaya PB, Silva GA, Reis RL. Natural-origin polymers as carriers and scaffolds for biomolecules and cell delivery in tissue engineering applications. Adv Drug Deliv Rev 2007; 59(4-5): 207-33. [http://dx.doi.org/10.1016/j.addr.2007.03.012] [PMID: 17482309]

[316] Gümüşderelioğlu M, Aday S. Heparin-functionalized chitosan scaffolds for bone tissue engineering. Carbohydr Res 2011; 346(5): 606-13. [http://dx.doi.org/10.1016/j.carres.2010.12.007] [PMID: 21333274]

[317] Park H, Choi B, Nguyen J, *et al.* Anionic carbohydrate-containing chitosan scaffolds for bone regeneration. Carbohydr Polym 2013; 97(2): 587-96. [http://dx.doi.org/10.1016/j.carbpol.2013.05.023] [PMID: 23911489]

[318] Wang Y, Wang X, Li L, Gu Z, Yu X. Controlled drug release from a novel drug carrier of calcium polyphosphate/chitosan/aldehyde alginate scaffolds containing chitosan microspheres. RSC Advances 2014; 4: 24810. [http://dx.doi.org/10.1039/c4ra03566f]

[319] Mattioli-Belmonte M, Gigante A, Muzzarelli RA, *et al.* N,N-dicarboxymethyl chitosan as delivery agent for bone morphogenetic protein in the repair of articular cartilage. Med Biol Eng Comput 1999; 37(1): 130-4.

[http://dx.doi.org/10.1007/BF02513279] [PMID: 10396855]

[320] Lee JE, Kim KE, Kwon IC, *et al.* Effects of the controlled-released TGF-β 1 from chitosan microspheres on chondrocytes cultured in a collagen/chitosan/glycosaminoglycan scaffold. Biomaterials 2004; 25(18): 4163-73. [http://dx.doi.org/10.1016/j.biomaterials.2003.10.057] [PMID: 15046906]

[321] Taravel MN, Domard A. Collagen and its interactions with chitosan, III some biological and mechanical properties. Biomaterials 1996; 17(4): 451-5. [http://dx.doi.org/10.1016/0142-9612(96)89663-3] [PMID: 8938241]

[322] Kim SE, Park JH, Cho YW, *et al.* Porous chitosan scaffold containing microspheres loaded with transforming growth factor-β1: implications for cartilage tissue engineering. J Control Release 2003; 91(3): 365-74. [http://dx.doi.org/10.1016/S0168-3659(03)00274-8] [PMID: 12932714]

[323] Tong J-C, Yao S-L. Novel scaffold containing transforming growth factor-β1 DNA for cartilage tissue engineering. J Bioact Compat Polym 2007; 22: 232-44. [http://dx.doi.org/10.1177/0883911507076460]

[324] Deng J, She R, Huang W, Dong Z, Mo G, Liu B. A silk fibroin/chitosan scaffold in combination with bone marrow-derived mesenchymal stem cells to repair cartilage defects in the rabbit knee. J Mater Sci Mater Med 2013; 24(8): 2037-46. [http://dx.doi.org/10.1007/s10856-013-4944-z] [PMID: 23677433]

[325] OShea TM, Miao X. Bilayered scaffolds for osteochondral tissue engineering. Tissue Eng Part B Rev 2008; 14(4): 447-64. [http://dx.doi.org/10.1089/ten.teb.2008.0327] [PMID: 18844605]

[326] Chen J, Chen H, Li P, *et al.* Simultaneous regeneration of articular cartilage and subchondral bone *in vivo* using MSCs induced by a spatially controlled gene delivery system in bilayered integrated scaffolds. Biomaterials 2011; 32(21): 4793-805. [http://dx.doi.org/10.1016/j.biomaterials.2011.03.041] [PMID: 21489619]

[327] Liu H, Fan H, Cui Y, Chen Y, Yao K, Goh JC. Effects of the controlled-released basic fibroblast growth factor from chitosan-gelatin microspheres on human fibroblasts cultured on a chitosan-gelatin scaffold. Biomacromolecules 2007; 8(5): 1446-55. [http://dx.doi.org/10.1021/bm061025e] [PMID: 17375950]

[328] Mahmoud AA, Salama AH. Norfloxacin-loaded collagen/chitosan scaffolds for skin reconstruction: Preparation, evaluation and in-vivo wound healing assessment. Eur J Pharm Sci 2016; 83: 155-65. [http://dx.doi.org/10.1016/j.ejps.2015.12.026] [PMID: 26733072]

[329] Ahmed S, Ikram S. Chitosan based scaffolds and their applications in wound healing. Achievements in the Life Sciences 2016; 10: 27-37. [http://dx.doi.org/10.1016/j.als.2016.04.001]

[330] Radhakumary C, Antonty M, Sreenivasan K. Drug loaded thermoresponsive and cytocompatible chitosan based hydrogel as a potential wound dressing. Carbohydr Polym 2011; 83: 705-13. [http://dx.doi.org/10.1016/j.carbpol.2010.08.042]

[331] Martins AM, Eng G, Caridade SG, Mano JF, Reis RL, Vunjak-Novakovic G. Electrically conductive chitosan/carbon scaffolds for cardiac tissue engineering. Biomacromolecules 2014; 15(2): 635-43. [http://dx.doi.org/10.1021/bm401679q] [PMID: 24417502]

[332] Wang K, Kievit FM, Florczyk SJ, Stephen ZR, Zhang M. 3D porous chitosan-alginate scaffolds as an *in vitro* model for evaluating nanoparticle-mediated tumor targeting and gene delivery to prostate cancer. Biomacromolecules 2015; 16(10): 3362-72. [http://dx.doi.org/10.1021/acs.biomac.5b01032] [PMID: 26347946]

[333] Wers E, Oudadesse H, Lefeuvre B, Merdrignac-Conanec O, Barroug A. Evaluation of the kinetic and relaxation time of gentamicin sulfate released from hybrid biomaterial Bioglass-chitosan scaffolds.

Appl Surf Sci 2015; 353: 200-8.
[http://dx.doi.org/10.1016/j.apsusc.2015.06.146]

[334] Neut D, van Horn JR, van Kooten TG, van der Mei HC, Busscher HJ. Detection of biomaterial-associated infections in orthopaedic joint implants. Clin Orthop Relat Res 2003; (413): 261-8.
[http://dx.doi.org/10.1097/01.blo.0000073345.50837.84] [PMID: 12897618]

[335] Dorati R, De Trizio A, Genta I, Merelli A, Modena T, Conti B. Formulation and *in vitro* characterization of a composite biodegradable scaffold as antibiotic delivery system and regenerative device for bone. J Drug Deliv Sci Technol 2016; 35: 124-33.
[http://dx.doi.org/10.1016/j.jddst.2016.04.008]

[336] Silva MC, Fideles TB, Oliveira SS, Fook MV. Development and characterization of scaffolds obtained by aggregation of balls of chitosan and chitosan/curcumin. J Chitin Chitosan Sci 2016; 4(1): 14-21.
[http://dx.doi.org/10.1166/jcc.2016.1094]

[337] Ardeshirzadeh B, Anaraki NA, Irani M, Rad LR, Shamshiri S. Controlled release of doxorubicin from electrospun PEO/chitosan/graphene oxide nanocomposite nanofibrous scaffolds. Mater Sci Eng C Mater Biol Appl 2015; 48: 384-90.
[http://dx.doi.org/10.1016/j.msec.2014.12.039] [PMID: 25579938]

CHAPTER 9

Application of Chitosan Based Materials as Edible Films and Coatings

Vanderlei C. Souza[1,*] and **Guilherme L. Dotto**[2]

[1] *Mechanical Engineering Department, Laboratory of Porous Media and Thermophysical Properties, Federal University of Santa Catarina-UFSC, Florianópolis-SC, Brazil*

[2] *Chemical Engineering Department, Federal University of Santa Maria-UFSM, Santa Maria-RS, Brazil*

Abstract: Chitosan is the only pseudo-natural cationic polymer and, owing to its biocompatibility, biodegradability, non-toxicity, and antimicrobial activity, combined with its cationic character, is considered as the greatest potential polymer for obtaining food packaging, especially in the form of edible films and coatings. The successful manufacture of edible films and coatings for food applications is based on specific functional properties that the biopolymer should meet for each type of food, due to the different properties of the various components (nanoparticles, lipids, plasticizers, proteins, *etc.*). The addition of these compounds can be beneficial in the films production, with superior functional properties and thus extend their applications. In this chapter, we highlight the application of chitosan-based materials as edible films and coatings for food protection. The manufacture, and mechanical properties and the water vapor permeability (WVP) of the films and coatings are also discussed.

Keywords: Mechanical properties, Packaging materials, Polysaccharide, Water vapor permeability.

INTRODUCTION

Owing to concerns about limited natural resources and environmental impacts caused by the use of synthetic polymers and rigid packaging, there is a renewed interest in developing biopolymers based materials used as food packaging [1]. The edible films and coatings are formed from the natural polymers of animal or vegetable origin, and when released into the environment, they are converted into simple mineralized compounds. These in turn, are redistributed through the carbon, nitrogen and sulfur cycles, causing no harm to the biosystem. Films are

* **Corresponding author Vanderlei C. Souza:** Mechanical Engineering Department, Laboratory of Porous Media and Thermophysical Properties (LMPT), Federal University of Santa Catarina-UFSC, Florianópolis-SC, Brazil; Tel/Fax: +55 (48) 996108066; E-mail: vanderleilei@bol.com.br

Guilherme Luiz Dotto, Sergio Paulo Campana-Filho and Luiz Antonio de Almeida Pinto (Eds)

defined as stand-alone thin layers that are applied to the material surface. Films can form pouches, wraps, capsules, bags, or casings through further fabrication processes. Coatings are a particular form of films directly applied to the surface of materials. The removal of coating layers may be possible; however, coatings are typically not intended for disposal separately from the coated materials. Therefore, coatings are regarded as a part of the final product [2 - 4].

Besides the possibility of returning to the biological cycle after use, biopolymer based films can be developed in order to obtain properties that meet the demanding applications, such as mass transfer barrier (water vapor and gas), transport of food additives (pigments, aromas, flavors) and controlled release of active ingredients (antioxidants and mycotic agents) [5]. Moreover, they should prevent physical damage to the material during the stages of handling, transport and storage. Thus, they must have the capability of deformation and mechanical strength [6]. Bio-based films offer advantages such as edibility, non-toxicity, biocompatibility, non-polluting and low cost [7]. Two major promising applications of such films are the replacement of short-life plastic in food packaging and use as edible food coatings [8].

In the past decades, several natural polymers have received increased attention for their applications in food, biomedical, and chemical industries [9, 10]. The main biopolymers used to obtain films are carbohydrates (polysaccharides, pectins and starches) [11 - 13], proteins [14], lipids [15] or combinations thereof [16, 1]. In particular, chitosan, a polysaccharide, is the only pseudonatural cationic polymer and, owing to its biocompatibility, biodegradability, non-toxicity, and antimicrobial activity, combined with its cationic character, is considered as the greatest potential biopolymer for obtaining food packaging, especially in the form of edible films and coatings [17]. In this chapter, we highlight the application of chitosan based materials as edible films and coatings for food protection. The manufacture, mechanical properties and water vapor permeability (WVP) of the films and coatings are also discussed.

MANUFACTURE METHODS OF EDIBLE FILMS AND COATINGS

The formulation of edible films and coatings requires using at least one component (macromolecule) capable of forming a structural matrix with a sufficient cohesiveness. Only high molecular weight polymers, owing to their sufficient cohesive strength and capacity for coalescence, can produce such film or coating structure [18]. In addition, in general, one needs a solvent and a plasticizer [19]. The formation of these materials usually involves associations or inter and intramolecular crosslinking of polymer chains. Plasticizers increase the space between the chains, causing a reduction of intermolecular forces along the matrix [20, 21], decreasing the glass transition temperature and increasing flexibility [22]. In structural terms, a semi rigid three-dimensional network that traps the solvent is formed [23].

The formation of edible films and coatings may involve one of the following mechanisms [24]:

- Simple coacervation: achieved by the precipitation of a hydrocolloid that is dispersed in aqueous solution. This precipitation can be obtained by precipitation from a solvent, by adding a non-electrolyte solute and in which the hydrocolloid is not soluble, by the addition of an electrolyte which induces the precipitation of the components or intersection, or by altering the solution pH;
- Complex coacervation: achieved by obtaining the precipitation by mixing two solutions of hydrocolloids with opposite electrical charges, which interact to form the complex polymer;
- Thermal gelation or coagulation: achieved by heating the macromolecule, leading to a denaturation and gel formation or precipitation. The gel can also be obtained by cooling the hydrocolloid solution.

The films are obtained from a solution or dispersion of film forming agent, followed by the formation of a thin layer using techniques such as casting, extrusion, thermoforming, blow molding, among others [18, 22]. The casting technique is widely used in laboratory studies; it does not require special equipment and large quantities of raw materials. This technique involves spreading the film forming solution (macromolecules, solvent, and plasticizer) in a container, which is then subjected to drying for solvent evaporation [19]. Extrusion has the advantage of low cost, in addition to having a system of versatile and continuous production, suitable for industrial scale [25].

Coatings are formed directly on the food product using either liquid film forming solutions (or dispersions) or molten compounds (*e.g.* lipids). They can be applied by different methods: with a paint brush or by spraying, dipping-dripping, fluidizing, *etc.* [26].

CHITOSAN FILMS

Chitosan, a high molecular weight polysaccharide consisting of (1,4)-linked 2-amino-deoxy-β-d-glucan, is a deacetylated derivative of chitin, which is the second most abundant polysaccharide found in nature after cellulose [27, 28]. The cationic nature of chitosan is owed to the free amino groups left by partial remotion of chitin acetyl groups. Chitosan is a linear polyelectrolyte at acidic pH with good film forming capacity [29]. Chitosan has been compared with other biomolecule based films used as packaging materials and the reported results showed that chitosan has more advantages because its antibacterial activity and cationic character [27, 30, 31].

The antimicrobial action of chitosan and their derivatives is proposed by three mechanisms. In the first one, the positive charges present in the polymeric chain of chitosan (due to its amino group), interacts with the negatively charged microbial cell membranes, leading to the leakage of proteinaceous and other intracellular constituents of the microorganisms [32]. Chitosan acted mainly on the outer surface of bacteria. At a lower concentration (0.2 mg/mL), the polycationic chitosan does probably bind to the negatively charged bacterial surface to cause agglutination, while at higher concentrations, the larger number of positive charges may have imparted a net positive charge to the bacterial surfaces to keep them in suspension [33 - 35]. The second mechanism proposes that chitosan acts as a chelating agent, creating compounds from traces of metals essential to the cell [36]. The third mechanism establishes that chitosan of low molecular weight is capable of entering the cell's nucleus itself, interacting with the DNA, interfering with the messenger RNA synthesis, affecting the synthesis of proteins and inhibiting the action of various enzymes [37].

Although chitosan films/coatings have recognized properties that allow them to be applied in food systems, they also have some drawbacks that restrict their use under specific conditions, owing to their low resistance to water vapor diffusion and relatively low tensile strength, especially in moist environments [1, 38]. As a result, there are a number of strategies that have been used in the literature such as crosslinking or blending with biopolymers (lipids, polysaccharides, proteins, cellulose nanocrystals, *etc.*), to reduce its water sensitivity [1, 11, 16, 17, 39 - 41].

Chitosan-lipids Films

The lipid based films and coatings provide better barrier to water vapor than the polysaccharides due to their hydrophobic character [42, 43]. Aiming to combine the benefits of lipids and polysaccharides, films involving these biopolymers are prepared by emulsion or bilayer, where the lipid helps to reduce the transmission of water vapor, and polysaccharides provides the needed strength and integrity to the polymer matrix [44, 45]. Bilayer films consist of the formation of an additional lipid layer on the matrix of the previously formed film, requiring two stages of drying. Whereas the emulsified need one drying step after obtaining a film forming solution, which means that they receive more attention from both, researchers and industries [45 - 47].

Emulsified films are manufactured from the dispersion of lipid fraction in the filmogenic matrix through homogenization, forming an emulsion that is subsequently subjected to drying for solvent evaporation [48]. The lipids are used due to their polar nature. Their efficiency, mechanical properties and permeability depend on their physical state (liquid, solid or crystalline), length of fatty acid chains, degree of saturation, polarity, particle size distribution in dispersed phase and chemical structure, among others [43, 49, 50]. The chemical nature of lipids, such as chain length, changes the properties of the films, since the greater the number of carbons in the lipid structure, the more difficult is the incorporation of lipid compounds in the film forming dispersion. In general, lipid emulsions are heterogeneous dispersions of two immiscible liquids and are subject to various instability processes such as aggregation, flocculation and coalescence, hence a possible phase separation. To obtain emulsion stability and mixing of the polysaccharide/lipid interface, the addition of a surfactant is necessary. Surfactants are used because they are amphiphilic substances that interact simultaneously with the discontinuous (lipid) and continuous phases (polysaccharide), stabilizing and forming the emulsion through an electrosteric effect [51, 52]. The emulsion stability is influenced by pH, ionic strength and viscosity of the continuous phase, density and size of lipid globules [53, 54].

The mass transfer through emulsified film is described by two models. The first called "microvoid" where Ukai and co-workers [55] suggested that the mass transfer of gases and vapors occurs by microvoids, which are formed between the microparticles of the hydrocolloid matrix and the hydrophobic material in the emulsion during drying. According to the "micropathway model" proposed by Krochta [56], mass transfer occurs through the polymer matrix itself. This may be because the polysaccharides are often compatible with the moisture and gases and can offer little resistance to their transference.

In order to improve the water barrier properties, the main lipid fractions incorporated into the chitosan films are fatty acids [51, 57, 58], vegetable oils [59], essential oils [60] and waxes [61]. For example, Souza and co-workers [1] added stearic acid, oleic acid, rice refined oil and refined fish oil to chitosan based films, obtaining a higher resistance to water vapor diffusion (reduction in WVP=65%) when refined fish oil has been used. At the same time, films with lipid fraction showed poor mechanical strength, compared to chitosan film. According to the authors, the low transfer of water vapor between the polymer chains of the composite (chitosan-refined oil) films may be due to low interaction of highly polar molecules such as hydroxyl and carboxyl groups of fatty acids present in the oil (Table **1**). In this sense, the obtained results indicate that the predominance of long chain lipid molecules, especially EPA and DHA, may have been responsible for the higher resistance to water vapor diffusion of the chitosan-fish oil composite films, where EPA + DHA contents in fish and rice oils were 5.3% and 0.1%, respectively (Table **1**).

Srinivasa and co-workers [58] modified the chitosan film by blending with polyols and fatty acids (stearic and palmitic acids) and studied the WVP and mechanical properties. The tensile strength of the blended films decreased with the addition of polyols and fatty acids, whereas the percent elongation has been increased in polyol blend films, but fatty acid blend films showed no significant differences. Glycerol blend films showed decrease, whereas sorbitol and polyethyleneglycol blend films showed increase in the WVP values.

Table 1. Fatty acid (%) profiles and lipid classes (% of total fatty acids) of the carp and rice refined oils [1].

Fatty acids and lipid classes	Refined carp oil	Refined rice oil
C16:0	15.47 ± 0.02	20.16 ± 0.01
C18:0	3.09 ± 0.02	1.18 ± 0.02
C16:1 ω-7	7.59 ± 0.04	0.22 ± 0.02
C18:1 ω-9 cis	26.52 ± 0.03	41.39 ± 0.02
C18:1 ω-9 trans	0.33 ± 0.01	0.46 ± 0.01
C18:2 ω-6 trans 9,12	0.44 ± 0.01	0.60 ± 0.03
C18:2 ω-6 cis	9.32 ± 0.04	28.00 ± 0.02
C18:3 ω-6	0.39 ± 0.02	0.27 ± 0.01
C18:3 ω-3	7.61 ± 0.04	1.24 ± 0.03
C20:5 ω-3 (EPA)	3.99 ± 0.03	–
C22:6 ω-3 (DHA)	1.30 ± 0.01	0.10 ± 0.01
∑ ni	5.24 ± 0.19	1.08 ± 0.24
∑ SFA	25.25 ± 0.15	23.98 ± 0.08
∑ MUFA	41.81 ± 0.14	43.80 ± 0.11
∑ PUFA	27.69 ± 0.20	31.11 ± 0.18
ω-3	14.29 ± 0.10	1.48 ± 0.08
ω-6	13.48 ± 0.13	28.99 ± 0.09

Mean value ± standard error (in replicate). EPA, eicosapentaenoic acid; DHA, docosahexaenoic acid; ∑ ni, sum of unidentified fatty acids; ∑ SFA, sum of unsaturated fatty acids; ∑ MUFA, sum of monounsaturated fatty acids; ∑ PUFA, sum of polyunsaturated fatty acids.

Vargas and co-workers [51] modified the chitosan film by blending with oleic acid and, studied the impact of the incorporation of oleic acid into the chitosan matrix. Results showed that the higher the oleic acid content, the lower the WVP. According to the authors, in general, the addition of oleic acid into the chitosan matrix leads to a significant increase in gloss and translucency and a decrease in the tensile strength, elongation at break and Young's modulus of the composite films.

Wong and co-workers [62] incorporated saturated fatty acids and fatty acid esters to chitosan, obtaining a significant decrease in WVP when butyric and lauric acids were used. Similar results with decreasing WVP of chitosan film have been achieved with the incorporation of different essential oils (caraway, citronella and cedarwood) [16, 60, 63] palm oil and margarine into the chitosan starch matrix obtaining a significant decrease in the WVP and a reduction in the tensile strength of the composite films; The waxes are lipids commonly used in films based on polysaccharides. Studies show that waxes are more effective than saturated fatty acids in improving the WVP, and this is because of its high hydrophobicity [42, 43].

Chitosan-cellulose Nanocrystals Films

In order to improve the mechanical properties, the main nanofillers incorporated into the base of polymer films are carbon nanotubes, nanoclays, inorganic nanoparticle and graphene [64]. The use of nanometer-sized (1 to 100 nm) single crystals of cellulose ("cellulose nanocrystals", CN) with non-toxicity character, non-abrasive nature, biodegradability, highly rigid crystalline, high specific area and good compatibility with several biopolymers attracted both, industrial and academic attention [65]. When CN and a biopolymer matrix are combined, they may interact favorably, resulting in materials with improved structural and functional properties [66, 67]. The performance as a reinforcing agent of the CN mainly depends on the source of cellulosic material and the extraction process [68].

Some studies have been published on the incorporation of CNs in chitosan matrices (Table **2**). In general, the presence of CNs in the chitosan matrix provides superior performances such as mechanical and barrier properties. This may be attributed to the high crystallinity of the CNs and interactions among their anionic groups with the cationic groups present in the chitosan polymer chains [69, 70].

Pereda and co-workers [77] reported a study on composite films based on chitosan/glycerol/olive and CNs by the casting technique. The results showed that although their physical, mechanical and barrier properties have been strongly related with the CNs concentration and the presence (or absence) of olive oil, the films developed in this work can be used as food packaging, but also, due to the edible, non-toxic and nutritional characteristics of the different constituents, they can also be considered as edible food packaging. Due to cellulose-glycerol-chitosan interactions, the composite films appeared less opaque as the cellulose concentration increases (up to 7 wt.% CN), which is an advantage from the consumer viewpoint. The tensile strength has been significantly increased by CN addition, while the utilization of olive oil moderates the reduction of the elongation at break; this synergistic effect is an improvement over the expected response (*i.e.* modifications that lead to an increase in the stiffness, usually also lead to a decrease in the deformation capability).

Table 2. Research works related to chitosan-based films with cellulose nanocrystals.

Polymer Matrix	Source of Cellulose Nanocrystals	Reference
Chitosan	Wood pulp	Fernandes and co-workers [71]
	Bacterial cellulose	Fernandes and co-workers [72]
	Softwood kraft pulp	Khan and co-workers [73]
	Cotton	Li and co-workers [74]
	Cotton	Li and co-workers [75]
	Eucalyptus wood cellulose	Mesquita and co-workers [76]
	Cotton	Pereda and co-workers [77]
	Corn cobs	Souza and co-workers [78]
	Corn cobs	Souza and Quadri [11]

The effect of introduction of CN (5 to 30 wt%) prepared from cotton linter on the properties of a chitosan film has been studied by Li and co-workers [75]. The results indicated that the CN have been well dispersed within the chitosan matrix and strong interactions occurred between fillers and matrix. The tensile strength of the composite films in dry and wet states significantly increased from 85 to 120 MPa and 9.9 to 17.3 MPa, respectively, with an increase of CN content from 0 to 15-20 wt %. Moreover, the nanocomposite films displayed excellent thermal stability and water resistance with the incorporation of fillers.

New nanocomposite films based on different chitosan matrices (HCH: high molecular weight chitosan; LCH: low molecular weight chitosan) and bacterial cellulose (BC) (Fig. **1**) have been reported by Fernandes and co-workers [72]. The films have been prepared by a simple and green procedure based on casting water (or 1% acetic solutions) suspensions of chitosan with different contents of bacterial cellulose. The obtained materials have been transparent, flexible and presented significantly better mechanical properties than the corresponding unfilled chitosan films. They also have a reasonable thermal stability and low O_2

permeability. Their notable properties are promising for applications in transparent biodegradable and anti-bacterial packaging.

Fig. (1). Images of CH and CHBC nanocomposite films placed in front of a green plant [72].

Cross-linked Chitosan Films

Filmogenic network can be modified by crosslinking agents. Crosslinking serve to improve the functionalities of the chitosan based films [79]. Reports in the literature describe the properties of chitosan based films upon the addition of crosslinking agents such as vanillin, transglutaminase, genipin, ferulic and tannic acids [79 - 85]. For example, Pierro and co-workers [86] reported the preparation of chitosan-ovalbumin films through a TGase-catalyzed reaction. The improved mechanical and solubility properties of these new materials showed that this enzymatic approach could be a useful tool for preparing edible films for food coating applications. Several researchers [87, 88] reported the physicochemical properties of chitosan-gelatine films as well as their improvement upon the incorporation of crosslinking agents. The improvements in crosslinking have been attributed to the formation of polyelectrolyte complexes between the ammonium groups of chitosan and the negatively charged side chain groups in the gelatine through electrostatic interaction [89]. Plasticized and unplasticized chitosan films with tannic acid addition have been developed by Rivero and co-workers [79]. The crosslinking capacity of tannic acid has been studied by evaluating both the structural modification produced on chitosan films and its effect on physicochemical, barrier and mechanical properties. Likewise the changes of these properties during the storage have been analyzed; formulations containing tannic acid increased 29% tensile strength whereas decreased 24% WVP. Film solubility decreased at 100 °C, confirming that exposure at high temperatures facilitates the crosslinking process. The presence of tannic acid and glycerol simultaneously showed a synergic effect on the studied properties, all of which presented intermediate behavior in relation to those determined with the addition of either acid or plasticizer. The effect of the storage has been also relevant since the films evidenced their tendency to a more stable structure due to the reorganization toward an anhydrous conformation, as has been demonstrated by X-ray diffraction and FTIR analysis. Ferulic acid has been used [90] to crosslink starch-chitosan blend films, which finds possible application as an edible film or coating. The blend exhibited good film forming property due to the presence of high density of amino groups and hydroxyl groups and inter and intra molecular hydrogen bonding. The various physicochemical analyses indicated the interaction between starch and chitosan molecules and ferulic acid. Incorporation of ferulic acid has been found to improve the barrier properties and tensile strength of starch-chitosan blend films and significantly enhanced the lipid peroxide inhibition capacity.

Use as Food Packaging

One of the best attributes of a good packaging film is the prevention of microbial spoilage due to pathogens during extended storage [35]. Direct incorporation of active agents into food results in an immediate but short-term reduction of bacterial populations, while the antimicrobial films can maintain their activity for a long period of time [91 - 93]. Moreover, the bioactive packagings potentially serve as a vehicle to incorporate and enhance the food value along with other additives such as antioxidant, flavoring, antimicrobial agents and coloring [35, 94]. Over the years authors [95 - 98] have established that chitosan possesses significant antimicrobial activity against bacteria such *L. monocytogenes*, *E. coli*, *S. aureus*, *S. dysenteriae* and *P. aeruginosa*, and fungi such as *A. niger*, *P. italicum*, *M. fructicola* and *S. sclerotium*.

The use of antimicrobial chitosan films at an experimental level on food such fruits, meats, vegetables, dairy and eggs have been subject of a great number of studies. In addition, some patents have been obtained. For example, Sanchez-Gonzalez and co-workers [99] studied the preparation and characterization of chitosan (CH)-tea tree essential oil (TTO) composite films. Film forming dispersions have been characterized in terms of rheological properties, particle size distribution and ζ-potential. The WVP and mechanical properties of the films have been evaluated. Furthermore, the antimicrobial effectiveness of CH-TTO composite films against L. *monocytogenes* and P. *italicum* has been studied. The physical properties of the initial emulsions reveal that the CH molecules adsorb on the oil-water interface, thus contributing to the emulsion stability. The incorporation of TTO, at a ratio of CH:TTO of 1:2 in the polysaccharide matrix, reduced the WVP by 40%. Likewise, the films resistance notably reduced by TTO incorporation due to the presence of discontinuities in the film matrix that affect its mechanical response. Only the composite films with TTO:CH ratios higher than 1 showed a limited antifungal effectiveness against *P. italicum*, which has been notably reduced after 3 days of storage. Nevertheless, CH films presented a significant antimicrobial activity against *L. monocytogenes* and the incorporation of TTO in the CH:TTO ratio of 1:2 improved the antibacterial properties of these films, showing a complete inhibition of the microbial growth during the first fifth days at 10 °C.

Chitosan-pectin based films incorporated with tea polyphenols, citric acid and natamycin have been proposed to protect vegetables and fruits against microbial enzymatic browning and microbial growth (CN 103719267A patent) [100]. Chitosan edible films have been used on precooked beef patties [101]. After 3 days of refrigerated storage, no difference has been found in the moisture loss between CH film-wrapped patties and unpackaged patties (control-A). Chitosan film decreased moisture loss but not as effective as control-B. Chitosan film reduced thiobarbituric acid-reactive substances and hexanal values compared to control-A. Chitosan films were not effective in controlling lipid oxidation. These films afford potentials as oxygen barriers for precooked meat products. Further research needs to be conducted to improve moisture barrier properties and the application of such films on precooked beef patties.

Chitosan-sodium alginate based film has been prepared and applied to prolong the shelf-life of meat (CN 102977414B patent) [102]. This patent discloses an edible composite preservative film applied for the preservation of fresh pork in a supermarket. The film contains lactic acid of which the volume concentration is 0.1%, ascorbic acid of which the mass concentration is 1%, nisin of which the mass concentration is 0.25%, chitosan of which the mass concentration is 0.5%, sodium alginate of which the mass concentration is 1%, and glycerol of which the volume concentration is 1%, wherein the pH value is 3.5-4.0. The edible film disclosed by the patent is effective, safe and innocuous, economical and suitable for practical application of the supermarket, and the shelf life of the fresh pork packaged on a supermarket tray stored at 5-8 °C can be prolonged to over 7 days from 1-3 days.

Souza and co-workers [78] studied the preparation, characterization and application of edible films formed by CN uniformly dispersed in a matrix of chitosan. The developed bionanocomposite films have been applied as interleaving of sliced cheese, and the efficiency has been assessed by the investigation of adhesion between the surfaces and by comparing its properties with two commercial interleaving products (polyethylene (PE), and Greasepel paper (GP)). The performances of the chitosan films have been improved by CN reinforcement; the elongation, Young's modulus, and WVP values have been improved by up to 136, 224, and 70%, respectively, after CN addition to the chitosan matrix, showing good mechanical reinforcing capacity and barrier activity of such nanocrystals in the chitosan matrix. These properties are superior to those of the evaluated commercial interleaving films (PE and GP). Peel-test results showed that the bionanocomposite films had lower peel strengths, indicating lower adhesion to the sliced cheese surfaces compared to pure chitosan, PE, and GP films. These results suggest that the developed films may be exploited as food packaging, for example, as an interleaving film, owing to its excellent mechanical properties, permeability, hydrophobicity, and low surface adhesion.

Bhale and co-workers [103] studied the internal and sensory quality of eggs coated with chitosan. Three chitosans with high (HMw, 1100 KDa), medium (MMw, 746 KDa), and low (LMw, 470 KDa) molecular weight have been used to prepare coating solutions. The results demonstrated that chitosan coating is effective in pre serving the interior quality of eggs. Chitosan coating increased the shelf life of eggs by at least 3 weeks at 25 °C compared with that of noncoated eggs. No drastic differences in weight loss, Haugh unit, and yolk index have been observed among 6 chitosan treatments (3 molecular weights, each with 2 concentrations) after 5 weeks of storage. However, the use of 2% LMw chitosan (470 kDa) as a coating material is advantageous in reducing weight loss and obtaining more desirable albumen (higher Haugh unit) and yolk (higher yolk index) quality during a 5 weeks storage period. This study also shows that chitosan coating on eggs has not been detectable by consumers and that no difference in acceptability among noncoated, commercial, and chitosan-coated eggs has been observed. Further studies are needed to compare the chitosan-coating effect on washed and unwashed eggs as well as to address the effect of chitosan coating on the eggshell strength and microbial contamination. The beneficial aspects of chitosan coated eggs, as demonstrated in this study, clearly warrant scale-up trials under the large production volume typical of commercial conditions.

Khan and co-workers [80] prepared genipin crosslinked chitosan-CN films to prevent the surface growth of non-pathogenic and pathogenic bacteria in fresh pork meats. Combination of low-dose gamma irradiation and genipin crosslinking has been found to be very useful to protect the antimicrobial activity of nisin-disodium ethylenediaminetetraacetate (EDTA) formulation in the films during storage at high RH. The gamma irradiated cross-linked films increased the microbiological shelf life of pork meats by more than 5 weeks by restricting the growth of psychrotrophs, mesophiles and *Lactobacillus* spp. below 6 log CFU/g. The films also inhibited the growth of E. *coli* and L. *monocytogenes* in meat samples.

PROPERTIES OF BIO-BASED PACKAGING

The methods used to evaluate the properties of bio-based packaging are adapted from traditional methods applied to synthetic materials, taking into account the high sensitivity, relative humidity and temperature, which are involved in biopolymer films/coatings [104]. The different characteristics and properties depend of the components and the process used to obtain the films/coatings. In general the main evaluated characteristics are thickness, transparency, mechanical properties and WVP.

Thickness

In order to give uniformity to the film, the thickness control is of paramount importance. This control ensures a proper repeatability of the measurements of material properties [105]. The thickness should be set according to the required application of the packaging, which depends on the food to be packed. In addition, the characteristics of the formulations must be taken into consideration when one wants to control the thickness of a film/coating. Bravin and co-workers [41] and Bertuzzi and co-workers [106], studying bio-packagings, observed that the WVP increases linearly with increasing thickness of the films, which has been attributed to structural changes in films, resulting from the drying of film-forming dispersions. According to Cuq and co-workers [107], the tensile strength may vary with the thickness of the film, the greater the amount of dry matter per surface, the greater the number of intermolecular interactions.

Transparency

When incident light passes through a material with minimal absorption or reflection, the material is said to be transparent. A material is said to be opaque when it absorbs and/or reflects all light that strikes it, without light transmission. Transparency can be evaluated by both transmittance and absorbance, depending on the type of material [108]. Among the various types of equipment available to assess the transparency, a spectrophotometer can be used [16].

Mechanical Properties

The mechanical properties of packaging are one of the most studied properties because they are directly related to their application. These properties determine the response to external mechanical influences, being associated with the ability to develop reversible and irreversible deformations and exhibit resistance to fracture.

Packaging obtained should be resistant to breakage and abrasion, protecting the food to be packed from losing its integrity and from handling and transportation damage risks. Moreover, this must have enough flexibility to adapt to deformations in the product without mechanical damage [109].

The main mechanical properties of bio-packagings are tensile strength and elongation at break. Young's modulus has also been analyzed. Tensile strength is the resistance of the material at the point of rupture. The elongation is the ratio between the elongation of the specimen in the test and its original length. Young's modulus is the relationship between tensile strength and elastic deformation in the region, where the response of the specimen elongation is increasing and linearly proportional to the imposed traction [110].

Mechanical relaxation is another mechanical property, though little studied, from relaxation curves it is possible to determine viscoelastic properties. According to Ferry [111], the viscoelasticity is important because it provides information related to the conformation of macromolecules and molecular relaxation phenomenon.

Water Vapor Permeability

The American Society for Testing and Materials Standard E96-95 [112] defines the WVP and gases as the transmission rate of water vapor or gas per unit area and thickness of the material, per unit difference in partial pressure between two surfaces, under specified conditions of temperature and relative humidity [24, 113].

Permeability is a process of dissolution and diffusion rather than transport through pores. The solute dissolves into one side of the packaging and spreads to the opposite side due to gradient of chemical potential. According to Roy and co-workers [114], the transfer of water vapor through biopolymers films involves steps of adsorption of water vapor on the film surface, solubility of the vapor in the polymer matrix, diffusion of vapor through the film and desorption of vapor in the other surface of the film.

Many foods are susceptible to deterioration due to high moisture content, so the WVP is of great importance in the development of packaging in order to obtain the desired shelf life [115]. The choice of a suitable packaging depends on the permeability properties. A package can be considered to have high resistance to water vapor, when the values are less than 8 g/m^2 day and very high resistance when it is less than 0.8 g/m^2 day^1 [116]. The gravimetric method is the most popular in the determination of WVP. A cell scheme used in permeability assays is shown in Fig. (**2**) [6].

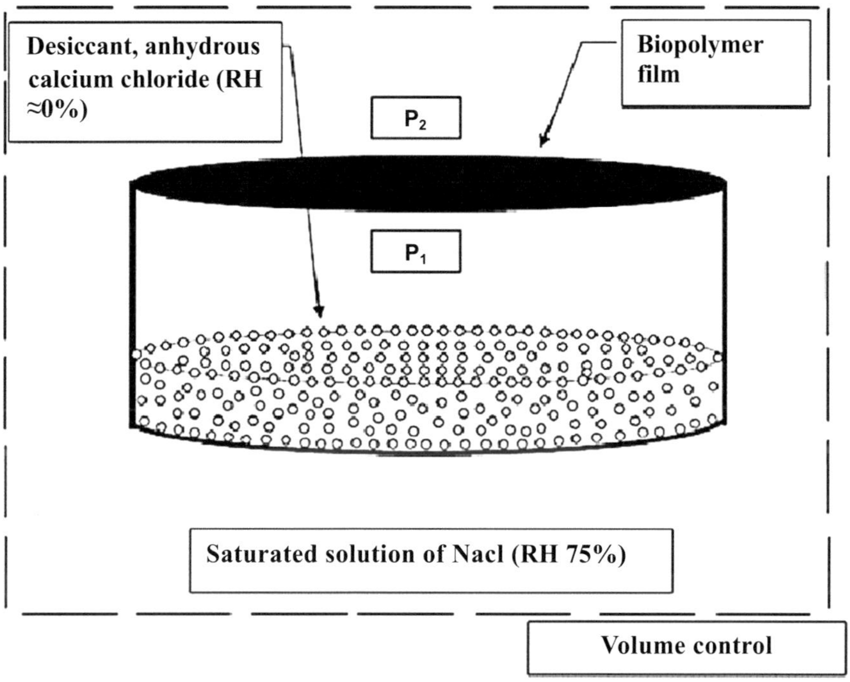

Fig. (2). Scheme of a diffusion cell used to determine the water vapor permeability of biopolymeric films [6].

As shown in Fig. (**2**), the disc shaped film is attached to the cell. At the cell bottom, there is a desiccant media with relative humidity (RH) assumed to be 0%. The cell is stored in an environment whose relative humidity is 75% [112]. The vapor pressure on the inner surface of the film, which is the same as the surface of the desiccant is defined as P_1 and the vapor pressure on the outer surface of the film, which is the same as the inside of the environment produced by saturated solution of sodium chloride, is defined as P_2. The WVP can be determined according to Eq. (**1**).

$$WVP = \frac{w_{ma}}{t}\frac{L}{A(\Delta P)} \tag{1}$$

where w_{ma} is the mass of absorbed moisture (g), t is the duration of the test (days), L is the thickness of the film (mm), A is the surface area of exposed film (m^2), and ΔP is the water vapor pressure difference between the two sides of the film (Pa).

CONCLUSIONS

Chitosan has a great potential as a packing material due to its biodegradability, biocompatibility, antimicrobial activity, cationic character and film-forming properties. Although chitosan films have recognized properties that allow them to be applied to food systems, they also have some drawbacks that restrict their use under specific conditions, owing to their low resistance to water-vapor diffusion and relatively low tensile strength, especially in moist environments. Thus, the blending of chitosan with other biopolymers (lipids, polysaccharides, proteins, cellulose nanocrystals, *etc*) has been a promising strategy to obtaining edible films with improved properties. Chitosan based edible films and coatings have been applied on fruits, vegetables, meats, eggs, and dairy over the last few years. Nevertheless, the several benefits to be afforded to food consumers and processors by effective edible film formulations justify further research in this field.

CONFLICT OF INTEREST

The authors confirm that they have no conflict of interest to declare for this publication.

ACKNOWLEDGEMENTS

The authors would like to express their thanks to the CAPES, CNPq, FAPERGS, Federal University of Santa Maria (UFSM) and Federal University of Santa Catarina (UFSC) for supporting this work.

REFERENCES

[1] Souza VC, Monte ML, Pinto LA. Preparation of biopolymer film from chitosan modified with lipid fraction. Int J Food Sci Technol 2011; 46: 1856-62. [http://dx.doi.org/10.1111/j.1365-2621.2011.02692.x]

[2] Chandra R, Rustgi R. Biodegradable polymers. Prog Polym Sci 1998; 23: 1273-335. [http://dx.doi.org/10.1016/S0079-6700(97)00039-7]

[3] Innovations in Food Packaging Part 4: Edible and biodegradable coatings and films 15 - Edible films and coatings: a review. Jung H. Han, Gennadios Aristippos 2005; pp. 239-62.

[4] Hu G, Chem J, Gao J. Preparation and characteristics of oxidized portao starch films. Carbohydr Polym 2009; 76: 291-8. [http://dx.doi.org/10.1016/j.carbpol.2008.10.032]

[5] Peressini D, Bravin B, Lapasin C, Sensidoni A. Starch-methylcellulose based edible films: rheological properties of film-forming dispersions. J Food Eng 2003; 59: 25-32. [http://dx.doi.org/10.1016/S0260-8774(02)00426-0]

[6] Souza VC. Produção de filmes biopoliméricos a partir de quitosana com incorporação de fração lipídica 2011.

[7] Han JH. Antimicrobial food packaging. Food Technol 2000; 54: 56-65.

[8] Adebiyi PA, Adebiyi AO, Jin DH, Ogawa T, Muramoto K. Rice bran protein-based edible films. Int J Food Sci Technol 2008; 43: 476-83. [http://dx.doi.org/10.1111/j.1365-2621.2006.01475.x]

[9] Geraldine RM, Soares NF, Botrel DA, Goncalves LA. Characterization and effect of edible coatings on minimally processed garlic quality. Carbohydr Polym 2008; 72: 403-9. [http://dx.doi.org/10.1016/j.carbpol.2007.09.012]

[10] Park SY, Lee BI, Jung ST, Park HJ. Biopolymer composite films based on k-carrageenan and chitosan. Mater Res Bull 2001; 36: 511-9. [http://dx.doi.org/10.1016/S0025-5408(01)00545-1]

[11] Souza VC, Quadri MG. Preparation and Properties of Cellulose Nanocrystals Derived from Corn Cobs: Application as a Reinforcing Material for "Green Nanocomposite" Films. J Biobased Mater Bioenergy 2014; 8: 495-505. [http://dx.doi.org/10.1166/jbmb.2014.1465]

[12] Lorevice MV, Otoni CG, Moura MR, Mattoso LH. Chitosan nanoparticles on the improvement of thermal, barrier, and mechanical properties of high- and low-methyl pectin films. Food Hydrocoll 2016; 52: 732-40. [http://dx.doi.org/10.1016/j.foodhyd.2015.08.003]

[13] Dang KM, Yoksan R. Morphological characteristics and barrier properties of thermoplastic starch/chitosan blown film. Carbohydr Polym 2016; 150: 40-7. [http://dx.doi.org/10.1016/j.carbpol.2016.04.113] [PMID: 27312611]

[14] Sharma L, Singh C. Sesame protein based edible films: Development and characterization. Food Hydrocoll 2016; 61: 139-47. [http://dx.doi.org/10.1016/j.foodhyd.2016.05.007]

[15] Nikoleli GP, Nikolelis DP, Evtugyn G, Hianik T. Advances in lipid film based biosensors. Trends Analyt Chem 2016; 79: 210-21. [http://dx.doi.org/10.1016/j.trac.2016.01.021]

[16] Bourtoom T, Chinnan MS. Improvement of water barrier property of rice starch-chitosan composite film incorporated with lipids. Food Sci Technol Int 2009; 15: 149. [http://dx.doi.org/10.1177/1082013208105993]

[17] Souza VC, Niehues E, Quadri MG. Development and characterization of chitosan bionanocomposites containing oxidized cellulose nanocrystals. J Appl Polym Sci 2016; 133: 43033. [http://dx.doi.org/10.1002/app.43033]

[18] Shellhammer TH, Rhim JW. Lipid-based films and coatings. Innov Food Packag 2005.

[19] Denavi G, Tapia-Blácido DR, Anon MC, Sobra PJ, Mauri AN, Menegalli FC. Effects of drying conditions on some physical properties of soy protein films. J Food Eng 2009; 90: 341-9. [http://dx.doi.org/10.1016/j.jfoodeng.2008.07.001]

[20] Souza BW, Cerqueira MA, Casariego A, Lima AM, Teixeira JA, Vicente AA. Effect of moderate electric fields in teh permeation properties of chitosan coatings. Food Hydrocoll 2009; 23: 2110-5. [http://dx.doi.org/10.1016/j.foodhyd.2009.03.021]

[21] Rodriguez M, Oses J, Ziani K, Mate JI. Combined effect of plasticizers and surfactants on the physical properties of starch based edible films. Food Res Int 2006; 39: 840-6. [http://dx.doi.org/10.1016/j.foodres.2006.04.002]

[22] Galdeano MC, Grossmann MV, Bello-Perez LA, Garcia MA, Zamudio-Flores PB. Effects of production process and palsticizers on stability of films and sheets of oat starch. Mater Sci Eng 2009;

29: 492-8.
[http://dx.doi.org/10.1016/j.msec.2008.08.031]

[23] Karbowiak T, Debeaufort F, Voilley A. Influence of thermal process on structure and functional properties of emulsion-base edible films. Food Hydrocoll 2007; 21: 879-88.
[http://dx.doi.org/10.1016/j.foodhyd.2006.07.017]

[24] Kester JJ, Fennema OR. Edible films and coatings. Food Technol 1986; 40: 47-59.

[25] Sothornvit T, Pitak N. Oxygen permeability and mechanical properties of banana films. Food Res Int 2007; 40: 365-70.
[http://dx.doi.org/10.1016/j.foodres.2006.10.010]

[26] Cuq B, Gontard N, Guilbert S. Edible films and coatings as active layers. In: Rooney ML, Ed. Glasgow, UK: Active Food Packaging, Blackie Academic and Professional 1995; pp. 111-42.

[27] Aider M. Chitosan application for active bio-based films production and potential in the food industry: review. LWT -. Food Sci Technol (Campinas) 2010; 43: 837-42.

[28] Rinaudo M. Chitin and chitosan: Properties and applications. Prog Polym Sci 2006; 31: 603-32.
[http://dx.doi.org/10.1016/j.progpolymsci.2006.06.001]

[29] Sanford P, Hutchings G. Chitosan—a natural, cationic biopolymer: Commercial applications. In: Yalpani M, Ed. Industrial polysaccharides: Genetic engineering, structure/property relations and applications. Amsterdam: Elsevier 1987.

[30] No HK, Meyers SP, Prinyawiwatkul W, Xu Z. Applications of chitosan for improvement of quality and shelf life of foods: a review. J Food Sci 2007; 72(5): R87-R100.
[http://dx.doi.org/10.1111/j.1750-3841.2007.00383.x] [PMID: 17995743]

[31] Chen XG, Zheng L, Wang Z, Lee CY, Park HJ. Molecular affinity and permeability of different molecular weight chitosan membranes. J Agric Food Chem 2002; 50(21): 5915-8.
[http://dx.doi.org/10.1021/jf020151g] [PMID: 12358459]

[32] Shahidi F, Arachchi JK, Jeon YJ. Food application of chitin and chitosans. Trends Food Sci Technol 1999; 10: 37-51.
[http://dx.doi.org/10.1016/S0924-2244(99)00017-5]

[33] Papineau AM, Hoover DG, Knorr D, Farkas DF. Antimicrobial effect of water-soluble chitosans with high hydrostatic pressure. Food Biotechnol 1991; 5: 45-57.
[http://dx.doi.org/10.1080/08905439109549790]

[34] Sudarshan NR, Hoover DG, Knorr D. Antibacterial action of chitosan. Food Biotechnol 1992; 6: 257-72.
[http://dx.doi.org/10.1080/08905439209549838]

[35] Dutta PK, Tripathi S, Mehrotra GK, Dutta J. Perspectives for chitosan based antimicrobial films in food applications. Food Chem 2009; 114: 1173-82.
[http://dx.doi.org/10.1016/j.foodchem.2008.11.047]

[36] Roller S, Covill N. The antifungal properties of chitosan in laboratory media and apple juice. Int J Food Microbiol 1999; 47(1-2): 67-77.
[http://dx.doi.org/10.1016/S0168-1605(99)00006-9] [PMID: 10357275]

[37] Rabea EI, Badawy ME, Stevens CV, Smagghe G, Steurbaut W. Chitosan as antimicrobial agent: applications and mode of action. Biomacromolecules 2003; 4(6): 1457-65.
[http://dx.doi.org/10.1021/bm034130m] [PMID: 14606868]

[38] Ferreira CO, Nunes CA, Delgadilo I, Silva JA. Characterization of chitosan-whey protein films at acid pH. Food Res Int 2009; 42: 887-13.
[http://dx.doi.org/10.1016/j.foodres.2009.03.005]

[39] Fernandez-Saiz P, Lagaron JM, Hernandez-Muñoz P, Ocio MJ. Characterization of the antimicrobial properties against S. aureus of novel renewable blends chitosan and gliadins of interest in active food

packaging and coating applications. Int J Food Microbiol 2008; 124: 13-20. [http://dx.doi.org/10.1016/j.ijfoodmicro.2007.12.019] [PMID: 18353476]

[40] Fernandez-Saiz P, Lagaron JM, Ocio MJ. Optimization of the biocide properties of chitosan for its application in the design of active films of interest in the food area. Food Hydrocoll 2009; 23: 913-21. [http://dx.doi.org/10.1016/j.foodhyd.2008.06.001]

[41] Bravin B, Peressini D, Sensidoni A. Development and application of polysaccharide-lipid edible coating to extend shelf-life of dry bakery products. J Food Eng 2006; 76: 280-90. [http://dx.doi.org/10.1016/j.jfoodeng.2005.05.021]

[42] Hambleton A, Debeaufort F, Bonnotte A, Voilley A. Influence of alginate emulsion-based films structure on its barrier properties and on the protection of microencapsulated aroma compound. Food Hydrocoll 2009; 23: 2116-24. [http://dx.doi.org/10.1016/j.foodhyd.2009.04.001]

[43] Fabra MJ, Talens P, Chiralt A. Tensile properties and water vapour permeability of sodium caseinate films containing oleic acid-beeswax mixtures. J Food Eng 2008; 85: 393-400. [http://dx.doi.org/10.1016/j.jfoodeng.2007.07.022]

[44] Rivero S, García MA, Pinotti A. Composite and bi-layer films based on gelatin and chitosan. J Food Eng 2009; 90: 531-9. [http://dx.doi.org/10.1016/j.jfoodeng.2008.07.021]

[45] Giancone T, Torrieri E, Di Pierro P, *et al.* Role of constituents on the network formation of hydrocolloids edible films. J Food Eng 2008; 89: 195-203. [http://dx.doi.org/10.1016/j.jfoodeng.2008.04.017]

[46] Fabra MJ, Talens P, Chiralt A. Influence of calcium on tensile, optical and water vapour permeability properties of sodium caseinate edible films. J Food Eng 2010; 96: 356-64. [http://dx.doi.org/10.1016/j.jfoodeng.2009.08.010]

[47] Kristo E, Biliaderis CG, Zampraka A. Water vapour barrier and tensile properties of composite caseinate-pullulan films: Bio polymer compositions effects and impact of beeswax lamination. Food Chem 2007; 101: 753-64. [http://dx.doi.org/10.1016/j.foodchem.2006.02.030]

[48] Debeaufort F, Voilley A, Luu D. Influence of hydrocolloid nature on the structure and functional properties of emulsified edible films. Food Hydrocoll 2009; 23: 691-9. [http://dx.doi.org/10.1016/j.foodhyd.2008.05.006]

[49] Bertan LC, Tanada-Palmu PS, Siani AC, Grosso CR. Effect of fatty acids and Brazilian elemi on composite filmes based on gelatin. Food Hydrocoll 2005; 19: 73-82. [http://dx.doi.org/10.1016/j.foodhyd.2004.04.017]

[50] Villalobos R, Chanona J, Hernández P, Guitiérrez G, Chiral A. Gloss and transparency of hydroxypropyl methylcellulose films containing surfactants as affected by their microstructure. Food Hydrocoll 2005; 19: 53-61. [http://dx.doi.org/10.1016/j.foodhyd.2004.04.014]

[51] Vargas M, Albors A, Chiralt A, González-Martínez C. Characterization of chitosan-oleic acid composite films. Food Hydrocoll 2009; 23: 536-47. [http://dx.doi.org/10.1016/j.foodhyd.2008.02.009]

[52] Narsimhan G. Stability of thin emulsion film between two oil phases with a viscoelastic liquid-liquid interface. J Colloid Interface Sci 2009; 330(2): 494-500. [http://dx.doi.org/10.1016/j.jcis.2008.10.080] [PMID: 19041985]

[53] Rangsansarid J, Fukada K. Factors affecting the stability of O/W emulsion in BSA solution: stabilization by electrically neutral protein at high ionic strength. J Colloid Interface Sci 2007; 316(2): 779-86. [http://dx.doi.org/10.1016/j.jcis.2007.07.040] [PMID: 17897667]

[54] Petersson M, Stading M. Water vapour permeability and mechanical properties of mixed starch-monoglyceride films and effect of film forming conditions. Food Hydrocoll 2005; 19: 123-32. [http://dx.doi.org/10.1016/j.foodhyd.2004.04.021]

[55] Ukai N, Ishibashi S, Tsutsumi T, Marakami K. Preservation of agricultural products. US Patent n, 3, 997, 674, 1976.

[56] Krochta JM. Emulsion films on food products to control mass transfer. In: Wan PJ, Cavallo JL, Saleeb FZ, McCarthy MJ, Eds. Food Emulsion and Foams; Theory and Practice, AIChE Symposium Series No 277. American Institute of Chemical Engineers: New York 1990; 86: pp. 57-61.

[57] Souza VC, Monte ML, Pinto LA. Effect of carp (Cyprinus carpio) oil incorporation on water vapour permeability, mechanical properties and transparency of chitosan films. Int J Food Sci Technol 2013; 48: 1309-17. [http://dx.doi.org/10.1111/ijfs.12093]

[58] Srinivasa PC, Ramesh MN, Tharanathan RN. Effect of plasticizers and fatty acids on mechanical and permeability characteristics of chitosan films. Food Hydrocoll 2007; 21: 1113-22. [http://dx.doi.org/10.1016/j.foodhyd.2006.08.005]

[59] Khalifa I, Barakat H, El-Mansy HA, Soliman SA. Improving the shelf-life stability of apple and strawberry fruits applying chitosan-incorporated olive oil processing residues coating. Food Packag Shelf Life 2016; 9: 10-9. [http://dx.doi.org/10.1016/j.fpsl.2016.05.006]

[60] Hromiš NM, Lazić VL, Markov SL, *et al.* Optimization of chitosan biofilm properties by addition of caraway essential oil and beeswax. J Food Eng 2015; 158: 86-93. [http://dx.doi.org/10.1016/j.jfoodeng.2015.01.001]

[61] Velickova E, Winkelhausen E, Kuzmanova S, Alves VD, Moldão-Martins M. Impact of chitosan-beeswax edible coatings on the quality of fresh strawberries (Fragaria ananassa cv Camarosa) under commercial storage conditions. Food Sci Technol (Campinas) 2013; 52: 80-92.

[62] Wong D, Gastineau F, Gregorski KS, Tillin SJ, Pavlath AE. Chitosan-Lipid films. Microstructure and surface energy. J Agric Food Chem 1992; 40: 540-4. [http://dx.doi.org/10.1021/jf00016a002]

[63] Shen Z, Kamdem DP. Development and characterization of biodegradable chitosan films containing two essential oils. Int J Biol Macromol 2015; 74: 289-96. [http://dx.doi.org/10.1016/j.ijbiomac.2014.11.046] [PMID: 25542174]

[64] Hussain F, Hojjati M, Okamoto M, Gorga RE. Polymer-matrix Nanocomposites, Processing, Manufacturing, and Application: An Overview. J Compos Mater 2006; 40: 1511-75. [http://dx.doi.org/10.1177/0021998306067321]

[65] Lin N, Huang J, Dufresne A. Preparation, properties and applications of polysaccharide nanocrystals in advanced functional nanomaterials: a review. Nanoscale 2012; 4(11): 3274-94. [http://dx.doi.org/10.1039/c2nr30260h] [PMID: 22565323]

[66] Dufresne A. Cellulose-based composites and nanocomposites. In: Belgacem MN, Gandini A, Eds. Monomers, Polymers and Composites from Renewable Resources. Amsterdam: Elsevier 2008; pp. 401-18. [http://dx.doi.org/10.1016/B978-0-08-045316-3.00019-3]

[67] Dufresne A. Polysaccharide nanocrystals reinforced nanocomposites. Can J Chem 2008; 86: 484-94. [http://dx.doi.org/10.1139/v07-152]

[68] Peng BL, Dhar N, Liu HL, Tam KC. Chemistry and applications of nanocrystalline cellulose and its derivatives: a nanotechnology perspective. Can J Chem Eng 2011; 89: 1191-206. [http://dx.doi.org/10.1002/cjce.20554]

[69] Mesquita JP. Nanocristais de celulose para preparacão de bionanocompositos com quitosana e carbonos nanoestruturados para aplicacões tecnologicas e ambientais 2012.

[70] Cao X, Ding B, Yu J, Al-Deyab SS. Cellulose nanowhiskers extracted from TEMPO-oxidized jute fibers. Carbohydr Polym 2012; 90(2): 1075-80.
[http://dx.doi.org/10.1016/j.carbpol.2012.06.046] [PMID: 22840042]

[71] Fernandes SC, Freire CS, Silvestre AJ, *et al.* Transparent chitosan films reinforced with a high content of nanofibrillated cellulose. Carbohydr Polym 2010; 81: 394-401.
[http://dx.doi.org/10.1016/j.carbpol.2010.02.037]

[72] Fernandes SC, Oliveira L, Freire CS, *et al.* Novel transparent nanocomposite films based on chitosan and bacterial cellulose. Green Chem 2009; 11: 2023-9.
[http://dx.doi.org/10.1039/b919112g]

[73] Khan A, Khan RA, Salmieri S, *et al.* Mechanical and barrier properties of nanocrystalline cellulose reinforced chitosan based nanocomposite films. Carbohydr Polym 2012; 90(4): 1601-8.
[http://dx.doi.org/10.1016/j.carbpol.2012.07.037] [PMID: 22944422]

[74] Li F, Biagioni P, Finazzi M, Tavazzi S, Piergiovanni L. Tunable green oxygen barrier through layer-by-layer self-assembly of chitosan and cellulose nanocrystals. Carbohydr Polym 2013; 92(2): 2128-34.
[http://dx.doi.org/10.1016/j.carbpol.2012.11.091] [PMID: 23399267]

[75] Li Q, Zhou JP, Zhang LN. Structure and properties of the nanocomposite films of chitosan reinforced with cellulose whiskers. J Polym Sci, B, Polym Phys 2009; 47: 1069-77.
[http://dx.doi.org/10.1002/polb.21711]

[76] de Mesquita JP, Donnici CL, Pereira FV. Biobased nanocomposites from layer-by-layer assembly of cellulose nanowhiskers with chitosan. Biomacromolecules 2010; 11(2): 473-80.
[http://dx.doi.org/10.1021/bm9011985] [PMID: 20055503]

[77] Pereda M, Dufresne A, Aranguren MI, Marcovich NE. Polyelectrolyte films based on chitosan/olive oil and reinforced with cellulose nanocrystals. Carbohydr Polym 2014; 101: 1018-26.
[http://dx.doi.org/10.1016/j.carbpol.2013.10.046] [PMID: 24299870]

[78] Souza VC, Niehues E, Quadri MG. Development and characterization of chitosan bionanocomposites containing oxidized cellulose nanocrystals. J Appl Polym Sci 2016; 133: 43033.
[http://dx.doi.org/10.1002/app.43033]

[79] Rivero S, García MA, Pinotti A. Crosslinking capacity of tannic acid in plasticized chitosan films. Carbohydr Polym 2010; 82: 270-6.
[http://dx.doi.org/10.1016/j.carbpol.2010.04.048]

[80] Khan A, Gallah H, Riedl B, Bouchard J, Safrany A, Lacroix M. Genipin cross-linked antimicrobial nanocomposite films and gamma irradiation to prevent the surface growth of bacteria in fresh meats. Innov Food Sci Emerg Technol 2016; 35: 96-102.
[http://dx.doi.org/10.1016/j.ifset.2016.03.011]

[81] Zhang ZH, Han Z, Zeng XA, Xiong XY, Liu YJ. Enhancing mechanical properties of chitosan films *via* modification with vanillin. Int J Biol Macromol 2015; 81: 638-43.
[http://dx.doi.org/10.1016/j.ijbiomac.2015.08.042] [PMID: 26314906]

[82] Benbettaïeb N, Karbowiak T, Brachais CH, Debeaufort F. Coupling tyrosol, quercetin or ferulic acid and electron beam irradiation to cross-link chitosan-gelatin films: A structure-function approach. Eur Polym J 2015; 67: 113-27.
[http://dx.doi.org/10.1016/j.eurpolymj.2015.03.060]

[83] Cao N, Fu Y, He J. Mechanical properties of gelatin films cross-linked, respectively, by ferulic acid and tannin acid. Food Hydrocoll 2012; 21: 575-84.
[http://dx.doi.org/10.1016/j.foodhyd.2006.07.001]

[84] Liu Y, Kim HI. Characterization and antibacterial properties of genipin-crosslinked chitosan/poly(ethylene glycol)/ZnO/Ag nanocomposites. Carbohydr Polym 2012; 89(1): 111-6. [http://dx.doi.org/10.1016/j.carbpol.2012.02.058] [PMID: 24750611]

[85] Kołodziejska I, Piotrowska B. The water vapour permeability, mechanical properties and solubility of fish gelatin-chitosan films modified with transglutaminase or 1-ethyl-3-(3-dimethylaminopropyl) carbodiimide (EDC) and plasticized with glycerol. Food Chem 2007; 103: 295-300. [http://dx.doi.org/10.1016/j.foodchem.2006.07.049]

[86] Pierro PD, Chico B, Villalonga R, Mariniello L, Masi P, Porta R. Transglutaminase-catalyzed preparation of chitosan-ovalbumin films. Enzyme Microb Technol 2007; 40: 437-41. [http://dx.doi.org/10.1016/j.enzmictec.2006.07.017]

[87] Kołodziejska I, Piotrowska B, Bulge M, Tylingo R. Effect of transglutaminase and 1-ethyl-3-(3-dimethylaminopropyl) carbodiimide on the solubility of fish gelatinechitosan films. Carbohydr Polym 2006; 65: 404-9. [http://dx.doi.org/10.1016/j.carbpol.2006.01.025]

[88] Liu Z, Ge X, Lu Y, Dong S, Zhao Y, Zeng M. Effects of chitosan molecular weight and degree of deacetylation on the properties of gelatine-based films. Food Hydrocoll 2012; 26: 311-7. [http://dx.doi.org/10.1016/j.foodhyd.2011.06.008]

[89] Yin Y, Li Z, Sun Y, Yao K. A preliminary study on chitosan/gelatine polyelectrolyte complex formation. J Mater Sci Lett 2005; 40: 4649-52. [http://dx.doi.org/10.1007/s10853-005-3929-9]

[90] Mathew S, Abraham TE. Characterisation of ferulic acid incorporated starch-chitosan blend films. Food Hydrocoll 2008; 22: 826-35. [http://dx.doi.org/10.1016/j.foodhyd.2007.03.012]

[91] Appendini P, Hotchkiss JH. Review of antimicrobial food packaging. Innov Food Sci Emerg Technol 2002; 3: 113-26. [http://dx.doi.org/10.1016/S1466-8564(02)00012-7]

[92] Hanušová K, Dobiáš J, Klaudisová K. Effect of packaging films releasing antimicrobial agents on stability of food products. Czech J Food Sci 2009; 27: 347-9.

[93] Lucera A, Costa C, Conte A, Del Nobile MA. Food applications of natural antimicrobial compounds. Front Microbiol 2012; 3: 287. [http://dx.doi.org/10.3389/fmicb.2012.00287] [PMID: 23060862]

[94] Tharanathan RN. Biodegradable films and composite coatings: Past, present and future. Trends Food Sci Technol 2003; 14: 71-8. [http://dx.doi.org/10.1016/S0924-2244(02)00280-7]

[95] Coma V, Deschamps A, Martial-Gros A. Bioactive packaging materials from edible chitosan polymer: antimicrobial activity assessment on diary-related contaminants. J Food Sci 2003; 68: 2788-92. [http://dx.doi.org/10.1111/j.1365-2621.2003.tb05806.x]

[96] Tsai GJ, Su WH, Chen HC, Pan CL. Antimicrobial activity of shrimp chitin and chitosan from different treatments and applications of fish preservation. Fish Sci 2002; 68: 170-7. [http://dx.doi.org/10.1046/j.1444-2906.2002.00404.x]

[97] Vargas M, Albors A, Chiralt A, Gónzalez-Martinez C. Quality of cold stored strawberries as affected by chitosan-oleic acid edible coatings. Postharvest Biol Technol 2006; 41: 164-71. [http://dx.doi.org/10.1016/j.postharvbio.2006.03.016]

[98] Li H, Yu T. Effect of chitosan on incidence of brown rot, quality and physiological attributes of postharvest peach fruit. J Sci Food Agric 2001; 81: 69-274. [http://dx.doi.org/10.1002/1097-0010(20010115)81:2<269::AID-JSFA806>3.0.CO;2-F]

[99] Sánchez-González L, González-Martínez C, Chiralt A, Cháfer M. Physical and antimicrobial properties of chitosan-tea tree essential oil composite films. J Food Eng 2010; 98: 443-52. [http://dx.doi.org/10.1016/j.jfoodeng.2010.01.026]

[100] Preparation method of novel edible film-coated fruit and vegetable fresh-keeping agent. Patent CN 103719267A, 2014.

[101] Wu Y, Rhim JW, Weller CL, Hamouz F, Cuppett S, Schnepf M. Moisture loss and lipid oxidation for precooked beef patties stored in edible coatings and films. J Food Sci 2000; 65: 300-4. [http://dx.doi.org/10.1111/j.1365-2621.2000.tb15997.x]

[102] Edible composite preservative film applied to preservation of fresh pork in supermarket. Patent CN 102977414B, 2015.

[103] Bhale S, No HK, Prinyawiwatkui W, Farr AJ, Nadarajah K, Meyers SP. Chitosan coating improves shelf life of eggs. J Food Sci 2003; 68: 2378-83. [http://dx.doi.org/10.1111/j.1365-2621.2003.tb05776.x]

[104] Guilbert S, Biquet B. Películas y envolturas comestibles de los alimentos de gran consume. Ed. Acríbia S. A., 22, 331-371

[105] Gennadios A, Weller CL, Testin RF. Modification of physical and barrier properties of edible wheat gluten-based films. Cereal Chem 1993; 70: 426-9.

[106] Bertuzzi MA, Vidaurre EF, Armada M, Gottifredi JC. Water vapor permeability of edible starch based films. J Food Eng 2007; 80: 972-8. [http://dx.doi.org/10.1016/j.jfoodeng.2006.07.016]

[107] Cuq B, Gontard N, Cuq JL, Guilbert S. Stability of myofibrillar protein based biopackagings during storage. LWT-Food Sci Technol 1996; 29: 334-48. [http://dx.doi.org/10.1006/fstl.1996.0052]

[108] Veiga-Santos P, Suzuki CK, Cereda MP, Scamparini AR. Microstructure and color of starch-gum films: Effect of gum deacetylation and additives. Part 2. Food Hydrocoll 2005; 19: 1064-73. [http://dx.doi.org/10.1016/j.foodhyd.2005.02.007]

[109] Gontard N, Duchez C, Cuq JL, Guilbert S. Edible composite films of wheat and lipids: water vapour permeability and other physical properties. Int J Food Sci Technol 1994; 29: 39-50. [http://dx.doi.org/10.1111/j.1365-2621.1994.tb02045.x]

[110] Van De Velde K, Kiekens P. Biopolymer: Overview of several properties and consequences on their applications. Polym Test 2002; 21: 433-42. [http://dx.doi.org/10.1016/S0142-9418(01)00107-6]

[111] Ferry JD. Viscoelastic properties of polymers. John Wiley & Sons 1980; p. 672.

[112] Standard test methods for water vapor transmission of materials. Designation E96-95. Annual book of American standard testing methods. Philadelphia: American Society for Testing and Materials 1995; pp. 406-13.

[113] Debeaufort F, Quezada-Gallo JA, Voilley A. Edible films and coatings: tomorrows packagings: a review. Crit Rev Food Sci Nutr 1998; 38(4): 299-313. [http://dx.doi.org/10.1080/10408699891274219] [PMID: 9626488]

[114] Roy S, Gennadios A, Weller CL, Testin RF. Water vapor transport parameters of a cast wheat gluten film. Ind Crops Prod 2000; 22: 43-50. [http://dx.doi.org/10.1016/S0926-6690(99)00032-1]

[115] Feldman D. Polymer Barrier Films. J Polym Environ 2001; 9: 49-55. [http://dx.doi.org/10.1023/A:1020231821526]

[116] Garcia EE, Padula M, Sarantópulos CI. Embalagens Plásticas: propriedades de barreira. Inst de Tecnol de Alim 1989; 44: 80-290.

CHAPTER 10

Application of Chitosan Membranes for Permeation and Pervaporation

Gabriela Dudek* and **Roman Turczyn**

Department of Physical Chemistry and Technology of Polymers, Faculty of Chemistry, Silesian University of Technology, Strzody 9, 44100 Gliwice, Poland

Abstract: This chapter provides a comprehensive overview on the application of chitosan membranes in pervaporation and permeation processes. Pervaporation is a widely used membrane process for volatile liquid mixture separation. The separation of solvents is of extreme importance in a variety of industries ranging from chemical to food and pharmaceutical. The main thrust of research efforts has been concentrated on the polymeric membranes: investigation of new polymeric materials, modeling of component transport through dense polymers and description of their application. Chitosan membranes, owing to their hydrophilicity, biocompatibility, and ease of modification have been extensively applied for numerous pervaporation processes. The general overview of the preparation and application of different kinds of chitosan membranes *e.g.* crosslinked, blended, hybrid, grafted and chitosan membranes on porous support, in the separation of various solvents is presented. The progress of research, challenges and opportunities, as well as the prospect of pervaporation technique are also discussed.

Keywords: Blended chitosan membranes, Chitosan membranes on porous support, Crosslinked chitosan membranes, Grafted chitosan membranes, Hybrid chitosan membranes, Pervaporation.

INTRODUCTION

Pervaporation is recognized as a separation process that has elements in common with reverse osmosis and membrane gas separation. The concept of pervaporation separation dates from the early twentieth century. In 1917, the phenomenon of pervaporation was first discovered by Kober, who introduced the term in a paper reporting selective permeation of water through cellulose nitrate membranes from albumin and toluene aqueous solutions [1 - 9].

* **Corresponding author Gabriela Dudek:** Department of Physical Chemistry and Technology of Polymers, Faculty of Chemistry, Silesian University of Technology, Strzody 9, 44-100 Gliwice, Poland; Tel/Fax: +48 32 237-1067; Email: gabriela.maria.dudek@polsl.pl

Guilherme Luiz Dotto, Sergio Paulo Campana-Filho and Luiz Antonio de Almeida Pinto (Eds)

Pervaporation is a widely used membrane separation process for the separation of liquid mixtures. During pervaporation, the separated mixture is in direct contact with one side of membrane, whereas the permeate vapor is removed from the opposite side in a vapor state under a vacuum. The permeate vapor can be condensed and collected or released as desired. The driving force for the mass transport of permeants is a gradient of chemical potential, which is established by an existing difference in the partial pressures of permeants across the membrane. The difference in partial pressures can be created either by reducing the total pressure by using a vacuum pump system or by sweeping an inert gas on the permeate side of the membrane [1 - 9].

Pervaporation transport is usually described by the generally accepted solution-diffusion (dual sorption) mechanism of mass transport through the nonporous membranes. According to this mechanism, the transport process can essentially be visualized to occur by the following three steps: (a) selective sorption of the components on the membrane surface; (b) diffusion of the components through the membrane; and (c) desorption on the permeate side. The sorption of the permeate component can be controlled by the affinity for the membrane material. The diffusion can be described as the permeation rate of the component through the membrane. Desorption is a very rapid step when permeate pressures are low. However as the permeate pressures approach the partial vapor pressure of the permeating species, the permeate flux becomes dependent on the rate of vaporization [1 - 9].

Pervaporation has substantial advantages in the separation of azeotropes, *i.e.* close boiling mixtures, and thermally sensitive compounds, and removing species present at low concentrations. Only a fraction of a mixture is vaporized during pervaporation, and lower temperatures than those required in distillation are usually used. Pervaporation can be mainly applied in the three fields classified as follows: (a) dehydration of aqueous organic mixtures; (b) removal of low concentrated organic compounds from organic aqueous mixtures and (c) separation of liquid organic mixtures [1 - 9].

The dehydration of organic solvents is the most important application of pervaporation. The first commercial application was the dehydration of ethanol, using a hybrid system with a distillation column. The ability to selectively separate ethanol and other alcohols from water by pervaporation has attracted significant attention over the years as alternatives to conventional separation processes. Considering that, it is the most widespread distillation, characterized as high energy and high capital cost, currently, in chemical engineering, membrane based pervaporation (PV) is widely used as one of the most effective methods for the separation of azeotropic mixtures, isomeric mixtures and temperature sensitive solutions. In the pervaporation process, water vapor diffuses faster than alcohol, and consequently leaves the alcohol to be dehydrated (purified) to the desired level of purity. In contrast, distillation enables the purification only up to the level of azeotrope (98.6% of ethyl alcohol) and needs heating [10 - 13].

For the dehydration of organic solvents, hydrophilic membranes are generally used because of the favorable solubility and diffusivity for selective permeation of water. Many researchers are making continuous efforts to develop new polymeric membranes. Recently, several reports have been published on using different types of hydrophilic polymers including poly(vinyl alcohol) (PVA) [14, 16], sodium alginate [11, 17, 19] and chitosan [19, 20]. Furthermore, hydrophilic polymers modified by grafting [21, 22], blending [18, 23, 24] and filler incorporated membranes [25 - 27] have been recently more promising in pervaporation dehydration of aqueous organic mixtures than plain membranes.

The second field, of enormous importance, is the water purification. Wastewater generated from various chemical products including commonly used refrigerants, plastics, adhesives, paints, petroleum products, and pharmaceutics, as well chemical plants is a serious environmental problem [28 - 32]. The wastewater mainly contains volatile organic compounds (VOCs), which are particularly halogenated hydrocarbons known for their negative effects on the environment and human health, with the most common ones being the depletion of stratospheric ozone, formation of the ground level smog, origination of the annoying odors and induction of the chronic toxicity [33 - 36]. Traditional VOCs control technologies such as conventional distillation process [37, 38], adsorption [39, 40] and biological treatment [41, 42] do not always provide a complete and economically reasonable solution, because the concentration of VOCs in the wastewater is usually very low. The pervaporation process is appropriate for the removal of VOC traces from the wastewater [34 - 36]. Membranes with high organic permeability and low water permeability are required in these processes. A typical organic permselective material for pervaporative membranes is poly(dimethylsiloxane) (PDMS) also known as silicone rubber [43, 44] or zeolite membranes [45, 46].

The third field of application of pervaporative separation is in the separation of the aromatic/aliphatic mixtures, such as benzene/cyclohexane [47 - 50], toluene/n-heptane [51 - 54] or toluene/octane [55 - 57]. This kind of separation is one of the most important and difficult process in the chemical and petrochemical industries, including the production of cyclohexane [47, 50], removal of benzene from gasoline [58, 59], and naphtha reforming [60, 61]. Many of these mixtures are difficult to separate because they contain close boiling point components that generally have similar physical and chemical properties, and/or the existence of azeotropes in the mixtures [47, 61]. The conventional separation technologies, such as liquid-liquid extraction, extractive distillation and azeotropic distillation have problems with the process complexity, large equipment investment and high energy consumption. Furthermore, no feasible processes are available for the separation of aromatic and aliphatic hydrocarbons in the range below 20 wt% of aromatics in the feed mixture. Compared with traditional technologies, pervaporation is considered to be a promising alternative due to its economical, safe, and eco-friendly nature [47, 61]. For the separation of aromatic/aliphatic mixtures by membrane pervaporation, various polar polymeric materials have been investigated including polyurethane [47], aromatic polyimides [56, 62, 63], polybenzimidazole [13, 55, 57], poly(methyl acrylate) [64], and poly(vinyl alcohol) [22, 23].

In this chapter, the application of chitosan based membranes in several pervaporation processes will be presented. Chitosan,, has along the years become one of the most studied biodegradable polymeric material for pervaporation membranes owing to its significant hydrophilicity, excellent chemical resistance and good film-forming properties. Chitosan is also a multifunctional polymer containing large number of reactive primary amine groups together with hydroxyl groups, which make the polymer a hydrophilic material and simultaneously capable of reacting with other groups, *e.g.* epoxy. These hydrophilic groups are considered to play a major role in preferential water sorption and diffusion through the chitosan membrane. In view of this feature, chitosan membranes have been studied extensively for the dehydration of organic liquids [19, 20, 26].

One of the major drawbacks of the hydrophilic polymer membranes in the pervaporative dehydration application is the relatively poor membrane stability, especially for the feeding solutions possessing high water contents. In order to increase the membrane stability, many efforts, including crosslinking the membranes [65 - 75], preparation of polymer blends [76 - 91], formation of hybrid organic - inorganic composites [48, 49, 92 - 103], chemical surface modification of hydrophilic polymers by grafting [50, 104 - 107], and formation a thin and dense separation layer on top of a porous substrate [108 - 117] have been made. The following chapter has been divided into subsections with each of them containing the contemporary data on the various chemical or physical modifications of chitosan membranes mentioned. Moreover, the achieved improvement in the separation properties are presented and deeply discussed.

PERVAPORATION EXPERIMENTS

Pervaporation experiments are carried out using typical pervaporation apparatus [94, 95] presented on (Fig. **1**). The feed is poured to the feed tank. Through the circulating pump, the solution is supplied into the high pressure top side of membrane placed inside the permeation cell, which is assembled from two half-cells usually made of stainless steel and fastened with bolted clamps. The membrane is placed on a finely porous plate support, *e.g.* stainless steel. During pervaporation experiments, the retentate is continuously recirculated to the feed tank and the permeate is received and frozen in the cold trap condenser unit. The reduced pressure in the system is produced by a vacuum pump and controlled with a vacuum gauge.

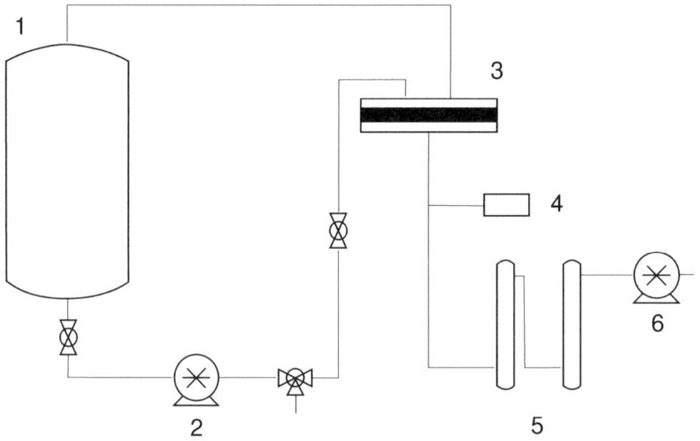

Fig. (1). Typical pervaporation apparatus. (**1**) - feed tank, (**2**) - circulating pump, (**3**) - permeation chamber combined from two half-cells, (**4**) - vacuum gauge, (**5**) - cold trap condenser unit, (**6**) - vacuum pump.

Firstly, the performance of tested membranes is characterized by evaluation of the permeation rate or flux and selectivity. Flux was calculated from the measured weight of cold trap liquid collected during certain time intervals at steady-state condition. The feed, permeate and retentate composition is analyzed by gas chromatography.

The permeation flux of component *i* is calculated using the following equation [18]:

$$J_i = \frac{m_i}{At} \quad \textbf{(1)}$$

where m_i - weight of component *i* in permeate, A - effective membrane area, t - permeation time.

For a homogeneous membrane, the flux could be normalized to a thickness of 1μm and calculated as the normalized flux of component *i* [18]:

$$J_{Ni} = \frac{J_i}{d} \tag{2}$$

where d - thickness of membrane.

Knowing the normalized flux and partial vapor pressure of component *i* in a feed, it is possible to estimate the permeability coefficient of component *i* [18]:

$$P_i = \frac{J_{Ni}}{p_{if}} \tag{3}$$

where p_{if} - partial vapor pressure of component *i* in a feed.

To describe the separation effectiveness of the membrane, two parameters are used: separation factor α_{AB} and selectivity coefficient Sc_{AB}. Separation factor is defined as follows [18]:

$$\alpha_{AB} = \frac{y_A / y_B}{x_A / x_B} \tag{4}$$

where x_A, x_B - weight fraction of components *A* and *B* in the feed [wt%], and y_A, y_B - weight fraction of components *A and* *B* in the permeate [wt%].

On the other hand, selectivity coefficient is equal to the ratio of the permeability coefficient of both separated components [18]:

$$S_{cAB} = \frac{P_A}{P_B} \tag{5}$$

The difference between these two parameters is related to their physical meanings, *i.e.* separation factor determines the separation efficiency of the mixture through the changes of component concentration in the feed and permeate, while the selectivity coefficient indicates the component which is preferentially transported through the membrane.

In order to compare the separation efficiency of different investigated membranes, pervaporation separation index expressed by following equation [18] is widely used:

$$PSI = J(\alpha_{AB} - 1) \qquad (6)$$

where J - total permeate flux, α_{AB} - separation factor.

PREPARATION METHODS OF CHITOSAN MEMBRANES

Typically, chitosan membranes are prepared using cast drying method. Chitosan powder or flakes are dissolved in a glacial acetic acid solution. Next, the prepared solution is filtered in order to remove any undissolved chitosan and debris, and then cast onto a Petri dish or other planar plate. The clear, transparent or yellowish membranes are obtained after drying in air or in a vacuum dryer for an appropriate time.

In order to increase the membrane stability, many modifications are used. One of the most popular is the membrane crosslinking process. By crosslinking, it is possible to decrease the polymer crystallinity and shrinkage of crystallite size. Crosslinking could also improve the selectivity of the membranes but the reduction in permeation flux is often accompanied. Researchers explored various crosslinking agents including miscellaneous organic species [55, 60] and inorganic acids [57, 61, 64]. The effect of crosslinking on the separation performance varies enormously with the crosslinking degree, the nature of interaction with chitosan matrix, the chemical structure of the crosslinking agents as well as the process conditions, etc [55, 64].

During the crosslinking process, chitosan membranes are usually immersed in an appropriate bath of crosslinking agent for adequate time and subsequently washed with sodium hydroxide and distilled or deionized water to neutralize and remove any excess crosslinking agent, until neutral pH is obtained [83, 84].

Another promising approach for the preparation of new materials is the combination of chitosan with another component to fabricate polymer blends [65, 80]. The addition of another substance, with suitable properties, can greatly affect the overall membrane properties, especially their transport parameters, leading to the improvement of the selectivity of the chitosan membranes. Additionally, blending allows to improve the mechanical properties, *e.g.* strength and flexibility, and thus enhances the process of pervaporation. Researchers typically added the following components to the chitosan: poly(vinyl alcohol) [65, 68], alginate [69, 72], poly(acrylic acid) [73, 74], cellulose [75, 76] and poly(vinyl pyrrolidone) [77, 79]. The preparation procedure of blended chitosan membranes relies on the mixing of chitosan solution with the solution of another polymer dissolved in a suitable compatible solvent. The rest of the preparation of the membranes is the same as that of pure chitosan membranes [65, 67].

In recent years, the organic/inorganic hybrid chitosan membranes composed of polymer matrix and inorganic nanoparticle fillers have become very popular [81, 90]. The results achieved showed that the addition of inorganic materials into the polymer matrix develop effective transport channels, and enhance its thermal stability. These factors could intensify the separation process and increase the ion conductivity. As the most popular inorganic additives, metal oxide nanoparticles [92 - 96], zeolites [97, 98], silica [99, 100] and carbon nanotubes are used. The broad spectrum of metal oxide nanoparticles introduced to polymer membranes contains Fe_3O_4, ZrO_2 TiO_2, MgO, CaO, Al_2O_3,CuO, SiO_2 and ZnO. Each of them could be incorporated into most of the polymeric materials used for membrane fabrication, showing the fine synergism properties between the polymeric materials and nanoparticles.

Zeolites are microporous crystalline aluminosilicates with uniform molecular sized pores and have unique features that are very attractive for a large number of industrial applications. The crystalline nature of zeolites offers the opportunity to obtain membranes with a regular 3D network of micro pores at the molecular scale, and they are therefore able to separate mixtures of substances on the basis of differences in the molecular size and shape, such as, isomers, compounds with similar molecular weight as well as azeotropic mixtures [45, 46, 97, 98].

Carbon nanotube loaded membranes are composed of cylindrical graphene sheets rolled up in a tube like structure with the appearance of latticework fence. Depending on their synthesis methods, CNTs could be produced as single walled CNTs (SWCNTs) or multiwalled CNTs (MWCNTs). The synthesis and applications have been extensively examined and published. Among different additives, CNTs have gathered considerable attention as nanofillers for water filtration applications. They also have unique reinforcement capabilities due to their exceptional thermal, mechanical and permeation properties. Exclusively dispersed CNTs provide enhanced interfacial adhesion between polymer chains [49, 101, 103].

Another promising approach to the modification of membrane properties is the formation of an active layer through the grafting of covalently bonded polymer layer on the surface of the porous ceramic substrate. The thin polymer layer provides effective reduction in transport resistance without compromising the mechanical strength. Compared to conventional single layer membranes, the thickness of the active separation layer is greatly reduced, leading to significant increase of the permeation flux. In general, the process of radiation induced grafting can be summarized in two steps: (1) creating active sites in the pre-existing polymer using physical or chemical techniques; and (2) initiating a polymerization reaction with a monomer bearing new tailored properties according to a designed requirement. Changing the chemical nature of the grafted monomer, membranes with desired properties can be fabricated through this process. Moreover, if the monomer possesses ionic groups, the grafted membrane acquires an ionic character with only limited influence from its inherent characteristics. The grafting process can be controlled to yield desirable architectures. To prepare the chitosan layer on the ceramic support, many techniques such as spray coating, grafting, spin coating, self-assembly, dip coating and vapor deposition have been reported. Among these methods, dip coating is simple and inexpensive, and thus most desirable choice for industrial purpose [21, 22, 50, 104, 107].

CROSSLINKED CHITOSAN MEMBRANES

In general, the advantages of chitosan as a material are a high swell ability, high permeability for water vapor and gases, good fluid transport *via* the membrane, and, high selectivity for the transport of voluminous and polar substances. Unfortunately, membranes prepared only with chitosan pose a lack of sufficient mechanical strength and stability, mainly due to excessive swelling in aqueous solutions. This disadvantage could be overcome by chemical or physical stabilization of the membrane structure. The most popular modification used for chitosan membranes is crosslinking. As cross-linkers, various organic chemicals, inorganic acids and metal salts are applied. Chemical crosslinking agents create covalent bonds with the polysaccharide backbone. In order to block amino groups and turn chitosan structures more inert and resistant to the acidic media, researchers widely explored the use of glutaraldehyde. With its hydrophobic character, the glutaraldehyde molecule when crosslinked with this agent, introduces more hydrophobic fragments to the chitosan membranes [65]. Jiangnan Shen *et al.* [66] studied the separation of water/isopropanol mixture through glutaraldehyde crosslinked carboxymethyl chitosan/polysulfone hollow fiber composite membranes. The authors concluded that the permeation flux decreased and separation factor increased with increasing glutaraldehyde content. The decrease in permeation flux was explained as a result of the increase in the crosslinking density of the chitosan membrane. Uragami *et al.* [67] found that the chitosan membrane had many intermolecular hydrogen bonds between their hydrophilic groups (NH_2 and OH), and the amount of free hydrophilic groups were quite low. At the time of crosslinking, some of these hydrogen bonds were broken and free hydrophilic groups could be generated. Therefore, the chain mobility and swelling ability of chitosan decreases. The thermal mobility of polymeric chains apparently drops and the available absolute free volume that diffuses the permeants through the membrane decreases due to the crosslinking. However, water molecules with relatively small molecular size can diffuse through the crosslinked membrane rather easily than isopropanol molecules with larger molecular size. Thus, the glutaraldehyde crosslinked chitosan membranes are effective pervaporation membranes for the separation of isopropanol/water mixtures. The achieved permeation flux and separation factor for 87.5 wt% isopropanol/water solution at 45°C was 38.6 $g/m^2 \cdot h$ and 3238.5 $g/m^2 \cdot h$, respectively.

Gnus *et al.* [68] applied glutaraldehyde crosslinking chitosan membranes in process of dehydration of ethanol. In this case, the overall membrane hydrophilicity was reduced by Schiff base formation reaction, *i.e.* formation of imine bonds between glutaraldehyde and chitosan backbone. The results showed that the hydrophobicity increased with increasing glutaraldehyde content and crosslinking time. Mechanical test showed that the strength of glutaraldehyde crosslinked chitosan membranes is much higher compared to the homogenous one. By comparison, the modified membranes showed lower degree of swelling than that of homogenous membranes. For both mechanical and swelling tests, it was found that the optimum crosslinking time was 15 min. The normalized permeation flux and separation factor for ethanol/water was 4740 $\mu m \cdot g/m^2 \cdot h$ and 16.9, respectively using 93 wt% ethanol solutions at room temperature.

Jian Hua Chen *et al.* [69] crosslinked chitosan membranes using 3-aminopropyl-triethoxysilane (CSHM). It was found that the amorphous region in the CSHMs crosslinked membranes increased with increasing crosslinking agent content. Likewise hydrophilicity of the CSHMs membranes change in the same way and arrived at the maximum when crosslinking agent content was equal to 10 wt%. The sorption experiments showed that the solubility selectivity and the diffusion selectivity, as well as the diffusion coefficient of water and ethanol were greatly influenced by the feed concentration and feed temperature. Compared to the unmodified chitosan membrane, the properties of CSHMs have been significantly improved. The permeation flux and the water permselectivity increased remarkably with increasing either 3-aminopropyltriethoxysilane content, and/or feed temperature.

Shude Xiao *et al.* [70] demonstrated the experimental data for chitosan membranes crosslinked by trimesoyl chloride (TMC, 1,3,5-benzenetricarboxylic acid chloride) in hexane solution used in pervaporative isopropanol/water mixture separation process. Trimesoyl chloride is a well-known monomer/crosslinking agent for interfacial polymerization, where it reacts at the liquid interface with different functional groups, *e.g.* amine. In chitosan, dissociated amino groups were formed from neutralization of the acetic acid containing membranes with sodium hydroxide solution. When the dried chitosan membrane was immersed in trimesoyl chloride/hexane, the solid surface was in direct contact with TMC. The acyl chloride groups of trimesoyl chloride reacted with the amino groups to form amide linkages between chitosan chains. Crosslinking might also occur between trimesoyl chloride and hydroxyl moieties when chitosan membranes are soaked in trimesoyl chloride/hexane. Hydroxyl groups are active enough to react with trimesoyl chloride, but this reaction is rather slow. Authors found that the trimesoyl chloride moieties changed the thermal properties of the chitosan membranes. Additionally, the relaxation of the segmental chains in chitosan matrix and the packing properties of the matrix were influenced by trimesoyl chloride presence, which led to changes in pervaporation flux and selectivity. The recognized separation factor for 10-40 wt% isopropanol/water solution was 1000 wt% at 60°C.

D. Anjali Devi *et al.* [71] used toluene-2,4-diisocyanate crosslinked chitosan membrane in isopropanol-water mixture separation process. Toluene-2, 4-diisocyanate appeared to be a useful and novel crosslinking agent, rendering chitosan highly selective to water without compromising heavily on the flux. The XRD results of the crosslinked chitosan showed a reduction in the effective spacing value, which is an indication of shrinkage in the cell size, which would in turn, improve the selective permeation properties of the membrane. Additionally, crosslinking did not bring about any unwanted major changes in the thermal and mechanical stability of the membrane.

The second largest group of crosslinking agents are substances used for physical stabilization of a membrane structure. Gnus *et al.* [68] explored the experimental data for chitosan membranes crosslinked by sulfuric and phosphoric acid in water/ethanol separation process. The results showed that physically crosslinked chitosan membranes had more diversified properties. For sulfuric acid crosslinked membrane, Coulombic interactions between amino groups in chitosan and sulfate ions are responsible for the crosslinking mechanism. These interactions strengthen the charge density in chitosan membranes, leading to an increase in the affinity of the membrane toward water molecules, and high water selectivity was observed [72]. The phosphoric acid reacts with the amine groups of chitosan in a similar way as sulfuric acid, resulting in strong ionic bonding between the amine group of the polymer and phosphate ion. Observed interactions were confirmed by Sunitha *et al.* [73]. The normalized fluxes of components for membranes crosslinked by sulfuric or phosphoric acid were significantly different between themselves. However, the normalized flux of ethanol for both examined membranes was higher, than for water.

Lee *et al.* [74] prepared phosphorylated chitosan membranes by reacting ortho-phosphoric acid and urea on the surface of chitosan membrane. They used them for separation of 90 wt% ethanol/water mixtures. The authors observed that the phosphorylated chitosan membranes showed higher water selectivity, up to 541, and flux of 182 g/m^2·h at 70 °C. However, the greatest surface modification of membranes could not improve the flux and the separation factor simultaneously.

Jegal *et al.* [75] crosslinked chitosan membranes with sulfosuccinic acid (SSA, 2-sulfobutanedioic acid) and used them for the pervaporative separation of water/alcohol mixtures. The sulfosuccinic acid has three crosslinkable sites (two carboxylic acid and one sulfuric acid groups) and a relatively bulky chemical structure. The three acidic groups are convenient for ionic crosslinking with chitosan and also good for keeping the membrane hydrophilicity. In this study, the permselective properties were estimated for three different kinds of alcohol solutions: methanol, ethanol, and isopropanol. It was found that when the temperature rose, the total fluxes also increased, whereas separation factors decreased. The observed behaviors of fluxes and separation factors as a function of temperature are usual in the case of hydrophilic membranes for the separation of water/alcohol mixtures. The high fluxes also confirmed indirectly that the free volume of the membrane was relatively large, due to the bulky SSA structure. The authors also studied the total flux and separation factor at 40°C as a function of the composition of the water/ethanol feed mixture. They observed that as the ethanol content in feed solution increases, the total flux decreases gradually, whereas the separation factor shows a maximum at 80 wt%. From this result, it was verified that this membrane showed a good performance for the wide range of feed compositions, not only at high ethanol content (such as 80 wt%), but also when it is low (10-40 wt%).

BLENDED CHITOSAN MEMBRANES

Blended polymer membranes are widely studied for pervaporative dehydration. Polymer blending proves to be an effective way to fabricate membranes with a number of desired properties. Successful blending can improve separation performance by altering polymer chain mobility out of intermolecular interactions between polymers. One of the most common polymer blends is a mixture of chitosan with poly(vinyl alcohol). Wu *et al.* [76] used such blend in pervaporative dehydration of ethanol. They stated that the total flux increased gradually, but the selectivity of the investigated system decreased sharply when the chitosan content in the blend increased. The sharp decrease observed was explained according to the structure of chitosan and poly(vinyl alcohol). Since poly(vinyl alcohol) has semi crystalline structure while chitosan has less compact structure and relatively larger free volume, blending of poly(vinyl alcohol) with chitosan can reduce the crystallinity, and as a result, flux can be improved. In increasing the membrane permeation flux, chitosan played an important role. With blending of chitosan into poly(vinyl alcohol), intra molecular hydrogen bonds are substituted by the intermolecular hydrogen bonds between chitosan and poly(vinyl alcohol), so the membrane structure becomes less compact and the free volume in the membrane is increased, and thus the permeation flux is also increased. Meanwhile, as the hydrophilic groups in poly(vinyl alcohol) and chitosan form intermolecular hydrogen bonds, hydrophilicity of the membranes decreased, thus the separation factor decreased slightly. The same tendency of total flux *versus* chitosan composition for 96 wt% isopropanol feed content at 50 °C was observed in the work of Ariyaskul *et al.* [77]. Additionally, it was found that chitosan membranes blended with poly(vinyl alcohol) at a ratio of 3/1 had the best pervaporation performance, high flux and excellent selectivity. Water content in the permeate from this membrane remained constant upon the change of either feed concentration or feed temperature and this was around 100 wt% water in the permeate.

Chitosan membranes blended with poly(vinyl alcohol) for dehydration of 1,4-dioxane were also studied by Devi *et al.* [78]. Pervaporation performance of the crosslinked poly(vinyl alcohol)/chitosan blends were investigated by shifting the feed compositions from 2 to 28.53 wt% of water. The permeation rate of 1,4-dioxane is much smaller compared to water, thus showing that the membrane was permselective to water. The developed membranes were able to break the azeotrope containing 18 wt% of water. With increasing water content in the feed compositions, the membrane performance exhibited a reduction in selectivity with an improvement in flux due to increased swelling. However, with increasing membrane thickness, selectivity improved, but the flux decreased. Higher permeate pressure caused a reduction in both flux and selectivity. Pervaporation through crosslinked poly(vinyl alcohol)/chitosan blend membrane for the feed mixture containing 9.35 wt% of water could dehydrate up to 97.04 wt%.

Nawawi *et al.* [79] studied separation of methanol/methyl *t*-butyl ether (MTBE) mixture using similar chitosan/poly(vinyl alcohol) blended membranes. The results showed that the unmodified membrane exhibited the highest permeation flux at 259 $g/m^2 \cdot h$ for 70 wt% feed solution. A rise of methanol feed concentration resulted in an increase of flux. This was very typical in the separation of methanol/methyl tbutyl ether mixture by pervaporation. It was because of the factor of the polarity in methanol compositions. The permeation fluxes of the permeating components increased with the increase in chitosan content in the membrane. The permeation flux of methanol was higher than that of MTBE for the whole range of the composition, which indicated that the membrane was more selective to methanol as compared to MTBE. The membrane containing 20 wt% chitosan had the lowest total permeation flux. This was attributed to the high composition of more condensed polymer network of PVA. The insertion of small amount of chitosan could have made certain parts of the poly(vinyl alcohol) compact network distorted and lead to higher free volume in the network.

Therefore, further increase of chitosan in the membrane prefers its higher permeation rate. However, at high concentrations of methanol in the feed, the permeation flux decreased as the poly(vinyl alcohol) ratios in the membrane increased. It could have happened owing to the structural properties of poly(vinyl alcohol) and chitosan. PVA is a linear aliphatic hydrophilic polymer while chitosan is a cycloaliphatic hydrophilic polymer. Therefore, the poly(vinyl alcohol) structure is usually regarded as more compact than that of chitosan. The separation factor decreased as both the chitosan composition in the membrane and methanol feed concentration increased. The highest separation factor was achieved for chitosan content in the blend between 20-40 wt% and 30 wt% methanol concentration in the feed. This may be attributed to the decrease in density of the membrane as higher dosage of chitosan was added to the blend. As the membrane becomes relatively less dense and the polymer network loosens, the membrane also becomes less selective. The flux obtained for 30 wt% methanol feed solution was between 52.28 and 66.92 $g/m^2 \cdot h$, and the corresponding separation factor was between 53.22 and 81.00.

Another very important polymer mixture applied to preparation of membrane material is the chitosan and sodium alginate blend. This polymer combination has specific properties such as high polarity, bulkiness and a strong structure with long time stability. Blending of the polymers resulted in the spontaneous formation of polyion complex membranes due to internal ionic crosslinking. This could be followed by covalent crosslinking of the blend with glutaraldehyde for further swelling reduction and introduction of specific properties such as structural strength along with thermal and mechanical stability. In fact, the polyion complex membranes made from anionic κ-carrageenan and cationic poly{1,3-bis[4-alkyl pyridinium]propane bromide} have been reported to have an excellent selectivity (more than 45 000) and permeability (150 $g/m^2 \cdot h$) as well as good stability for the separation of alcohol and water [80].

Kanti *et al.* [81] used chitosan/sodium alginate membranes for dehydration of ethanol. The membrane was crosslinked in an isopropanol-water bath (90/10 v/v) containing 5 vol.% of glutaraldehyde as a crosslinking agent. They reported that with increasing water concentration in the feed, the membrane performance was found to be substantially affected to the extent of swelling of the polymer, which resulted in a rise in flux but a reduction in selectivity. The increase in the flux for a corresponding increase in the concentration of water was due to the availability of more water molecules for sorption and diffusion. The preferential interaction with water caused the membrane to swell due to absorption of the liquid in the polymer matrix leading to plasticization and unrestricted transport of both the feed components through the upstream layer. Due to the relatively high selectivity obtained at low feed water concentrations, the PSI was found to follow the same trend as the selectivity and varied from 95.9 to 14.1 over the studied feed concentration range. The crosslinked blends were found to show promising performance for dehydration of ethanol containing small amounts of water ($\leq$ 26.35 mol%).

Kim *et al.* [82] studied the process of pervaporation separation of methyl t-butyl ether (MTBE) and methanol mixtures through polyion complex composite membranes consisting of sodium alginate (SA) and chitosan (CS). With increasing sodium alginate content, the separation factor increased from 133 to infinite value and the permeation rate decreased from 999 to 245 $g/m^2 \cdot h$. Particularly, the membranes made from 1.5 wt% SA solution exhibited only methanol permeation, with a flux high than 240 $g/m^2 \cdot h$. Consequently, the overall membranes prepared showed a very excellent pervaporation performance in the separation of MTBE/methanol mixtures. It could be deduced that the Coulomb interactions between the carboxylic group of sodium alginate and the protonated amino group of chitosan made the charge density stronger within the membrane, leading to an increase in the affinity of a membrane toward polar component and thus resulting in high selectivity in case of such components.. Therefore, the polyion complex composite membranes absorb and diffuse methanol molecules very selectively compared to the other component.

During the dehydration of 1,4-dioxane, Subba Reddy *et al.* [83] developed membranes by blending alginate with chitosan followed by crosslinking with maleic anhydride (MA-CA/CS). The pervaporation performance of MA-CA/CS membrane was investigated by varying feed composition between 5 - 30 wt% of water. A rise in the feed concentration of water produced an increase in the water normalized flux from 0.53 to 0.81 $kg/m^2 \cdot h$. In addition, the sorption data for binary feed mixtures showed a high degree of sorption. At 100% water composition, the sorption was very high and equal to 133.84%. This showed the hydrophilic nature of the membrane, which has high affinity for water and was capable of being selective towards the same during separation. An increased swelling had a negative impact on membrane selectivity. Therefore, the water concentration in the permeate was found to reduce from 96.5 to 79.3% indicating a distinct drop in selectivity from 512 to as low as 8.8 over the same water feed concentration range of 5-30%.

One of the effective way to enhance the pervaporation performance of chitosan is the preparation of other polyelectrolyte complex membranes composed of chitosan and poly(acrylic acid) (PAAc). Nam and Lee [84] systematically investigated the mechanism of polyelectrolyte complexes formation and its pervaporation performance in the separation of water-ethanol, water-isopropanol and methanol-methyl t-butyl ether mixtures. The formation of polyelectrolyte complexes is essentially due to the electrostatic interaction between oppositely charged polyions, *i.e.* crosslinking between protonated amino group NH_3^+ of chitosan and carboxylic anion COO^- coming from poly(acrylic acid), hydrogen bonding between carboxylic group of poly(acrylic acid) and -OH of chitosan. The formed polyelectrolyte complexes have increased charge density, and exhibit higher affinity toward water molecules than pure chitosan membranes. The same membranes were investigated in [85]. Changlai Hu *et al.* employed the obtained polyelectrolyte complex membrane with different blending ratio for dehydration of 80 wt% ethylene glycol aqueous solution. The permeation flux decreased with increase in PAAc content in polyelectrolyte complex and reached minimum when PAAc content was about 40 wt%. The separation factor appeared to reverse the change tendency, reaching maximum value when PAAc content was 40 wt%. Among all investigated membranes, the membrane containing 60 wt% of chitosan and 40 wt% of poly(acrylic acid) exhibited the best separation performance with a separation factor of 105 and a permeation flux of 216 $g/m^2 \cdot h$ at 70 °C, and 80 wt% ethylene glycol content in aqueous solution.

A promising approach for the improvement of the transport properties in the pervaporation process results from a blend of chitosan and hydroxyethylcellulose (CS/HEC). Hydroxyethylcellulose is a water soluble cellulose ether. It is a non-ionic polymer that is compatible with a wide range of other water soluble polymers and it can also form homogeneous blends with chitosan acetate salt [86]. Chanachai *et al.* [86] investigated such membranes crosslinked with urea formaldehyde sulfuric acid solution for the dehydration of ethanol-water and isopropanol-water mixtures. They demonstrated that the total permeation flux increased with an increase in hydroxyethylcellulose content. While the separation factor increased initially with HEC content, it reached maximum at 25 wt%, and decreased afterwards. It was explained by the molecular structure and hydrophilic property of hydroxyethylcellulose. Since an increase in the HEC content leads to a more flexible polymer network, the membrane has more chain mobility and larger free volume, resulting in increased permeation. This caused a decrease in the separation factor when the hydroxyethylcellulose content was higher than 25 wt%. For the HEC content range of 10-25 wt%, the increase in separation factor with hydroxyethylcellulose content was due to the increase in the hydrophilicity of the membrane. The membrane for ethanol/water mixture separation with HEC content of 25 wt%, showed the maximum overall pervaporation efficiency in terms of PSI, indicating that the polymer density and hydrophilicity were well balanced. On the other hand, for the separation of isopropanol/water mixture, a membrane with 10 wt% HEC content exhibited the highest PSI value. On account of the higher water activity compared to ethanol/water mixture, the higher water sorption in the membrane was observed for this solution, manifested in the larger degree of swollen.

Veerapur *et al.* [87] investigated chitosan/hydroxypropyl cellulose blend membranes crosslinked with the same crosslinking species, *i.e.* urea formaldehyde sulfuric acid solution for pervaporative dehydration of isopropanol. Crosslinked chitosan/hydroxypropyl cellulose blend membranes were found to be more effective than the plain chitosan membrane. The presence of the hydrophilic HEC in another natural polymer such as chitosan, converted the resultant blend to a more hydrophilic material, and allowed to achieve deeper dehydration of isopropanol. The authors showed that the permeation flux increased with increasing HEC content in the blend, and that the more flexible polymer network induced higher free volume in such membrane. Selectivity also increased and reached optimum value for 20 wt% HEC content. In case of a larger amount of hydroxypropyl cellulose, the selectivity was reduced, which could be connected with the molecular structure and hydrophilic nature of the hydroxypropyl cellulose. In any case, the chitosan membrane with HPC content of 20 wt% offered optimum pervaporation separation efficiency, and for 10 wt% water in the feed mixture, the highest selectivity of 11.241 and a permeation flux of 0.132 $kg/m^2 \cdot h$ were reported.

Nowadays, there are very popular blends of chitosan with poly(vinyl pyrrolidone) (PVP). PVP is an excellent biopolymer used in dehydration processes, and it is highly compatible with chitosan, forming a homogeneous blend matrix. Devi *et al.* [88] studied the separation of the dehydration of tetrahydrofuran using this type of membranes. As the crosslinking agents glutaraldehyde and sulfuric acid were used. Pervaporation performance was investigated for various feed compositions ranging from 2.28 to 50.77 wt% of water. A growth of water concentration in the feed produced an increase in water flux from 0.0902 to 0.2936 $kg/m^2 \cdot h$, with a slight drop in selectivity from 23.53 to 22.71. Such a variation in membrane performance was due to the preferential transport of feed components through the membrane matrix, which produces stronger interactions with the membrane material and excessive swelling, producing a negative impact on the membrane selectivity due to a plasticizing effect. To avoid the above mentioned problems, these types of membranes are best suitable for the final dehydration process of tetrahydrofuran, when the concentration of water in the feed is low.

Zhang *et al.* [89] investigated chitosan/poly(vinyl pyrrolidone) (CH/PVP) blend membranes in the pervaporative dehydration of ethyl acetate/ethanol/water mixture process. It was found that the permeation flux increased with increasing PVP content, whereas the separation factor decreased at the same time. On the one hand, introduction of PVP into chitosan could disorder the arrangement of the chitosan chains and decrease the degree of crystallinity, and in consequence increased the free volume in the membrane. On the other hand, introduction of

PVP could significantly enhance the hydrophilicity of the membranes because of its strong polarity and hydrophilic character. The formation of hydrogen bonds between the membranes and water molecules was hence promoted. Therefore the affinity between water and the membranes increased. All these factors caused an increase in the sorption of water in CH/PVP blend membranes. Addition of silica to the previous system allowed Zhang *et al.* to fabricate chitosan/poly(vinyl pyrrolidone)silica hybrid membranes suitable for the removal of methanol from the ethylene glycol azeotrope [90]. With the introduction of PVP, the disordered packing of chitosan chains were favorable and the increased volume of the amorphous region promoted the diffusion of the penetrant through the membrane. As a result, increased flux and decreased separation factor were observed. On the other hand, the PSI increased until the 5.34 wt% PVP content, and later decreased. The crosslinking of the membrane with tetraethoxysilane reduced the mobility of the polymer chains. Larger ethylene glycol molecules were found to have more difficulty in passing through the membranes than methanol. Further, selectivity of the methanol sorption increased with the increasing tetraethoxysilane content. Consequently, the permeation flux of the hybrid membrane decreased rapidly, whereas the separation factor increased.

Konjac glucomannan (KGM) is an uncommon material which has been blended with chitosan for pervaporative separation of caprolactam from its water solution by Lin *et al.* [91]. Intermolecular interaction between KGM and chitosan formed firm hydrogen bonds. It was found that KGM/CS membranes had similar flux but exhibited a higher separation factor under the same condition compared to the poly (vinyl alcohol) and poly (vinyl alcohol)/chitosan blend membranes. During blending of konjac glucomannan and chitosan, their molecule structures were disturbed. The interaction of the functional groups in both polysaccharides formed the three-dimensional networks. Due to the high molecular weight, KGM is characterized by gel behavior, and formed stable inter chains with chitosan. This type of membranes had a higher flux and better separation factor.

HYBRID CHITOSAN MEMBRANES

A useful strategy for improving separation efficiency of polymeric membranes is the formation of hybrid materials through the incorporation of various inorganic materials, typically oxides or metal nanoparticles, into the polymer matrix. In hybrid membranes, inorganic components are dispersed in a polymer matrix. These membranes benefit from the selectivity of active components and the simplicity of polymeric membrane processing. The effectiveness of this kind of membranes depends mainly on the interaction between these two components and with the penetrants, their compatibility and homogeneous dispersion of inorganic component, into the polymer matrix [92, 93].

Dudek *et al.* [94, 95] studied the pervaporation and permeation processes of water/ethanol mixture through the hybrid chitosan membranes containing ferrous ferric oxide nanoparticles (Fe_3O_4) crosslinked by sulfuric acid and glutaraldehyde. Iron oxide particle content in the studied membranes ranged from 0 to 15 wt%. It was demonstrated that pure chitosan membranes crosslinked by sulfuric acid without addition of iron oxide nanoparticles were not suitable for the pervaporative separation of the investigated mixture. Mainly because evaluated permeation coefficients for water and ethanol were almost the same. The presence of the magnetite nanoparticles significantly changed the membrane transport properties. Starting from the 2 wt% magnetite content, permeation of water was gradually increased with the reduction of ethanol permeation. Regarding the transport mechanism, addition of the iron oxide nanoparticles into the sulfuric acid crosslinked membranes changed the contribution of diffusion and solubility in the total flux. Thus, the evaluated diffusion coefficient of ethanol slightly increased, whilst diffusion coefficient of water increased up to more than seventy times. At the same time, solubility coefficients of both water and ethanol significantly decreased.

Diffusion and solubility coefficients of water, evaluated for glutaraldehyde crosslinked pure chitosan membrane, were higher than corresponding values estimated for ethanol, thus the permeation of water was about 6 times higher. The presence of iron oxide nanoparticles in a membrane structure further enhanced the difference between the permeation efficiency of both components. Since the solubility coefficient slightly decreased for both species, the observed increase in water permeation was connected to its faster diffusion in the presence of the iron oxide nanoparticles, with only a minor increase in ethanol diffusion. Comparing the chitosan membranes crosslinked by two different agents, it was demonstrated that glutaraldehyde crosslinking promoted superior separation properties over the sulfuric acid crosslinked membranes and presence of iron oxide particles in the polymer matrix further enhanced the difference in the separation properties.

Yang *et al.* [96] investigated chitosan/TiO_2 nanocomposite membranes for pervaporative dehydration of ethanol. The results showed that the permeation flux increased with the increase in TiO_2 content up to 6 wt%. The formation of TiO_2 particles inside the polymer matrix alter the packing order of chitosan chains, which resulted in the increase of the amorphous region. Therefore, the permeating molecules could more easily diffuse through the hybrid membranes containing titanium dioxide. When the TiO_2 content was greater than 6 wt%, the permeation flux decreased. Yang explained this phenomenon by the fact that the hydrophilicity of hybrid membranes was changed. When the TiO_2 content was higher than 6 wt%, the total amount of hydroxyl groups on the surface of TiO_2 particles decreased due to the aggregation process. The separation factor of chitosan/TiO_2 hybrid membranes firstly decreased for 0-4 wt% TiO_2 content, then promptly increased in the range of 4-8 wt%, and decreased again after reaching the maximum at about 8 wt%. It was found that at low TiO_2 content, nanometer scale voids were formed considerably at the chitosan/TiO_2 interface, which led to the decrease in the separation factor. With the increase in TiO_2 content, the size selective effects of TiO_2 particles became more pronounced, resulting in the increased separation factor. Due to the aggregation of TiO_2 particles for high loaded membranes, their separation factor decreased again. Considering the permeation flux and separation factor totally, investigated chitosan/TiO_2 nanocomposite membranes exhibited the best pervaporation performance among all the prepared membranes until the 6 wt% TiO_2 content. The permeation flux and separation factor reported for this membrane were 0.340 $kg/m^2 \cdot h$ and 196, respectively, and were evaluated at 80 °C for the 90 wt% aqueous feed solution of ethanol.

Premakshi *et al.* [97] modified the blocked di-isocyanate crosslinked chitosan membranes by incorporating to the polymer matrix NaY zeolite in different amounts, and used them for pervaporative separation of water-isopropanol mixtures. It was found that an increase in the NaY zeolite content caused an increase in both permeation flux and selectivity. This was explained on the basis of significant enhancement of the membranes hydrophilicity, selective adsorption and establishment of molecular sieving action by the creation of pores in the polymer matrix. Both the total flux and flux of water almost overlapped each other, meaning that the modified membranes were highly selective toward water. The highest separation selectivity was found to be 11.241 with a flux of $11.37 \cdot 10^{-2}$ $kg/m^2 \cdot h$ and was obtained for the membrane having 40 wt% of NaY zeolite at 30 °C, and for 10 wt% content of water in the feed. With regard to temperature effect, the permeation rate increased. The modified membranes exhibited lower activation energy values compared to the crosslinked one, suggesting that the permeants require less energy during the separation process owing to molecular sieving action attributed to the presence of sodalite and super cages in the framework of NaY zeolite.

Sun *et al.* [98] studied the pervaporation process of water/ethanol mixture separation through HZSM5 filled chitosan membranes. HZSM5 is a kind of hydrophilic zeolite, in which the most important acid type center forms the bridged framework of hydroxyl groups with Si-OH-Al configuration. Due to the contribution of Al atoms framework to the hydroxyl groups of the Si-OH-Al structure, the ratio of Si atoms to Al atoms influenced the hydrophilicity of HZSM5 zeolite. The results showed that the incorporation of HZSM5 increased the permeation flux. The separation factor did not increase substantially in the most HZSM5 content due to the formation of non-selective voids at the interface.

By increasing Si/Al ratio in HZSM5, the hydrogen bonding interaction between organic/inorganic phases increased and the interface morphology of filled membranes improved, leading to the increase in separation factor in the whole range of the HZSM5 content. In addition, HZSM5 exhibited desirable size, selectivity and affinity effects for aqueous ethanol solution. Compared with the chitosan membrane (permeation flux 54.18 $g/m^2 \cdot h$ and separation factor 158.02 for 90 wt% aqueous ethanol solution at 80°C), the HZSM5 filled chitosan membranes containing 8 wt% of zeolite, exhibited better separation properties under identical experimental condition with permeation flux equal to 230.96 $g/m^2 \cdot h$ and separation factor 152.82.

Another chitosan silica hybrid membrane was investigated by Liu *et al.* [99]. These membranes were loaded with the sulfonic acid functionalized silica nanoparticles and were used for the pervaporative dehydration of ethanol water solutions. It was noticeable that the formation of complex membranes significantly increased the values of the separation factor, *i.e.* enhanced the selectivity of membranes for ethanol water separation. In his work, the crosslinked hybrid chitosan silica membranes exhibited a similar or even better permeation rate as compared to the original plain chitosan one. This property is highly appreciated in membranes for pervaporative dehydration, especially for the deep dehydration of the mixture with low water content, in which small fluxes are a critical problem. Authors explained this phenomenon by the fact that the addition of silica nanoparticles to chitosan matrix created extra free volumes in the polymer, and in consequence, offered space for water molecules to permeate easily through the membranes. The best result was obtained for the chitosan membrane containing 5 wt% of silica. The resulting membrane used for the pervaporative dehydration of 90 wt% aqueous ethanol solution at 70 °C exhibited a separation factor of 919 and a permeation flux of 410 $g/m^2 \cdot h$.

Varghese *et al.* [100] also developed hybrid membranes using chitosan and silica precursors and used them for pervaporative separation of water/isopropanol mixtures. It was found that the total permeation flux increased for all investigated membranes almost linearly with the increasing mass fraction of water in the feed and content of γ-glycidoxypropyltrimethoxysilane (GPTMS) in membrane. This was due to an increase of permeation driving force, degree of swelling, as well as the tuned hydrophilic character of the prepared membranes. The GPTMS is a well-known modified sol-gel precursor which comprises of three reactive alkoxide groups modified with a short organic moiety containing an extra epoxy ring, able to form an additional hydroxyl group so that the inorganic sol has higher hydrophilicity and better compatibility toward chitosan matrix. Both permeation flux and selectivity of examined membranes increased with increasing GPTMS content up to 25 wt%. However, when its content exceeded this value a decrease in selectivity was observed. This was explained on the basis of the prominent enhancement of hydrophilic character and increased crosslinking density. The PSI parameter clearly indicated that the membrane with 25 wt% of GPTMS showed an excellent pervaporation performance: highest separation selectivity of 18 981 accompanied by a normalized flux of 7.45×10^7 kg/m^2·h at 30 °C estimated for 0.05 mass fraction of water in the feed. With an increase in temperature, the permeation flux was further increased while a typical decrease in selectivity was noticed. The hybrid membranes containing higher amount of GPTMS showed significantly lower activation energy, indicating that the permeants consumed less energy, what could be connected with the increase of the hybrid membrane. The activation energy of the water permeation process was lower than that of isopropanol, meaning that the chitosan/GPTMS membranes demonstrated excellent separation efficiency toward water.

Lu *et al.* [48] fabricated graphite filled poly (vinyl alcohol)/chitosan (PVA/CS) hybrid membrane applied for the pervaporation of benzene/cyclohexane mixtures. Graphite is a suitable inorganic filler for the separation of such mixtures because of the repetition of a hexagonal carbon ring in its structure, similar to that of the benzene aromatic ring. This structural feature induced generation of an extended 2D structure and promoted the σ and π bonds interaction between the graphite planes and π electron clouds of the aromatic system. It was found that the separation factor increased when graphite content increased from 0 to 6 wt% and further decreased when the content exceeded 6 wt%. It resulted from the combined effect of sorption behavior and free volume property. Benzene can diffuse freely when adsorbed on the surface of graphite particles but when it is firmly immobilized on the graphite surface, the mobility of its molecules is strongly restricted and cannot diffuse freely. With low graphite loading, the contribution of benzene firmly immobilized on the graphite surface was considerably low, and the improvement effect that contributed to the free volume formation was dominated in the separation process, and an increase in the permeation flux was observed. With excess graphite loading, benzene molecules stayed immobilized on the graphite surface, and the permeation flux decreased. Among the all studied graphite filled PVA/CS membranes, the highest separation factor of 59.8 with a permeation flux of 124.2 g/m^2·h exhibited a membrane with a 60/40 mass ratio of PVA/CS applied for benzene/cyclohexane mixture pervaporative separation.

For further improvement of membrane performance, the incorporation of several nanofillers to form a "green" polymer-inorganic nanocomposite has been widely used. Among others, multiwalled carbon nanotubes (MWNTs) are of particular interest because of their relatively low cost and abundant availability. Carbon nanotubes with 1D structure and nanochannels are speculated to be an excellent candidate for substituting or complementing conventional nanofillers in the fabrication of nanocomposite pervaporation membranes.

Shen *et al.* [49] investigated membranes made of chitosan and Ag^+/carbon nanotubes for pervaporative separation of benzene/cyclohexane mixtures. The results showed that the total flux increased with increasing MWNTs-Ag^+ content. Due to the fact that the benzene molecule has π electrons, it may have a strong interaction with Ag^+ ion incorporated in the hybrid membrane. The more MWNTs-Ag^+ were fed in unit area, the more benzene permeated through the membranes. The selectivity also increased with an increase of the concentration of MWNTs-Ag^+ up to 1.5 wt%. For the higher content of MWNTs-Ag^+, the observed selectivity decreased because high MWNTs-Ag^+ concentration may destroy the intensity and integrity of the membrane. It was also stated that the selectivity increased with increasing benzene content in the feed mixture, proving that the MWNTs-Ag^+ membrane had high affinity towards benzene. The highest selectivity reached 7.89 when the content of benzene in the feed was 50 vol%. In addition, after introducing MWNTs and MWNTs-Ag^+, the Arrhenius activation coefficient of the total permeation flux decreased.

Ong *et al.* [101] studied poly(3-hydroxybutyrate)functionalized MWNTs/chitosan green nanocomposite membranes and their application for the pervaporative dehydration of 1,4-dioxane. Applied MWNTs were functionalized with the poly(3-hydroxybutyrate) (PHB), which is a bio based and biodegradable thermoplastic polymer produced by bacteria. PHB is miscible with chitosan. The functionalization of MWNTs was accomplished by dispersing the oxidized MWNTs in poly(3-hydroxybutyrate) solutions. Because both oxidized MWNTs and PHB contain hydroxyl groups, this provides a pathway for PHB to interact with MWNTs surface. Upon incorporation of PHB-MWNTs onto the chitosan matrix, the permeation flux and selectivity of the PHB-MWNTs/chitosan hybrid nanocomposite membrane exhibited similar trends as the pure chitosan membrane when subjected to the different 1,4-dioxane concentration in the feed. Decreasing the feed concentration of 1,4-dioxane from 95 to 30 wt% resulted in an increase in permeation flux from 16.8 to 220.7 $g/m^2{\cdot}h$ and decrease in the selectivity from 623.1 to 10.7. Compared to the plain chitosan membrane, the PHB-MWNTs/chitosan nanocomposite showed an increased permeation flux with only a small sacrifice in selectivity. These phenomena can be attributed to existence of the molecular transport channels in MWNTs, thus the presence of MWNTs in the hybrid membrane provided an additional transport pathway that enables the permeating molecules access through their inner cores, thereby enhancing the permeability of the membrane. Additionally, the incorporation of PHB-MWNTs was found to reduce the excessive swelling behavior of chitosan.

Peng *et al.* [102] fabricated nanocomposite pervaporation membranes composed of poly(vinyl alcohol) and chitosan rapped carbon nanotubes (PVA-CNT(CS)) for pervaporative separation of benzene/cyclohexane (50/50 w/w) mixtures at 50 °C. The results showed that the permeation flux and separation factor of plain PVA membrane were only 20.3 g/m^2·h and 9.6, respectively, while the corresponding values evaluated for PVA-CNT(CS) (CNT content: 2.0 wt%) nanocomposite membrane were 65.9 g/m^2·h and 53.4, respectively. Observed enhancement was explained by the modification of PVA chain packing both by the CNT and chitosan, leading to the further and tailored increase of free volume. Therefore, the CNT could control both the solubility selectivity and diffusion selectivity. The increased solubility selectivity should be contributed from the stronger affinity of CNT towards benzene due to the $\pi\pi$ stacking effect between the carbon nanotube and benzene ring. The diffusion selectivity was increased due to the size, concentration, and distribution of the network and aggregate pores. It was also found that PVA-CNT(CS) nanocomposite membrane exhibited very high permselectivity toward benzene in the wide feed composition range. The permeation flux rapidly increased as the weight fraction of benzene in the feed was increased, which was caused by the more pronounced membrane swelling in this solvent.

Yeang *et al.* [103] developed three-layer nanocomposite membranes synthesized by bulk aligning of the poly(vinyl alcohol) functionalized multiwalled carbon nanotubes (PVA-MWCNTs) on a poly(vinylidene fluoride) (PVDF) membrane and coating with a thin chitosan layer. The results of the pervaporative dehydration of acetone showed that water permeation increased with PVA-MWCNTs content until an optimum value of 87.52 g/m^2·h with separation factor value of 1505.9, at 0.06 wt% PVA-MWCNTs. When the PVA-MWCNTs content was increased beyond 0.06 wt%, a decline in the water permeation was observed. A decrease in selectivity was also exhibited with increasing PVA-MWCNTs content until a minimum value of 27.242 with 0.06 wt% PVA-MWCNTs. This finding suggested that the presence of multiwalled carbon nanotubes in the chitosan membrane provides an additional pathway for the mass transport of diffusing molecules *via* their inner cores. Both water and acetone molecules, with molecular diameters of 0.30 nm and 0.62 nm, respectively, may transport *via* the inner cores of the MWCNTs as their inner diameters are approximately 40-60 nm. Thus, water permeation was enhanced with the increasing content of PVA-MWCNTs, because more water molecules were able to diffuse and penetrate through the membrane. When PVA-MWCNTs content was above 0.06 wt%, the decrease in water permeation and increase in selectivity indicated the possibility of MWCNTs agglomeration due to their high concentration. At higher MWCNTs content, many layers of MWCNTs may stack together, entangle or agglomerate. As a result, the number of accessible diffusion pathways for the diffusing molecules provided by the inner cores of the MWCNTs are greatly reduced, hindering the transport of molecules across the membrane.

GRAFTED CHITOSAN MEMBRANES AND MEMBRANES ON POROUS SUPPORT

Sometimes developed membranes have a lack of mechanical strength and chemical resistance in aqueous solution mainly due to excessive swelling. Chemical modification of hydrophilic polymers through grafting is anticipated to be a promising method providing a wide variety of molecular design. Moreover, the properties of polymers can be properly tuned depending on the nature and concentration of grafted species. For these reasons, grafting became an important approach to enhance the separation performance of membranes.

Varghese *et al.* [104] investigated chitosan-g-polyaniline membranes for the dehydration of isopropanol. It was found that the grafted membranes exhibited better pervaporation performance than that of pure chitosan membranes. This was explained on the basis of increased hydrophobic character and increased degree of crystallinity. Among the grafted membranes, the membrane with 1:3 grafting ratio exhibited the highest separation selectivity of 2092 with a flux of 1.19×10^{-2} $kg/m^2 \cdot h$ at 30°C for 10 wt% of water in the feed. For higher grafting ratio the total flux and water flux were almost close to each other, suggesting that the developed membranes demonstrated an excellent selectivity towards water. With an increase in temperature, the permeation flux was increased while the selectivity slightly suffered, a fact attributed to the decreased viscosity of the permeating liquid and increased driving force. The grafted membranes also showed lower activation energy of water permeation than that of isopropanol, proving that grafted membranes had higher separation ability.

The separation efficiency of the same mixture, *i.e.* water/isopropanol was studied by Sajjan *et al.* [105]. The authors grafted the glycidyltrimethylammonium chloride (GTMAC) onto chitosan surface. The obtained results showed that with an increase of GTMAC content, both permeation flux and selectivity of the membrane increased. This was connected with the significant enhancement of membrane hydrophilicity and its induced amorphous nature. The best pervaporation performance was achieved for membranes incorporated with 40 wt% of GTMAC, which showed the highest separation selectivity of 2134 and a flux of $6.91 \cdot 10^{-2}$ $kg/m^2 \cdot h$ at 30°C for 10 wt% of water in the feed. Demonstrated experimental data also revealed that the total flux and flux of water were almost overlapping each other for all tested grafted membranes in the investigated range of water composition, suggesting high separation ability of these membranes. As in previous work, the temperature study indicated that permeation flux was increased with simultaneous loss in selectivity along the process temperature growth. The membrane showed significantly lower activation energy values for water permeation than that of isopropanol, what again suggests that rightly grafted chitosan membranes have high separation efficiency.

Han *et al.* [106] used grafted chitosan poly(benzoimidazole) (PBI) membranes for pervaporative dehydration of aqueous isopropanol solutions. The results showed that surface modification of the poly(benzoimidazole) membrane with chitosan, slightly increased the membrane selectivity and significantly increased the membrane permeability comparing with unmodified poly(benzoimidazole) membranes. The membrane permeability increased with the increase in the operation temperature and showed a value of about 250 g/m^2·h at 70 °C. The activation energy for water permeation was 10.5 and 5.7 kJ/mol for the pure poly(benzoimidazole) membranes and the grafted PBI-CS membrane, respectively. The results indicated that the energy barrier for water permeation through the PBI-CS membrane was lower than for the pure PBI membrane. Incorporation of the hydrophilic chitosan layer to PBI membrane surface enhances the surface hydrophilicity and increases the rate of water dissolution into the membrane. As a result, the water permeability of the PBI membrane has been significantly improved after chitosan grafting. Moreover, the high water dissolution rate into the PBICS membrane also diminished the dissolution of isopropanol, so as to increase the membrane selectivity especially for the feeding solutions with high water contents.

Liu *et al.* [107] developed chitosan/polysulfone (CS/PSf) composite hollow fiber membranes by casting a chitosan solution containing γ-glycidoxypropyltrimethoxysilane (GPTMS) on the poly(styrene sulfonic acid)-g-polysulfone hollow-fiber (PSSA) membranes. The composite hollow fiber membranes were used in the pervaporative dehydration process of 70 wt% aqueous isopropanol solution. The results showed that the surface modification on PSf hollow fiber membrane enhanced their compatibility with the chitosan coating solutions. A full and uniform coverage of chitosan layer on the PSf hollow fiber surface was achieved for membranes with 1.5 wt% chitosan content, making the composite membrane highly resistant to permeating water. That membrane showed a lower permeation flux and a higher separation factor.

The contemporary development trend for the preparation of pervaporation membranes relies on the use of a composite structure possessing a thin and dense separation layer on top of a porous substrate. These membranes are chemically and mechanically stable and usually display very low mass transport resistance. Owing to the unique, stable structure, they can withstand various kinds of organic solvents and higher temperatures.

Zhu *et al.* [108] investigated ceramic supported poly(vinyl alcohol)-chitosan blend membranes for pervaporative dehydration of alcohol/water and ester/water mixtures. It was found that the pore size of ceramic support directly affected the pervaporation performance of this membrane. When the pore size of ceramic support was equal to 0.2 μm, the pervaporation performance of the composite membrane was acceptable. For ceramic support with a smaller pore size of 0.05 μm, both flux and separation factors were really low. In case of pure PVA/ceramic and CS/ceramic membranes, the fluxes were low, while the estimated separation factors were high. In the blended CS/PVA membrane, intramolecular hydrogen bonding was substituted by intermolecular hydrogen bonding between both components, so the membrane structure became less compact and free volume was increased. For the pervaporative dehydration of alcohol/water mixtures it was stated, that with increase in process temperature and water concentration in the feed, the flux increased without a noticeable decrease in the separation factor. For the dehydration of ester/water mixtures, the best pervaporation performance was found when the water content in the feed was less than 5 wt%. In this condition, the achieved separation factor of ethyl acetate/water was larger than 10000. In addition, the used ceramic support was non-deformable, so the deformation of the organic top layer can be restrained and the membrane performance can be improved.

Huang *et al.* [109] applied the chitosan composite membranes having a microporous polysulfone substrate for the pervaporative dehydration of aqueous isopropanol solution. The results showed that the structural stability of the chitosan/polysulfone composite membrane was greatly increased as a result of the increased affinity between chitosan top layer and polysulfone substrate. It was believed that the porous substrate in the composite membrane acts not only as a mechanical support but also as part of an active layer affecting the overall pervaporation performance. In some other work, Huang *et al.* [110] studied the pervaporative dehydration of aqueous ethanol/isopropanol mixtures through alginate/chitosan two-ply composite membranes supported by poly(vinylidene fluoride) porous membrane. The results showed that porous support was useful in improving surface properties for adhesion of the alginate and chitosan top layers. The permeation flux of the alginic acid composite membrane was larger than that of the sodium alginate composite membrane over the whole feed concentration range. It was explained according to the possibility of hydrogen bonding formation which was larger for alginic acid membrane than the sodium alginate one. Additionally, it was found that the separation factors obtained for alginate composite membranes did not follow the typical tradeoff phenomenon which has been widely used to explain the high flux and low separation efficiency frequently encountered in pervaporation experiments.

Wu *et al.* [111] studied the process of dehydration of ethanol through chitosan/sulfonated polyethersulfone-polyethersulfone (CS/SPES-PES) composite membranes. It could be seen that the separation factor increased with the increase of sulfonated polyethersulfone (SPES) content in the support and the highest separation factor was obtained with SPES content of 50 wt%. It was explained by the enhanced interfacial interaction, which originated from the ionic interaction between the amino groups on chitosan and the sulfonic acid groups on sulfonated polyethersulfone. This enhanced interaction restricted the mobility of chitosan chains, thus increasing the separation factor. The high SPES content resulted in a high roughness and even pores as a result of severe phase separation between sulfonated polyethersulfone and polyethersulfone. These large pores caused structure heterogeneity or even defects in the very thin active layer, thus lowering the separation factor. The permeation flux reached the highest value (1700 $g/m^2{\cdot}h$) at sulfonated polyethersulfone content of 20 wt% and decreased thereafter. The variation of the flux was most likely related to the structure of interfacial region when sulfonated polyethersulfone content reached above 20 wt%. As the SPES content increased, the interfacial acid–base interaction became stronger, a denser interfacial region formed and resulted in reduced permeation flux.

Han *et al.* [112] developed hydrophilic chitosan modified poly(benzoimidazole) (PBI) membranes for pervaporative dehydration of isopropanol aqueous solutions. It was found that the surface modification of poly(benzoimidazole) membrane with hydrophilic chitosan chains could effectively enhance its ability to trap water from the feeding aqueous solution, so as to improve the membrane performance in the pervaporative dehydration process. The water permeability of the pure poly(benzoimidazole) membrane at 25 °C was equal to 158 and 28 $g/m^2{\cdot}h$ for the feeding solutions of 30 and 90 wt% of isopropanol, respectively. After functionalization with chitosan, the water permeability of the chitosan modified poly(benzoimidazole) membranes for the same concentration of feeding solutions increased to 247 and 133 $g/m^2{\cdot}h$, respectively. As a result, the water permeability of PBI-CS membranes were about 1.6-fold and 5-fold of the value recorded for the pure poly(benzoimidazole) membrane. Hydrophilic surface functionalization was an effective way for increasing the permeability of the poly(benzoimidazole) membrane. Moreover, introduction of chitosan layer on the surface of PBI membrane could also enhance the membrane selectivity toward water. For the feeding solution containing 30 wt% of isopropanol, the achieved water concentration of the permeate was 95.8 and 98.0 wt% for pure PBI and PBI-CS membranes, respectively. The selectivity factor of the PBI based membrane had increased from 10 to 21 after modification with chitosan. The modified PBICS membrane had gained a 3.1-fold increase in the pervaporation separation index PSI for the 30 wt% aqueous isopropanol feed solution.

Tsai *et al.* [113] studied the pervaporation performance of aqueous isopropanol solution through the chitosan/polyacrylonitrile (CS/mPAN) hollow fiber membrane. The chitosan layer of composite hollow fiber membrane was formed by dip coating chitosan solution onto the outer surface of NaOH hydrolyzed polyacrylonitrile hollow fibers membrane. The crosslinking agent γ-glycidoxy-propyltrimethoxysilane (GPTMS) was used. The results showed that the permeation rate decreased while water content in the permeate was increased with increasing chitosan concentration. These phenomena might be related to the thickness of chitosan coating layer. The average thickness of the chitosan coating layer obtained for 0.5, 1, 1.5, and 2 wt% chitosan solution were 0.69 μm, 1.24 μm, 1.86 μm, and 3.04 μm, respectively. An increase in the chitosan solution concentration led to a much higher chitosan concentration at the outer surface of composite hollow fiber membrane, resulting in an increase in the coating layer thickness. This implies that the volume fraction of polymer increases and, consequently, a lower porosity is obtained. Thus, the water content increase in the permeate and a decrease of the permeation rate arose from the increase of the coating layer thickness. For the pervaporative dehydration of 90 wt% aqueous isopropanol solution at 25 °C the best results were obtained for chitosan/mPAN composite hollow fiber membranes modified with 1.5 wt% chitosan solution. The resulting membrane exhibited permeation flux of 145 $g/m^2 \cdot h$ and 99.7 wt% water content in permeate.

Li *et al.* [114] applied chitosan/poly(vinyl alcohol)/poly(acrylonitrile) (CS/PVA/PAN) composite pervaporation membranes for the separation of ethanol-water solutions. The results showed that the separation factor of the CS/PVA/PAN composite membrane increased with an increase of PVA concentration in the CS/PVA blend from 0 to 40 wt%. With an increase of the membrane thickness from 12 to 18 μm, the separation factor of the CS/PVA/PAN composite membrane also increased and the permeation flux decreased. The separation factor of the CS/PVA/PAN composite membrane increased to a maximum value as PVA concentration in the CS/PVA blend increased to 40 wt%. With further increase of PVA concentration in the CS/PVA blend from 40 to 100 wt%, the observed separation factor decreased but the flux remained nearly the same. When the CS/PVA concentration in the CS/PVA/PAN composite membrane increased from 1 to 2 wt%, the separation factor of the CS/PVA/PAN membrane increased significantly but at the same time flux rapidly decreased. Therefore, 2 wt% CS-PVA (CS/PVA 60:40 weight ratios) concentrations were suitable for the fabrication of the CS/PVA/PAN composite membrane.

Liu *et al.* [115] used chitosan/poly(tetrafluoroethylene) (CS/PTFE) composite membranes for pervaporative dehydration processes. CS/PTFE composite membranes were obtained by the casting of chitosan dense layer on the surface modified with sulfonated poly(styrene) PTFE film. The sulfone groups existing on the modified PTFE surface could react with the chitosan amino groups, to provide interfacial linkages between chitosan and PTFE layers. It was found that this membrane was not effective for separation of solvents with small molecular radius, as very low separation factors of 14 and 81 for methanol and ethanol solutions were achieved respectively. On the other hand, the solvents with a larger molecular radius dissolved in the membrane and inhibited the permeation of water, so as to reduce its permeation flux. Therefore, the large molar volume of tetrafluoropropanol and N,N-dimethylformamide bring about much steric hindrance to water permeation, so as to reduce the total flux and permeability of tetrafluoropropanol and N,N-dimethylformamide. Consequently, both of the high separation factors for tetrafluoropropanol and N,N-dimethylformamide aqueous solutions were distinguished. Moreover, it was seen that the CS-5/PTFE membrane showed superior or comparable performance to other membranes, especially for the relatively critical solutions like isopropanol (70 wt%), tetrafluoropropanol (90 wt%), and N,N-dimethylformamide (90 wt%) aqueous solutions. Therefore, the chitosan/PTFE composite membranes, which possess a thin chitosan dense layer and stable PTFE supporting layer, exhibited high performance of pervaporative dehydrations for various aqueous solutions of organic compounds, including different hydrocarbon alcohols, tetrafluoro-propanol, and N,N-dimethylformamide.

Huang *et al.* [116] prepared composite chitin membranes supported by porous polyetherimide substrate for the pervaporative separation of ethanol/toluene and methanol/toluene mixtures. With the introduction of acetyl groups into the matrix of chitosan, the chemical structure of the chitosan membrane became bulky and stiff, and then the acetylated chitosan was mechanically brittle. Thus, the chain mobility of the membrane was substantially hindered due to the bulky acetyl groups, in addition, the free volume which continuously accommodates the permeating component, decreased. Studying ethanol/toluene mixture, a lower total flux was found for N-acetylated chitosan membranes than for pure chitosan membranes. It was explained by the fact that N-acetylated membranes were attributed to the strong inter or intra molecular bonding among acetyl functional groups resulting in less interaction between the feed mixture and membrane. In the case of N-acetylated membrane in 7 M acetic anhydride (90% degree of acetylation), permeation flux increased to the minimal extent through the whole investigated feed concentration range, probably due to its rigid structure, which was hardly swollen in the feed mixtures. In fact, 7 M acetic anhydride was the maximum concentration which acetylated membrane can function for the separation of ethanol/toluene mixture. 10 M acetic anhydride was proven to be excessive, because the membrane acetylated with this reagent was too rigid and brittle. In addition, there was almost no transport through this membrane. For the methanol/toluene mixture the total flux was much larger than that of ethanol/toluene mixture. The smaller molecular size of methanol compared to ethanol played a positive role in the increased flux. The total flux of methanol/ toluene mixture increased proportionally with the methanol content in the feed. In the case of the acetylated membrane in 4 M acetic anhydride, the flux increase trend was somewhat similar to that of pure chitosan. However, the total flux was still lower than that of pure chitosan membrane due to the introduction of the rigid acetyl functional group. Although the flux difference was less pronounced compared to that found for ethanol/toluene mixture.

Prasad *et al.* [117] synthesized composite chitosan membranes on porous and solvent resistant poly(ether-b-amide) (PEBA-2533) block copolymer substrate for pervaporative dehydration of the polar aprotic N-methyl-2-pyrolidone (NMP) "green" solvent. The process was performed over a wide range of feed compositions of N-methyl-2-pyrolidone/water mixtures at 30 °C and pressure on the permeate side of 2 mmHg. Moreover, the chitosan layer was crosslinked by tetraethylorthosilicate. The results showed that this kind of membrane exhibited remarkable flux and selectivity besides high chemical stability in the otherwise polar and aggressive solvent medium. The flux increased with increasing water concentration in the feed due to possible hydrogen bonding of water molecules with the residual OH and NH_2 groups of chitosan as well as NH moiety and polar terminal OH groups present in the amide block of PEBA-2533. Over an ascending feed, water concentration ranged between 1.33-18 wt%. The increase in N-methyl-2-pyrolidone flux was more profound, resulting in a fall in water selectivity. At higher water contents in the feed mixture, the hydrophilic membranes became swollen and plasticized due to greater accessibility of water molecules, which evoked reduction in selectivity from 1007.3 to 5.14 and 838.33 to 4.74 for tetraethylorthosilicate crosslinked and unmodified chitosan/poly(ether-b-amide) membranes, respectively. Therefore, the hydrophilic membrane was found to be suitable for the removal of small concentrations of preferentially sorbed water (≤10 wt%) from N-methyl-2-pyrrolidone, wherein swelling is under control.

CONCLUSION

The application of chitosan membranes in pervaporation process has played an important role in liquid mixture separation. They have been intensively studied and developed in numerous research groups, and widely used in a plain or modified form. The successful strategies applied for the improvement and tuning of chitosan membrane properties covers several modification methods such as crosslinking [65 - 75], formation of polymer blends [76 - 91], preparation of hybrid organic - inorganic composites [48, 49, 92 - 103], chemical modification by grafting [50, 104 - 107], as well as the use of porous support [108 - 117]. The most often used crosslinkers for the modification of chitosan membranes are several organic reagents and inorganic acids. Organic chemicals such as glutaraldehyde, 3-aminopropyltriethoxysilane, trimesoyl chloride and toluene-2, 4-diisocyanate were used in the membranes applied for separation of the water/isopropanol [66, 67, 70, 71] or water/ethanol [69] mixture. Chitosan membranes can also be modified by inorganic acid. In the process of water/ethanol separation, cross-linkers such as sulfuric acid, phosphoric acid or sulfosuccinic acid were used [69, 74]. Sulfosuccinic acid in turn [75], with three crosslinkable sites, was applied in the pervaporation separation of water/alcohol mixtures.

The second group of chitosan membranes modification involves the blending of chitosan with another polymeric material. Blending with another polymer allows for the increase of the free volume in the membrane, facilitating the permeation of mixture components, thus generally, much bigger flux can be obtained. On the other hand, growth in the flux is often accompanied with some sacrifice in selectivity. The most often investigated polymer for this purpose is poly(vinyl alcohol) [76 - 79]. The chitosan and poly(vinyl alcohol) blends were investigated in the dehydration of ethanol [76], isopropanol [77], 1,4-dioxane [78], and for the separation of methanol/methyl t-butyl ether mixture [79]. In this blends, the hydrophilic groups of poly(vinyl alcohol) and chitosan form intermolecular hydrogen bonds that influence the hydrophilic/hydrophobic properties of the final membrane materials. Another interesting polymer in this group, is sodium alginate that is blended with chitosan and could form the polyion complex membranes due to strong ionic crosslinking between both components. The polyion complex membranes have been reported to have an excellent selectivity and permeability while keeping good stability. Chitosan/alginate blends have been applied, like chitosan/poly(vinyl alcohol), for the dehydration of alcohols [80, 81], 1,4-dioxane [82] and separation of methyl t-butyl ether and methanol mixtures [83]. Poly(acrylic acid) chitosan is also applied to form polyelectrolyte complex membranes that is one of the effective way to enhance pervaporation performance. The formed polyelectrolyte complexes have increased charge density, and exhibit higher affinity toward water molecules than pure chitosan membranes. Such membranes were investigated in the separation of water-ethanol, water-isopropanol and methanol-methyl tbutyl ether mixtures [84] and in the dehydration of 80 wt% ethylene glycol aqueous solution [85]. A new promising approach for the improvement of the transport properties in the pervaporation process gives a blend of chitosan and hydroxyethylcellulose [86] or hydroxypropyl cellulose [87]. The blends were investigated in the dehydration of ethanol-water and isopropanol-water mixtures. The results showed that the addition of hydrophilic hydroxyethylcellulose or hydroxypropylcellulose gives blend membranes more hydrophilicity. It leads to a more flexible polymer network, and the membrane has more chain mobility and larger free volume, resulting in increased permeation ratio. An interesting example of a biopolymer compatible with chitosan is poly(vinyl pyrrolidone) (PVP), which forms a homogeneous blend matrix. Such blends were studied in the dehydration of tetrahydrofuran [88], in the separation of ethyl acetate/ethanol/water mixture [89] and in the removal of methanol from the ethylene glycol azeotrope [90]. The introduction of PVP into chitosan can disorder the arrangement of the chitosan chains, decreasing the degree of the crystallinity of chitosan and consequently increasing the free volume in the membrane. On the other hand, introduction of poly(vinyl pyrrolidone) could significantly enhance the hydrophilicity of the membranes because it has strong polarity and hydrophilic groups. This causes plasticizing effect into the membrane, thus this type of membranes are suitable for the final dehydration process, when the concentration of water in the feed is low.

The next hopeful direction of chitosan membranes property modification is the formation of the organic - inorganic composite hybrid membranes [48, 49, 92 - 103]. As an inorganic additive, various inorganic fillers such as metal oxide nanoparticles, zeolites, silica and carbon nanotubes are used. From several metal oxide nanoparticles, the most important are Fe_3O_4 and TiO_2. Chitosan membranes with iron oxide nanoparticles were studied in the pervaporation process of water/ethanol separation [94, 95]. Addition of the magnetite particles to the membrane matrix changed their transport properties, and enhancing the difference between the permeation of both components. The increase in water permeation was connected to its faster diffusion in the presence of the iron oxide nanoparticles, with simultaneous minor increase in the ethanol diffusion. Chitosan/TiO_2 nanocomposite membranes were investigated for ethanol dehydration [96]. The introduction of TiO_2 particles in the chitosan matrix caused the packing of chitosan chains, which resulted in the increase in the amorphous region in the chitosan matrix. Therefore, the permeating molecules could be more easily diffused through the hybrid membranes. Chitosan membranes filled with zeolite or silica were used in the separation of water-isopropanol [97, 100] and water/ethanol [98, 99] mixtures. Compared with pure chitosan membrane, the zeolite or silica filled chitosan membranes exhibited better separation properties. This was explained on the basis of significant enhancement of hydrophilicity, selective adsorption and establishment of molecular sieving action by the creation of pores in the polymer matrix. The addition of silica or zeolite nanoparticles to chitosan provided extra free volumes in the polymer matrixes, consequently to offer spaces for water molecules to permeate through the membranes. Another, popular chitosan membrane modification is the incorporation of nanofillers to form a "green" polymer-inorganic nanocomposite. Chitosan green nanocomposite membranes were applied for the separation of benzene/cyclohexane mixtures [49,

102] and dehydration of 1,4-dioxane [101], or acetone [103]. The obtained membranes exhibited better pervaporation performance than the pure chitosan membranes because carbon nanotubes provided an additional transport pathway in the chitosan membrane that enabled molecules access through their inner cores, thereby enhancing the permeability of the membrane. Additionally, the incorporation of carbon nanotubes reduced the excessive swelling behavior of chitosan.

The important approach to enhance the separation performance of the membrane is the chemical modification of their surface through grafting polymerization. For the dehydration of isopropanol, chitosan-g-polyaniline membranes [104], grafted chitosan glycidyltrimethylammonium chloride membranes [105], chitosan-g-poly(benzoimidazole) membranes [106], grafted chitosan polysulfone membranes [107], and others were used. The grafted membranes exhibited considerably better pervaporation performance than that of pure chitosan membranes. This was explained on the basis of increased hydrophobic character and increased degree of crystallinity.

The last group of modification method used for improving chitosan membrane properties involves laying a thin and dense chitosan layer on the top of porous substrate. In all cases, the support gives an increased chemical and mechanical stability, and higher temperature resistance than polymer membranes without porous support. Furthermore, the used support could play an important role in the tailoring of the membrane transport properties. The following materials are used as porous substrates: ceramic [108], polysulfone [109 - 111], poly (benzoimidazole) [112], polyacrylonitrile [113, 114], poly(tetrafluoroethylene) [115], polyetherimide [116], poly(ether-block-amide) [117], and other substrates.

CONFLICT OF INTEREST

The authors confirm that they have no conflict of interest to declare for this publication.

ACKNOWLEDGEMENTS

Declared none.

REFERENCES

[1] Kober PA. Pervaporation, perstillation and percrystallization. J. Am. Chem. Soc., 1917, 39, 944. J Membr Sci 1995; 100: 61-4. [http://dx.doi.org/10.1016/0376-7388(94)00234-P]

[2] Basile A, Figoli A, Khayet M. Pervaporation, vapour permeation and membrane distillation. 1st ed., Woodhead Publishing: Elsevier 2015.

[3] Huang RY. Pervaporation membrane separation processes. Amsterdam, The Netherlands: Elsevier Science Publishers 1991.

[4] WitcheyLakshmanan L.C.. Pervaporation membrane separation processes. Am Sci 1992; 80(6): 615.

[5] Baig FU. Pervaporation. In: Li NN, Fane AG, Ho WS, Matsuura T, Eds. Advanced Membrane Technology and Applications. USA: Wiley & Sons 2008. [http://dx.doi.org/10.1002/9780470276280.ch17]

[6] Baker RW. Membrane technology and applications. New York: Wiley&Sons 2004. [http://dx.doi.org/10.1002/0470020393]

[7] Wijmans JG, Baker RW. The solution diffusion model a review. J Membr Sci 1995; 107: 121. [http://dx.doi.org/10.1016/0376-7388(95)00102-I]

[8] Geankoplis CJ. Transport processes and separation process principles. 4th ed., USA: Prentice Hall Press Upper Saddle River, NJ 2013.

[9] Noble RD, Stern SA. Membrane separation technology: principles and applications. Amsterdam, The Netherlands: Elsevier Science Publishers 1995.

[10] Xu S, Wang Y. Novel thermally crosslinked polyimide membranes for ethanol dehydration *via* pervaporation. J Membr Sci 2015; 496: 142155. [http://dx.doi.org/10.1016/j.memsci.2015.08.055]

[11] Uragami T, Banno M, Miyata T. Dehydration of an ethanol/water azeotrope through alginate DNA membranes crosslinked with metal ions by pervaporation. Carbohydr Polym 2015; 134: 3845. [http://dx.doi.org/10.1016/j.carbpol.2015.07.054]

[12] Amirilargani M, Sadatnia B. Poly(vinyl alcohol)/zeolitic imidazolate frameworks (ZIF8) mixed matrix membranes for pervaporation dehydration of isopropanol. J Membr Sci 2014; 469: 110. [http://dx.doi.org/10.1016/j.memsci.2014.06.034]

[13] Shi GM, Yang T, Chung TS. Polybenzimidazole (PBI)/zeolitic imidazolate frameworks (ZIF8) mixed matrix membranes for pervaporation dehydration of alcohols. J Membr Sci 2012; 415-416: 577586. [http://dx.doi.org/10.1016/j.memsci.2012.05.052]

[14] Penkova AV, Acquah SF, Sokolova MP, Dmitrenko ME, Toikka AM. Polyvinyl alcohol membranes modified by low-hydroxylated fullerenol $C_{60}(OH)_{12}$. J Membr Sci 2015; 491: 2227. [http://dx.doi.org/10.1016/j.memsci.2015.05.011]

[15] Singha NR, Parya TK, Ray SK. Dehydration of 1,4-dioxane by pervaporation using filled and crosslinked polyvinyl alcohol membrane. J Membr Sci 2009; 340: 3544. [http://dx.doi.org/10.1016/j.memsci.2009.05.003]

[16] Yee KF, Ong YT, Abdul RM, Soon HT. Novel MWCNT-buckypaper/polyvinyl alcohol asymmetric membrane for dehydration of etherification reaction mixture: Fabrication, characterisation and application. J Membr Sci 2014; 453: 546555. [http://dx.doi.org/10.1016/j.memsci.2013.11.040]

[17] Wei Z, Yunxia X, Zhejun Y, Shui L, Xinping W. Separation of acetic acid/water mixtures by pervaporation with composite membranes of sodium alginate active layer and microporous polypropylene substrate. J Membr Sci 2014; 451: 135147. [http://dx.doi.org/10.1016/j.memsci.2013.09.027]

[18] Kuila SB, Ray SK. Dehydration of dioxane by pervaporation using filled blend membranes of polyvinyl alcohol and sodium alginate. Carbohydr Polym 2014; 101: 1154-65. [http://dx.doi.org/10.1016/j.carbpol.2013.09.086] [PMID: 24299887]

[19] Sajjan AM, Kumar BK, Kittur AA, Kariduraganavar MY. Novel approach for the development of pervaporation membranes using sodium alginate and chitosan wrapped multiwalled carbon nanotubes for the dehydration of isopropanol. J Membr Sci 2013; 425-426: 7788. [http://dx.doi.org/10.1016/j.memsci.2012.08.042]

[20] Sajjan AM, Premakshi HG, Kariduraganavar MY. Synthesis and characterization of GTMAC grafted chitosan membranes for the dehydration of low water content isopropanol by pervaporation. J Ind Eng Chem 2015; 25: 151161.
[http://dx.doi.org/10.1016/j.jiec.2014.10.027]

[21] Wang M. ArnalHerault, C.; Rousseau, C.; Palenzuela, A.; Babin, J.; David, L.; Jonquieres, A. Grafting of multiblock copolymers: A new strategy for improving membrane separation performance for ethyl tertbutyl (ETBE) biofuel purification by pervaporation. J Membr Sci 2014; 469: 3142.
[http://dx.doi.org/10.1016/j.memsci.2014.06.038]

[22] Khoonsap S, Supanchaiyamat N, Hunt AJ, Klinsrisuk S, Amnuaypanich S. Improving water selectivity of poly (vinyl alcohol) (PVA) - Fumed silica (FS) nanocomposite membranes by grafting of poly (2hydroxyethyl methacrylate) (PHEMA) on fumed silica particles. Chem Eng Sci 2015; 122: 373383.
[http://dx.doi.org/10.1016/j.ces.2014.09.049]

[23] Zhou K, Zhang QG, Han GL, Zhu AM, Liu QL. Pervaporation of water-ethanol and methanol-MTBE mixtures using poly (vinyl alcohol)/cellulose acetate blended membranes. J Membr Sci 2013; 448: 93101.
[http://dx.doi.org/10.1016/j.memsci.2013.08.005]

[24] Zhu M, Qian J, Zhao Q, An Q, Song Y, Zheng Q. Polyelectrolyte complex (PEC) modified by poly(vinyl alcohol) and their blend membranes for pervaporation dehydration. J Membr Sci 2011; 378: 233242.
[http://dx.doi.org/10.1016/j.memsci.2011.05.007]

[25] Zhao J, Zhu Y, Pan F, *et al.* Fabricating graphene oxide-based ultrathin hybrid membrane for pervaporation dehydration *via* layer-by-layer self-assembly driven by multiple interactions. J Membr Sci 2015; 487: 162172.
[http://dx.doi.org/10.1016/j.memsci.2015.03.073]

[26] Zhao J, Wang F, Pan F, *et al.* Enhanced pervaporation dehydration performance of ultrathin hybrid membrane by incorporating bioinspired multifunctional modifier and TiCl4 into chitosan. J Membr Sci 2013; 446: 395404.
[http://dx.doi.org/10.1016/j.memsci.2013.06.044]

[27] Gong L, Zhang L, Wang N, *et al. In situ* ultraviolet-light-induced TiO_2 nanohybrid superhydrophilic membrane for pervaporation dehydration. Separ Purif Tech 2014; 122: 3240.
[http://dx.doi.org/10.1016/j.seppur.2013.10.032]

[28] Fernández JG, Almeida CA. FernándezBaldo, M.A., Felici, E., Raba, J., Sanz, M.I. Development of nitrocellulose membrane filters impregnated with different biosynthesized silver nanoparticles applied to water purification. Talanta 2016; 146: 237243.
[http://dx.doi.org/10.1016/j.talanta.2015.08.060]

[29] Wang Z, Shi Y, Cheng M, He C, Zhuang J, Sheng L. Purification of small-scale stagnant water in urban regions: Human-powered water lift and bank-based constructed wetland. Ecol Eng 2015; 83: 108111.
[http://dx.doi.org/10.1016/j.ecoleng.2015.06.012]

[30] Li P, Wang Z, Qiao Z, *et al.* Recent developments in membranes for efficient hydrogen purification. J Membr Sci 2015; 495: 130168.
[http://dx.doi.org/10.1016/j.memsci.2015.08.010]

[31] An S, Lee MW, Joshi BN, Jo A, Jung J, Yoon SS. Water purification and toxicity control of chlorophenols by 3D nanofiber membranes decorated with photocatalytic titania nanoparticles. Ceram Int 2014; 40: 33053313.
[http://dx.doi.org/10.1016/j.ceramint.2013.09.104]

[32] Sun XF, Qin J, Xia PF, *et al.* Graphene oxide-silver nanoparticle membrane for biofouling control and water purification. Chem Eng J 2015; 281: 5359.
[http://dx.doi.org/10.1016/j.cej.2015.06.059]

[33] Ozturk B, Kuru C, Aykac H, Kaya S. VOC separation using immobilized liquid membranes impregnated with oils. Separ Purif Tech 2015; 153: 16. [http://dx.doi.org/10.1016/j.seppur.2015.08.032]

[34] Shah MR, Noble RD, Clough DE. Pervaporation-air stripping hybrid process for removal of VOCs from groundwater. J Membr Sci 2004; 241: 257263. [http://dx.doi.org/10.1016/j.memsci.2004.04.032]

[35] Bruggen BV, Luis P. Chapter four - pervaporation. Progress in Filtration and Separation 2015; p. 101154.

[36] Heymes F, Demoustier PM, Charbit F, Fanlo JL, Moulin P. Treatment of gas containing hydrophobic VOCs by a hybrid absorption-pervaporation process: The case of toluene. Chem Eng Sci 2007; 62: 25762589. [http://dx.doi.org/10.1016/j.ces.2007.02.001]

[37] Kiss AS. Advanced distillation technologies: design, control and applications. New York: Wiley & Sons 2013. [http://dx.doi.org/10.1002/9781118543702]

[38] Saidur R, Elcevvadi ET, Mekhilef S, Safari A, Mohammed HA. An overview of different distillation methods for small scale applications. Renew Sustain Energy Rev 2011; 15: 47564764. [http://dx.doi.org/10.1016/j.rser.2011.07.077]

[39] Liu Y, Zhou X, Wang D, Song C, Liu J. A prediction model of VOC partition coefficient in porous building materials based on adsorption potential theory. Build Environ 2015; 93: 221233. [http://dx.doi.org/10.1016/j.buildenv.2015.06.025]

[40] Campesi MA, Luzi CD, Barreto GF, Martínez OM. Evaluation of an adsorption system to concentrate VOC in air streams prior to catalytic incineration. J Environ Manage 2015; 154: 216-24. [http://dx.doi.org/10.1016/j.jenvman.2015.02.028] [PMID: 25734958]

[41] Padhi SK, Gokhale S. Biological oxidation of gaseous VOCs - rotating biological contactor a promising and eco-friendly technique. J Environ Chem Eng 2014; 2: 20852102. [http://dx.doi.org/10.1016/j.jece.2014.09.005]

[42] Raúl Muñoz R, Souza TS, Glittmann L, Pérez R, Quijano G. Biological anoxic treatment of O_2 free VOC emissions from the petrochemical industry: A proof of concept study. J Hazard Mater 2013; 260: 442450.

[43] Ramaiah KP, Satyasri D, Sridhar S, Krishnaiah A. Removal of hazardous chlorinated VOCs from aqueous solutions using novel ZSM-5 loaded PDMS/PVDF composite membrane consisting of three hydrophobic layers. J Hazard Mater 2013; 261: 362-71. [http://dx.doi.org/10.1016/j.jhazmat.2013.07.048] [PMID: 23959256]

[44] Kim HJ, Nah SS, Min BR. A new technique for preparation of PDMS pervaporation membrane for VOC removal. Adv Environ Res 2002; 6: 255264. [http://dx.doi.org/10.1016/S1093-0191(01)00056-9]

[45] Aguado S, Coronas J, Santamaría J. Use of zeolite membrane reactors for the combustion of VOCs present in air at low concentrations. Chem Eng Res Des 2005; 83: 295301. [http://dx.doi.org/10.1205/cherd.04208]

[46] Drobek M, Figoli A, Algieri C, Santoro S, Trotta A, Gaeta N. Preparation of novel MFI zeolite PVDF mixed matrix membranes for potential VOCs removal. Procedia Eng 2012; 44: 1806. [http://dx.doi.org/10.1016/j.proeng.2012.08.956]

[47] Dong Y, Guo H, Su Z, Wei W, Wu X. Pervaporation separation of benzene/cyclohexane through AAOMionic liquids/polyurethane membranes. Chem Eng Process: Process Intensification 2015; 89: 6269. [http://dx.doi.org/10.1016/j.cep.2015.01.006]

[48] Lu L, Sun H, Peng F, Jiang Z. Novel graphite-filled PVA/CS hybrid membrane for pervaporation of benzene/cyclohexane mixtures. J Membr Sci 2006; 281: 245-52. [http://dx.doi.org/10.1016/j.memsci.2006.03.041]

[49] Shen J, Chu Y, Ruan H, Wu L, Gao C, Bruggen BV. Pervaporation of benzene/cyclohexane mixtures through mixed matrix membranes of chitosan and Ag+/carbon nanotubes. J Membr Sci 2014; 462: 160-9. [http://dx.doi.org/10.1016/j.memsci.2014.03.040]

[50] Li Y, Liu L, Fang Y. Plasmainduced grafting of hydroxyethyl methacrylate (HEMA) onto chitosan membranes by a swelling method. Polym Int 2003; 52: 285-90. [http://dx.doi.org/10.1002/pi.1082]

[51] Zhang Y, Wang N, Ji S, Zhang R, Zhao C, Li JR. Metal-organic framework/poly(vinyl alcohol) nanohybrid membrane for the pervaporation of toluene/n-heptane mixtures. J Membr Sci 2015; 489: 144152. [http://dx.doi.org/10.1016/j.memsci.2015.04.012]

[52] Wang N, Ji S, Li J, Zhang R, Zhang G. Poly(vinyl alcohol)-graphene oxide nanohybrid "pore-filling" membrane for pervaporation of toluene/n-heptane mixtures. J Membr Sci 2014; 455: 113120. [http://dx.doi.org/10.1016/j.memsci.2013.12.023]

[53] Iravaninia M, Mirfendereski M, Mohammadi T. Pervaporation separation of toluene/n-heptane mixtures using a MSEmodified membrane: Effects of operating conditions. Chem Eng Res Des 2012; 90: 397408. [http://dx.doi.org/10.1016/j.cherd.2011.07.020]

[54] Aouinti L, Roizard D, Hu GH, Thomas F, Belbachir M. Investigation of pervaporation hybrid polyvinylchloride membranes for the separation of toluene-n-heptane mixtures — case of clays as filler. Desalination 2009; 241: 174181. [http://dx.doi.org/10.1016/j.desal.2007.12.049]

[55] Kung G, Jiang LY, Wang Y, Chung TS. Asymmetric hollow fibers by polyimide and polybenzimidazole blends for toluene/iso-octane separation. J Membr Sci 2010; 360: 303314. [http://dx.doi.org/10.1016/j.memsci.2010.05.030]

[56] Xu W, Paul DR, Koros WJ. Carboxylic acid containing polyimides for pervaporation separations of toluene/iso-octane mixtures. J Membr Sci 2003; 219: 89102. [http://dx.doi.org/10.1016/S0376-7388(03)00188-1]

[57] Kung G, Jiang LY, Wang Y, Chung TS. Asymmetric hollow fibers by polyimide and polybenzimidazole blends for toluene/iso-octane separation. J Membr Sci 2010; 360: 303314. [http://dx.doi.org/10.1016/j.memsci.2010.05.030]

[58] Qu H, Kong Y, Lv H, Zhang Y, Yang J, Shi D. Effect of crosslinking on sorption, diffusion and pervaporation of gasoline components in hydroxyethyl cellulose membranes. Chem Eng J 2010; 157: 6066. [http://dx.doi.org/10.1016/j.cej.2009.09.044]

[59] Matlack A. Introduction to Green Chemistry. 2nd ed., CRC Press 2010.

[60] Rahimpour MR, Jafari M, Iranshahi D. Progress in catalytic naphtha reforming process: A review. Appl Energy 2013; 109: 79-93. [http://dx.doi.org/10.1016/j.apenergy.2013.03.080]

[61] Wu T, Wang N, Li J, *et al.* Tubular thermal crosslinked PEBA/ceramic membrane for aromatic/aliphatic pervaporation. J Membr Sci 2015; 486: 19. [http://dx.doi.org/10.1016/j.memsci.2015.03.037]

[62] Ribeiro CP, Freeman BD, Kalika DS, Kalakkunnath S. Aromatic polyimide and polybenzoxazole membranes for the fractionation of aromatic/aliphatic hydrocarbons by pervaporation. J Membr Sci 2012; 390-391: 182193.

[http://dx.doi.org/10.1016/j.memsci.2011.11.042]

[63] Katarzynski D, Staudt-Bickel C. Separation of multi component aromatic/aliphatic mixtures by pervaporation with copolyimide membranes. Desalination 2006; 189: 8186. [http://dx.doi.org/10.1016/j.desal.2005.06.015]

[64] Matsui S, Paul DR. Pervaporation separation of aromatic/aliphatic hydrocarbons by crosslinked poly(methyl acrylate-co-acrylic acid) membranes. J Membr Sci 2002; 195: 229245. [http://dx.doi.org/10.1016/S0376-7388(01)00560-9]

[65] Beppu MM, Vieira RS, Aimoli CG, Santana CC. Crosslinking of chitosan membranes using glutaraldehyde: Effect on ion permeability and water absorption. J Membr Sci 2007; 301: 12630. [http://dx.doi.org/10.1016/j.memsci.2007.06.015]

[66] Shen J, Wu L, Qiu J, Gao C. Pervaporation separation of water/isopropanol mixtures through crosslinked carboxymethyl chitosan/polysulfone hollow-fiber composite membranes. J Appl Polym Sci 2007; 103: 1959-65. [http://dx.doi.org/10.1002/app.24910]

[67] Uragami T, Matsuda T, Okuno H, Miyata T. Structure of chemically modified chitosan membranes and their characteristics of permeation and separation of aqueous ethanol solutions. J Membr Sci 1994; 88(23): 243251.

[68] Gnus M, Dudek G, Turczyn R, Tórz A, Machoczek T. The influence of crosslinking agent on chitosan membranes separation properties, Membranes and Membrane Processes in Environmental Protection, Monographs of the Environmental Engineering Committee. Polish Academy of Sciences 2014; 118: 283292.

[69] Chen JH, Liu QL, Zhang XH, Zhang QG. Pervaporation and characterization of chitosan membranes cross-linked by 3-aminopropyltriethoxysilane. J Membr Sci 2007; 292: 125-32. [http://dx.doi.org/10.1016/j.memsci.2007.01.026]

[70] Xiao S, Feng X, Huang RY. Trimesoyl chloride crosslinked chitosan membranes for CO_2/N_2 separation and pervaporation dehydration of isopropanol. J Membr Sci 2007; 306: 36-46. [http://dx.doi.org/10.1016/j.memsci.2007.08.021]

[71] Devi DA, Smitha B, Sridhar S, Aminabhavi TM. Pervaporation separation of isopropanol/water mixtures through crosslinked chitosan membranes. J Membr Sci 2005; 262: 91-9. [http://dx.doi.org/10.1016/j.memsci.2005.03.051]

[72] Lee YM, Nam SY, Woo DJ. Pervaporation of ionically surface crosslinked chitosan composite membranes for water-alcohol mixtures. J Membr Sci 1997; 133: 103110. [http://dx.doi.org/10.1016/S0376-7388(97)00089-6]

[73] Sunitha S, Satyanarayana SV, Sridhar S. Phosphorylated chitosan membranes for the separation of ethanol-water mixtures by pervaporation. Carbohydr Polym 2012; 87: 15691574. [http://dx.doi.org/10.1016/j.carbpol.2011.09.054]

[74] Lee YM, Shin EM. Pervaporation separation of water-ethanol through modified chitosan membranes. IV. Phospohorylated chitosan membranes. J Membr Sci 1991; 64: 145-52. [http://dx.doi.org/10.1016/0376-7388(91)80085-K]

[75] Jegal J, Lee KH. Chitosan membranes crosslinked with sulfosuccinic acid for the pervaporation separation of water/alcohol mixtures. J Appl Polym Sci 1999; 71: 671-5. [http://dx.doi.org/10.1002/(SICI)1097-4628(19990124)71:4<671::AID-APP19>3.0.CO;2-T]

[76] Wu LG, Zhu CL, Liu M. Study of a new pervaporation membrane. Part 1. Preparation and characteristics of the new membrane. J Membr Sci 1994; 90: 199. [http://dx.doi.org/10.1016/0376-7388(94)80071-5]

[77] Svang-Ariyaskul A, Huang R.Y.M, Douglas P.L, Pal R, Feng X, Chen, P, Liu L. Blended chitosan and polyvinyl alcohol membranes for the pervaporation dehydration of isopropanol. J Membr Sci 2006; 280: 815-23.

[http://dx.doi.org/10.1016/j.memsci.2006.03.001]

[78] Devi DA, Smitha B, Sridhar S, Aminabhavi TM. Dehydration of 1,4-dioxane through blend membranes of poly(vinyl alcohol) and chitosan by pervaporation. J Membr Sci 2006; 280: 138-47. [http://dx.doi.org/10.1016/j.memsci.2006.01.006]

[79] Nawawi MGM, Zamrud Z, Idham Z, Hassan O, Sakri NM. Blended chitosan and polyvinyl alcohol membrane for pervaporation separation methanol/methyl tertButyl ether mixture. (II) Effect of operating parameters. Jurnal Teknologi (Sciences & Engineering) 2013; 65: 39-43.

[80] Jegal J, Lee JH. Development of polyion complex membranes for the separation of water&alcohol mixtures. III. Preparation of polyion complex membranes based on the kcarrageenan for the pervaporation separation of waterethanol. J Appl Polym Sci 1996; 60: 11771183. [http://dx.doi.org/10.1002/(SICI)1097-4628(19960523)60:8<1177::AID-APP10>3.0.CO;2-#]

[81] Kanti P, Srigowri K, Madhuri J, Smitha B, Sridhar S. Dehydration of ethanol through blend membranes of chitosan and sodium alginate by pervaporation. Separ Purif Tech 2004; 40: 259-66. [http://dx.doi.org/10.1016/j.seppur.2004.03.003]

[82] Kim SG, Lim GT, Jegal J, Lee KH. Pervaporation separation of MTBE (methyl *tert*butyl ether) and methanol mixtures through polyion complex composite membranes consisting of sodium alginate/chitosan. J Membr Sci 2000; 174: 1-15. [http://dx.doi.org/10.1016/S0376-7388(00)00339-2]

[83] Reddy AS, Kalyani S, Kumar NS, Boddu VM, Krishnaiah A. Dehydration of 1,4-dioxane by pervaporation using crosslinked calcium alginate-chitosan blend membranes. Polym Bull 2008; 61: 779-90. [http://dx.doi.org/10.1007/s00289-008-1003-x]

[84] Nam SY, Lee YM. Pervaporation and properties of chitosan-poly(acrylic acid) complex membranes. J Membr Sci 1997; 135: 161-71. [http://dx.doi.org/10.1016/S0376-7388(97)00144-0]

[85] Hu C, Li B, Guo R, Wu H, Jiang Z. Pervaporation performance of chitosan-poly(acrylic acid) polyelectrolyte complex membranes for dehydration of ethylene glycol aqueous solution. Separ Purif Tech 2007; 55: 327-34. [http://dx.doi.org/10.1016/j.seppur.2007.01.005]

[86] Chanachai A, Jiraratananon R, Uttapap D, Moon GY, Anderson WA, Huang RY. Pervaporation with chitosan/hydroxyethylcellulose (CS/HEC) blended membranes. J Membr Sci 2000; 166: 271-80. [http://dx.doi.org/10.1016/S0376-7388(99)00269-0]

[87] Veerapur RS, Gudasi KB, Aminabhavi TM. Pervaporation dehydration of isopropanol using blend membranes of chitosan and hydroxypropyl cellulose. J Membr Sci 2007; 304: 102-11. [http://dx.doi.org/10.1016/j.memsci.2007.07.006]

[88] Devi DA, Smitha B, Sridhar S, Aminabhavi TM. Novel crosslinked chitosan/poly(vinylpyrrolidone) blend membranes for dehydrating tetrahydrofuran by the pervaporation technique. J Membr Sci 2006; 280: 45-53. [http://dx.doi.org/10.1016/j.memsci.2006.01.003]

[89] Zhang XH, Liu QL, Xiong Y, Zhu AM, Chen Y, Zhang QG. Pervaporation dehydration of ethyl acetate/ethanol/water azeotrope using chitosan/poly (vinyl pyrrolidone) blend membranes. J Membr Sci 2009; 327: 274-80. [http://dx.doi.org/10.1016/j.memsci.2008.11.034]

[90] Zhang QG, Hu WW, Liu QL, Zhu AM. Chitosan/polyvinylpyrrolidone-silica hybrid membranes for pervaporation separation of methanol/ethylene glycol azeotrope. J Appl Polym Sci 2013; 129: 3178-84. [http://dx.doi.org/10.1002/app.39058]

[91] Lin W, Li Q, Zhu T. New chitosan/Konjac glucomannan blending membrane for application in

pervaporation dehydration of caprolactam solution. J Ind Eng Chem 2012; 18: 934-40. [http://dx.doi.org/10.1016/j.jiec.2011.09.008]

[92] Dudek G, Turczyn R, Strzelewicz A, Krasowska M, Rybak A, Grzywna ZJ. Studies of separation of vapours and gases through composite membranes with ferroferric oxide magnetic nanoparticles. Separ Purif Tech 2013; 109: 5563. [http://dx.doi.org/10.1016/j.seppur.2013.02.024]

[93] Dudek G, Turczyn R, Strzelewicz A, Rybak A, Krasowska M, Grzywna ZJ. Preparation and characterization of iron oxides - polymer composite membranes. Sep Sci Technol 2012; 47: 1390-4. [http://dx.doi.org/10.1080/01496395.2012.672519]

[94] Dudek G, Gnus M, Turczyn R, Strzelewicz A, Krasowska M. Pervaporation with chitosan membranes containing iron oxide nanoparticles. Separ Purif Tech 2014; 133: 8-15. [http://dx.doi.org/10.1016/j.seppur.2014.06.032]

[95] Dudek G, Strzelewicz A, Turczyn R, Krasowska M, Rybak A. The study of ethanol/water vapors permeation through sulfuric acid crossLinked chitosan magnetic membranes. Sep Sci Technol 2014; 49: 17611767. [http://dx.doi.org/10.1080/01496395.2014.906455]

[96] Yang D, Li J, Jiang Z, Lu L, Chen X. Chitosan/TiO_2 nanocomposite pervaporation membranes for ethanol dehydration. Chem Eng Sci 2009; 64: 3130-7. [http://dx.doi.org/10.1016/j.ces.2009.03.042]

[97] Premakshi HG, Ramesh K, Kariduraganavar MY. Modification of crosslinked chitosan membrane using NaY zeolite for pervaporation separation of water-isopropanol mixtures. Chem Eng Res Des 2015; 94: 3243. [http://dx.doi.org/10.1016/j.cherd.2014.11.014]

[98] Sun H, Lu L, Chen X, Jiang Z. Pervaporation dehydration of aqueous ethanol solution using HZSM5 filled chitosan membranes. Separ Purif Tech 2008; 58: 429-36. [http://dx.doi.org/10.1016/j.seppur.2007.09.012]

[99] Liu YL, Hsu CY, Su YH, Lai JY. Chitosan-silica complex membranes from sulfonic acid functionalized silica nanoparticles for pervaporation dehydration of ethanol-water solutions. Biomacromolecules 2005; 6: 368373. [http://dx.doi.org/10.1021/bm049531w]

[100] Varghese JG, Karuppannan RS, Kariduraganavar MY. Development of hybrid membranes using chitosan and silica precursors for pervaporation separation of water + isopropanol mixtures. J Chem Eng Data 2010; 55: 2084-92. [http://dx.doi.org/10.1021/je9003993]

[101] Ong YT, Ahmad AL, Zein SH, Sudesh K, Tan SH. Poly(3-hydroxybutyrate)-functionalised multi-walled carbon nanotubes/chitosan green nanocomposite membranes and their application in pervaporation. Separ Purif Tech 2011; 76: 419-27. [http://dx.doi.org/10.1016/j.seppur.2010.11.013]

[102] Peng F, Pan F, Sun H, Lu L, Jiang Z. Novel nanocomposite pervaporation membranes composed of poly(vinyl alcohol) and chitosanwrapped carbon nanotube. J Membr Sci 2007; 300: 13-9. [http://dx.doi.org/10.1016/j.memsci.2007.06.008]

[103] Yeang QW, Zein SH, Sulong AB, Tan SH. Comparison of the pervaporation performance of various types of carbon nanotube-based nanocomposites in the dehydration of acetone. Separ Purif Tech 2013; 107: 252-63. [http://dx.doi.org/10.1016/j.seppur.2013.01.031]

[104] Varghese JG, Kittur AA, Rachipudi PS, Kariduraganavar MY. Synthesis, characterization and pervaporation performance of chitosan-g-polyaniline membranes for the dehydration of isopropanol. J Membr Sci 2010; 364: 111-21. [http://dx.doi.org/10.1016/j.memsci.2010.08.007]

[105] Sajjan AM, Premakshi HG, Kariduraganavar MY. Synthesis and characterization of GTMAC grafted chitosan membranes for the dehydration of low water content isopropanol by pervaporation. J Ind Eng Chem 2015; 25: 151-61.
[http://dx.doi.org/10.1016/j.jiec.2014.10.027]

[106] Han YJ, Wang KH, Lai JY, Liu YL. Hydrophilic chitosan-modified polybenzoimidazole membranes for pervaporation dehydration of isopropanol aqueous solutions. J Membr Sci 2014; 463: 17-23.
[http://dx.doi.org/10.1016/j.memsci.2014.03.052]

[107] Liu YL, Yu CH, Ma LC, Lin GC, Tsai HA, Lai JY. The effects of surface modifications on preparation and pervaporation dehydration performance of chitosan/polysulfone composite hollow-fiber membranes. J Membr Sci 2008; 311: 243-50.
[http://dx.doi.org/10.1016/j.memsci.2007.12.040]

[108] Zhu Y, Xia S, Liu G, Jin W. Preparation of ceramicsupported poly(vinyl alcohol)-chitosan composite membranes and their applications in pervaporation dehydration of organic/water mixtures. J Membr Sci 2010; 349: 341-8.
[http://dx.doi.org/10.1016/j.memsci.2009.11.065]

[109] Huang RY, Pal R, Moon GY. Crosslinked chitosan composite membrane for the pervaporation dehydration of alcohol mixtures and enhancement of structural stability of chitosan/polysulfone composite membranes. J Membr Sci 1999; 160: 1730.
[http://dx.doi.org/10.1016/S0376-7388(99)00074-5]

[110] Huang RY, Pal R, Moon GY. Pervaporation dehydration of aqueous ethanol and isopropanol mixtures through alginate/chitosan two ply composite membranes supported by poly(vinylidene fluoride) porous membrane. J Membr Sci 2000; 167: 275-89.
[http://dx.doi.org/10.1016/S0376-7388(99)00293-8]

[111] Wu H, Li X, Zhao C, Shen X, Jiang Z, Wang X. Chitosan/Sulfonated Polyethersulfone–Polyethersulfone (CS/SPESPES) Composite Membranes for Pervaporative Dehydration of Ethanol. Ind Eng Chem Res 2013; 52: 5772-80.
[http://dx.doi.org/10.1021/ie303437r]

[112] Han YJ, Wang KH, Lai JY, Liu YL. Hydrophilic chitosan-modified polybenzoimidazole membranes for pervaporation dehydration of isopropanol aqueous solutions. J Membr Sci 2014; 463: 17-23.
[http://dx.doi.org/10.1016/j.memsci.2014.03.052]

[113] Tsai HA, Chen WH, Kuo CY, Lee KR, Lai JY. Study on the pervaporation performance and longterm stability of aqueous isopropanol solution through chitosan/polyacrylonitrile hollow fiber membrane. J Membr Sci 2008; 309: 146-55.
[http://dx.doi.org/10.1016/j.memsci.2007.10.018]

[114] Li BB, Xu ZL, Qusay FA, Li R. Chitosan-poly (vinyl alcohol)/poly (acrylonitrile) (CS-PVA/PAN) composite pervaporation membranes for the separation of ethanol-water solutions. Desalination 2006; 193(1): 171-81.
[http://dx.doi.org/10.1016/j.desal.2005.08.021]

[115] Liu YL, Yu CH, Lee KR, Lai JY. Chitosan/poly(tetrafluoroethylene) composite membranes using in pervaporation dehydration processes. J Membr Sci 2007; 287: 230-6.
[http://dx.doi.org/10.1016/j.memsci.2006.10.040]

[116] Huang RY, Moon GY, Pal R. N-acetylated chitosan membranes for the pervaporation separation of alcohol/toluene mixtures. J Membr Sci 2000; 176: 223-31.
[http://dx.doi.org/10.1016/S0376-7388(00)00443-9]

[117] Prasad NS, Moulik S, Bohra S, Rani KY, Sridhar S. Solvent resistant chitosan/poly(ether-block-amide) composite membranes for pervaporation of n-methyl-2-pyrrolidone/water mixtures. Carbohydr Polym 2016; 136: 1170-81.
[http://dx.doi.org/10.1016/j.carbpol.2015.10.037] [PMID: 26572460]

SUBJECT INDEX

A

Guilherme Luiz Dotto, Sergio Paulo Campana-Filho and Luiz Antonio de Almeida Pinto (Eds)

D

E

K

L

M

N

O

P

Q

R

S

T

U

V

W

X

Z

Made in the USA
Las Vegas, NV
04 November 2026

57425980R00200